The Minerals, Metals & Materials Series

Gisele Azimi · Takanari Ouchi · Kerstin Forsberg ·
Hojong Kim · Shafiq Alam ·
Alafara Abdullahi Baba · Neale R. Neelameggham
Editors

Rare Metal Technology 2021

Editors
Gisele Azimi
University of Toronto
Toronto, ON, Canada

Kerstin Forsberg
KTH Royal Institute of Technology
Stockholm, Sweden

Shafiq Alam
University of Saskatchewan
Saskatoon, SK, Canada

Neale R. Neelameggham
IND LLC
South Jordan, UT, USA

Takanari Ouchi
The University of Tokyo
Tokyo, Japan

Hojong Kim
Pennsylvania State University
University Park, PA, USA

Alafara Abdullahi Baba
University of Ilorin
Ilorin, Nigeria

ISSN 2367-1181 ISSN 2367-1696 (electronic)
The Minerals, Metals & Materials Series
ISBN 978-3-030-65488-7 ISBN 978-3-030-65489-4 (eBook)
https://doi.org/10.1007/978-3-030-65489-4

© The Minerals, Metals & Minerals Society 2021
This work is subject to copyright. All rights are solely and exclusively licensed by the Publisher, whether the whole or part of the material is concerned, specifically the rights of translation, reprinting, reuse of illustrations, recitation, broadcasting, reproduction on microfilms or in any other physical way, and transmission or information storage and retrieval, electronic adaptation, computer software, or by similar or dissimilar methodology now known or hereafter developed.
The use of general descriptive names, registered names, trademarks, service marks, etc. in this publication does not imply, even in the absence of a specific statement, that such names are exempt from the relevant protective laws and regulations and therefore free for general use.
The publisher, the authors and the editors are safe to assume that the advice and information in this book are believed to be true and accurate at the date of publication. Neither the publisher nor the authors or the editors give a warranty, expressed or implied, with respect to the material contained herein or for any errors or omissions that may have been made. The publisher remains neutral with regard to jurisdictional claims in published maps and institutional affiliations.

This Springer imprint is published by the registered company Springer Nature Switzerland AG
The registered company address is: Gewerbestrasse 11, 6330 Cham, Switzerland

Preface

Rare Metal Technology 2021 is the proceedings publication of the symposium on Rare Metal Extraction & Processing sponsored by the Hydrometallurgy and Electrometallurgy Committee of the TMS Extraction & Processing Division. The symposium has been organized to encompass the extraction of rare metals as well as rare extraction processing techniques used in metal production and mineral processing. This is the eighth symposium since 2014.

This symposium focuses on critical metals essential for critical modern technologies including electronics, electric motors, generators, energy storage systems, and specialty alloys. The rapid development of these technologies entails fast advancement of the resource and processing industry for their building materials. This symposium brings together researchers from academia and industry to exchange knowledge on developing, operating, and advancing extractive and processing technologies. In this proceedings publication, papers are presented on rare earth elements (magnets, catalysts, phosphors, and others), energy storage materials (lithium, cobalt, vanadium, and graphite), alloy elements (scandium, niobium, and titanium), and materials for electronic (gallium, germanium, indium, gold, and silver) commodities.

The papers cover various processing techniques in Mineral Beneficiation, Hydrometallurgy, Separation, Purification (Solvent Extraction, Ion Exchange, Precipitation, and Crystallization), Pyrometallurgy, Electrometallurgy, Supercritical Fluid Extraction, and Recycling (batteries, magnets, waste electrical, and electronic equipment). Papers address topics on process development and operations, Feed and Product Characterization, Critical Metals and the Environment, and Processing Plant Engineering Operations and Challenges.

We acknowledge the efforts of the symposium organizers and proceedings editors: Gisele Azimi, Takanari Ouchi, Kerstin Forsberg, Hojong Kim, Shafiq Alam, Alafara Abdullahi Baba, and Neale R. Neelameggham. The support from TMS staff members Matt Baker and Patricia Warren is greatly appreciated in assembling and publishing the proceedings. We sincerely thank all the authors, speakers, and participants and look forward to continued collaboration in the advancement of science and technology in the area of rare metal extraction and processing.

Gisele Azimi
Lead Organizer

Contents

Part I Li, Co, and Ni

Application of Eutectic Freeze Crystallization in the Recycling of Li-Ion Batteries 3
Yiqian Ma, Michael Svärd, James Gardner, Richard T. Olsson, and Kerstin Forsberg

Recovery of Valuable Metals from End-of-Life Lithium-Ion Battery Using Electrodialysis 11
Ka Ho Chan, Monu Malik, and Gisele Azimi

Lithium Adsorption Mechanism for Li_2TiO_3 19
Rajashekhar Marthi and York R. Smith

Study on the Production of Lithium by Aluminothermic Reduction Method 29
Huimin Lu and Neale R. Neelameggham

Effect of Synthesis Method on the Electrochemical Performance of $LiNi_xMnCo_{1-x-y}O_2$ (NMC) Cathode for Li-Ion Batteries: A Review 37
Monu Malik, Ka Ho Chan, and Gisele Azimi

Recovery of Cobalt as Cobalt Sulfate from Discarded Lithium-Ion Batteries (LIBs) of Mobile Phones 47
Pankaj Kumar Choubey, Archana Kumari, Manis Kumar Jha, and Devendra Deo Pathak

Part II Li, Co, Au, Ag, PGMs, Te, Na, W, and In

Environmental Aspects of the Electrochemical Recovery of Tellurium by Electrochemical Deposition-Redox Replacement (EDRR) 57
P. Halli, M. Rinne, B. P. Wilson, K. Yliniemi, and M. Lundström

Sodium Metal from Sulfate 65
Jed Checketts and Neale R. Neelameggham

Preparation of High Grade Ammonium Metatungstate (AMT) as Precursor for Industrial Tungsten Catalyst 73
Alafara A. Baba, Sadisu Girigisu, Mustapha A. Raji, Abdullah S. Ibrahim, Kuranga I. Ayinla, Christianah O. Adeyemi, Aishat Y. Abdulkareem, Mohammed J. Abdul, and Abdul G. F. Alabi

Industrial-Scale Indium Recovery from Various e-Waste Resources Through Simulation and Integration of Developed Processes 79
Basudev Swain, Jae Ryang Park, Kyung Soo Park, Chan-Gi Lee, Hyun Seon Hong, and Jae-chun Lee

Recovery of Lithium (Li) Salts from Industrial Effluent of Recycling Plant 91
Archana Kumari, Pankaj Kumar Choubey, Rajesh Gupta, and Manis Kumar Jha

Extraction of Platinum Group Metals from Spent Catalyst Material by a Novel Pyro-Metallurgical Process 101
Ana Maria Martinez, Kai Tang, Camilla Sommerseth, and Karen Sende Osen

Developed Commercial Processes to Recover Au, Ag, Pt, and Pd from E-waste 115
Rekha Panda, Manis Kumar Jha, Jae-chun Lee, and Devendra Deo Pathak

Part III   REEs

Innovative Reactors for Recovery of Rare Earth Elements (REEs) 129
Alison Lewis, Jemitias Chivavava, Jacolien du Plessis, Dane Smith, and Jody-Lee Smith

Recovery of Rare Earth Elements from Recycled Hard Disk Drive Mixed Steel and Magnet Scrap 139
Tedd E. Lister, Michelle Meagher, Mark L. Strauss, Luis A. Diaz, Harry W. Rollins, Gaurav Das, Malgorzata M. Lencka, Andre Anderko, Richard E. Riman, and Alexandra Navrotsky

Extraction Chromatography for Separation of Rare Earth Elements 155
Meher Sanku, Kerstin Forsberg, and Michael Svärd

Tool and Workflow for Systematic Design of Reactive Extraction for Separation and Purification of Valuable Components 163
Hana Benkoussas, David Leleu, Swagatika Satpathy, Zaheer Ahmed Shariff, and Andreas Pfennig

Rethinking Mineral Processing and Extractive Metallurgy Approaches to Ensure a Sustainable Supply of High-tech and Critical Raw Materials 173
Yousef Ghorbani, Glen T. Nwaila, Steven E. Zhang, and Jan Rosenkranz

Extraction of Rare Earth Metals: The New Thermodynamic Considerations Toward Process Hydrometallurgy 187
Ajay B. Patil, Rudolf P. W. J. Struis, Andrea Testino, and Christian Ludwig

Part IV REEs and Sc

Developing Feasible Processes for the Total Recycling of WEEE to Recover Rare Metals 197
Jae-chun Lee, Manis Kumar Jha, Rekha Panda, Pankaj Kumar Choubey, Archana Kumari, and Tai Gyun Kim

Rare Earth Elements Extraction from Coal Waste Using a Biooxidation Approach 211
Prashant K. Sarswat, Zongliang Zhang, and Michael L. Free

Scandium Extraction from Bauxite Residue Using Sulfuric Acid and a Composite Extractant-Enhanced Ion-Exchange Polymer Resin 217
Efthymios Balomenos, Ghazaleh Nazari, Panagiotis Davris, Gomer Abrenica, Anastasia Pilihou, Eleni Mikeli, Dimitrios Panias, Shailesh Patkar, and Wen-Qing Xu

Scandium – Leaching and Extraction Chemistry 229
Dag Øistein Eriksen

Preparation of Industrial Sodium Chromate Compound from an Indigenous Chromite Ore by Oxidative Decomposition 239
Alafara A. Baba, Kuranga I. Ayinla, Bankim Ch. Tripathy, Abdullah S. Ibrahim, Girigisu Sadisu, Daud T. Olaoluwa, and Mustapha A. Raji

Part V Recycling, Co, and REE

The Italian National Research Council Operations Within the EIT Raw Materials Framework 249
Paolo Dambruoso, Salvatore Siano, Armida Torreggiani, Ornella Russo, Stefania Marzocchi, and Vladimiro Dal Santo

Experimental Determination of Liquidus Temperature and Phase Equilibria of the $CaO–Al_2O_3–SiO_2–Na_2O$ Slag System Relevant to E-Waste Smelting 265
Md Khairul Islam, Michael Somerville, Mark I. Pownceby, James Tardio, Nawshad Haque, and Suresh Bhargava

How to Prepare Future Generations for the Challenges in the Raw Materials Sector 277
Armida Torreggiani, Alberto Zanelli, Alessandra Degli Esposti, Eleonora Polo, Paolo Dambruoso, Renata Lapinska-Viola, Kerstin Forsberg, and Emilia Benvenuti

Part VI V, Ce, Mo, Cr, and Fe

Transformation and Distribution of Vanadium Phases in Stone Coal and Combustion Fly Ash 291
Deng Zhi-gan, Tang Fu-li, Wei Chang, Fan Gang, Li Min-ting, Li Xing-bin, and Li Cun-xiong

Solvo-Chemical Recovery of Cerium from Sulfate Solution Using Cyanex 923 and Oxalate Precipitation 303
Sadia Ilyas, Hyunjung Kim, and Rajiv Ranjan Srivastava

Recovery of Molybdenum from Low Concentration Molybdenum-Containing Solution with Addition of Fe(III) 311
Bei Zhang, Bingbing Liu, Yuanfang Huang, Guihong Han, and Shengpeng Su

An Effective Way to Extract Cr from Cr-Containing Tailings 321
Jie Cheng, Hong-Yi Li, Shuo Shen, Jiang Diao, and Bing Xie

Study on the Enhancement of Iron Removal in the Becher Aeration by a Novel Tubular Reactor 327
Lei Zhou, Qiuyue Zhao, Mingzhao Zheng, Zimu Zhang, Guozhi Lv, and Tingan Zhang

Author Index 337

Subject Index 341

About the Editors

Gisele Azimi is an Associate Professor and Canada Research Chair in Urban Mining Innovations. She is jointly appointed by the Departments of Chemical Engineering and Applied Chemistry, and Materials Science and Engineering at University of Toronto. She is also a registered Professional Engineer in Ontario. Her research program is aligned well with the "Sustainability" and "Advanced Materials and Manufacturing" research themes. In her research program, she strives to achieve a sustainable future and mitigate the adverse effects of climate change through (1) advanced recycling and urban mining of waste electrical and electronic equipment (WEEE), utilizing innovative recycling processes based on supercritical fluids; (2) industrial solid waste reduction through waste valorization to produce strategic materials like rare earth elements; (3) development of innovative materials with unique properties with far-reaching applications in structural and energy materials sectors; and (4) energy storage focusing on the development of a new generation of rechargeable batteries made of aluminum. She received her Ph.D. in 2010 from the Department of Chemical Engineering and Applied Chemistry at University of Toronto. Before returning to University of Toronto as a faculty member in 2014, she completed two postdoctoral appointments at MIT in the Departments of Materials Science and Engineering and Mechanical Engineering. She has received a number of awards including the McCharles Prize for Early Career Research Distinction, TMS EPD Young Leaders Award, Emerging Leaders Award in Chemical Engineering, Dean's Spark Professorship, Early Researcher Award, TMS Light

Metals/Extraction and Processing Subject Award—Recycling, and Connaught New Researcher Award.

Takanari Ouchi is a Research Associate in the Institute of Industrial Science at The University of Tokyo. He received his Ph.D. in Nano-Science and Nano-Engineering from Waseda University in 2011. In this tenure, Dr. Ouchi developed electrochemical deposition processes to fabricate metal nano-structures with both well-controlled crystallinity and uniformity at the single nano-meter scale, and demonstrated the applicability of these processes for the fabrication of bit-patterned magnetic recording media for future hard disk drives. After completing his doctoral degree, Dr. Ouchi joined MIT, where he developed liquid metal batteries, which are, in principle, bi-directional electrolysis (electro-refining) cells, for application in grid-scale energy storage. As a research scientist, Dr. Ouchi led the systematic investigation of the electrochemical properties of liquid metal electrodes in molten salt electrolytes and developed novel lithium, calcium, and sodium liquid metal batteries. Since he began work as a research associate at The University of Tokyo in 2017, he has developed new recycling processes for rare metals and precious metals using pyrometallurgical and electrometallurgical methods. As a member of the Hydrometallurgy and Electrometallurgy Committee at The Minerals, Metals & Materials Society (TMS), Dr. Ouchi has contributed to the development of the vibrant field of metal extraction, organized technical symposia at TMS, solicited papers as a guest editor of *JOM*, and has earned several awards and honors, such as the TMS EPD Young Leaders Professional Development Award in 2015, based on his contributions to electrometallurgical processing.

About the Editors

Kerstin Forsberg is an Associate Professor in Chemical Engineering at KTH Royal Institute of Technology in Sweden. Her research program is focused on separation processes, in particular crystallization. This knowledge is often applied in projects concerning the recovery of resources from waste. Forsberg is the Program Director of the Master's program in Chemical Engineering for Energy and Environment at KTH. She is also the Deputy Director for the Research Platform for Industrial Transformation at KTH, and she represents the School of Engineering Sciences in Chemistry, Biotechnology and Health as a board member of the Water Centre and as a member of the management team for the Initiative in Circular Economy at KTH. Dr. Forsberg represents KTH as expert in the technical committee on circular economy at Swedish Institute for Standards (SIS) and with participation in the work within International Organization for Standardization (ISO).

Hojong Kim is an Associate Professor of Material Science and Engineering and Nuclear Engineering at Penn State University. He received a B.S. degree from Seoul National University and Ph.D. degree at MIT in the Uhlig Corrosion Laboratory. Dr. Kim worked as a senior researcher at Samsung Corning Precision Glass to improve the process yield for TFT-LCD glass manufacturing by engineering high-temperature materials. After 5 years of industrial experience, Dr. Kim returned to MIT as a postdoctoral researcher to contribute to the growing need for sustainable technology, with a research focus on molten oxide electrolysis for carbon-free iron production and liquid metal batteries for large-scale energy storage. His current research focuses on electrochemical processes for the separation of energy-critical elements and the development of corrosion-resistant materials. He is the recipient of the NSF CAREER award and the new doctoral new investigator award from American Chemical Society. He served as chair (2017–2019) and vice-chair (2015–2017) of the Hydrometallurgy and Electrometallurgy Committee of The Minerals, Metals & Materials Society.

Shafiq Alam is an Associate Professor at University of Saskatchewan, Canada. He is an expert in the area of mining and mineral processing with profound experience in industrial operations, management, engineering, design, consulting, teaching, research, and professional services. As a productive researcher, he has secured two patents, and has produced over 170 publications. He is the co-editor of eight books and an associate editor of *International Journal of Mining, Materials and Metallurgical Engineering*. He is the winner of the 2015 Technology Award from the Extraction & Processing Division of The Minerals, Metals & Materials Society (TMS), USA.

With extensive relevant industry experience as a registered professional engineer, Dr. Alam has worked on projects with many different mining industries. He is an Executive Committee Member of Hydrometallurgy Section of the Canadian Institute of Mining, Metallurgy and Petroleum (CIM). During 2015–2017, he served as the Chair of Hydrometallurgy and Electrometallurgy Committee of the Extraction & Processing Division (EPD) of TMS. He is a co-organizer of many symposia at international conferences through CIM and TMS. Dr. Alam is one of the founding organizers of the Rare Metal Extraction & Processing Symposium at TMS. He was involved in organizing International Nickel-Cobalt 2013 Symposium and TMS 2017 Honorary Symposium on applications of Process Engineering Principles in Materials Processing, Energy and Environmental Technologies. He was also involved in organizing the 9th International Symposium on Lead and Zinc Processing (PbZn 2020), co-located with TMS 2020 Annual Meeting and Exhibition in San Diego, California.

About the Editors

Alafara Abdullahi Baba is a Professor of Analytical/Industrial and Materials Chemistry in the Faculty of Physical Sciences, University of Ilorin, Nigeria. He holds a Ph.D. degree in Chemistry from University of Ilorin in 2008. His dissertation "Recovery of Zinc and Lead from Sphalerite, Galena and Waste Materials by Hydrometallurgical Treatments" was judged the best in the area of Physical Sciences at University of Ilorin in 2010. Until his current appointment as head of the Department of Industrial Chemistry in 2017, he was a deputy director—Central Research Laboratories, University of Ilorin (2014–2017). He is a fellow of the Chemical Society of Nigeria (CSN) and the Materials Science and Technology Society of Nigeria (MSN); is currently the secretary of the Hydrometallurgy and Electrometallurgy Committee of the Extraction & Processing Division (EPD) of The Minerals, Metals & Materials Society (TMS); is a co-organizer of the Rare Metal Extraction & Processing Symposium, and Energy Technologies and Carbon Dioxide Management Symposium at TMS Annual Meeting and Exhibition; and is on the TMS Materials Characterization, Education, and EPD Awards committees.

Dr. Baba has a keen interest in teaching, community services, and research covering solid minerals and materials processing through hydrometallurgical routes; reactions in solution and dissolution kinetic studies; and preparation of phyllosilicates, porous, and bio-ceramic materials for industrial value additions. He has more than 120 publications in nationally and internationally acclaimed journals of high impact, and has attended many national and international workshops, conferences, and research exhibitions to present his research breakthroughs. He is the recipient of several awards and honors including the 2015 Misra Award of Indian Institute of Mineral Engineers (IIME) for the best paper on Electro-/Hydro-Bio-Processing at the IIME International Seminar on Mineral Processing Technology—2014 held at Andhra University, Visakhapatnam, India; the 2015 MTN Season of Surprise Prize as Best Lecturer at University of Ilorin—Nigeria category; Award of Meritorious Service in recognition of immense contributions to Development of the Central Research Laboratories, University of Ilorin, Nigeria (2014–2017);

and the 2018 Presidential Merit Award in Recognition of Passion, Outstanding and Selfless Service to the Materials Science and Technology Society of Nigeria.

Neale R. Neelameggham is "The Guru" at IND LLC, involved in international technology and management consulting in the field of metals and associated chemicals, Thiometallurgy, energy technologies, soil biochemical reactor design, lithium-ion battery design, and agricultural uses of coal.

He has more than 38 years of expertise in magnesium production and was involved in the process development of its start-up company NL Magnesium to the present US Magnesium LLC, UT until 2011, during which he was instrumental in process development from the solar ponds to magnesium metal foundry. His expertise includes competitive magnesium processes worldwide and related trade cases.

In 2016, Dr. Neelameggham and Brian Davis authored the ICE-JNME award-winning paper "Twenty-First Century Global Anthropogenic Warming Convective Model." He is presently developing Agricoal® to greening arid soils. He authored the eBook *The Return of Manmade CO_2 to Earth: Ecochemistry*, published through Smashwords in November 2018.

Dr. Neelameggham holds 16 patents and patent applications and has published several technical papers. He has served in the Magnesium Committee of the TMS Light Metals Division (LMD) since its inception in 2000, chaired in 2005, and since 2007 has been a permanent co-organizer for the Magnesium Technology Symposium. He has been a member of Reactive Metals Committee, Recycling Committee, Titanium Committee, and Programming Committee for LMD and LMD council.

Dr. Neelameggham was the inaugural chair, when in 2008, LMD and the TMS Extraction and Processing Division (EPD) created Energy Committee, and has been a co-editor of Energy Technology Symposium through the present. He received LMD Distinguished Service Award in 2010. As chair of the Hydrometallurgy and Electrometallurgy Committee, he initiated Rare Metal Technology Symposium in 2014 and was a co-organizer for it through 2021. He organized the 2018 TMS Symposium on Stored Renewable Energy in Coal.

Part I
Li, Co, and Ni

Application of Eutectic Freeze Crystallization in the Recycling of Li-Ion Batteries

Yiqian Ma, Michael Svärd, James Gardner, Richard T. Olsson, and Kerstin Forsberg

Abstract The widespread and increasing use of Li-ion batteries has led to an impending need for recycling solutions. Consequently, recycling of spent Li-ion batteries with energy-efficient, environmentally sustainable strategies has become a research hotspot. In this work, eutectic freeze crystallization (EFC), which requires less energy input than conventional evaporative crystallization (EC), has been investigated as a method for the recovery of Ni and Co sulfates from synthetic acidic strip solution in the recycling of NMC or NCA Li-ion batteries. Two binary sulfate systems have been studied. Batch EFC experiments have been conducted. It is shown that, with suitable control of supersaturation, ice and salt crystals can be recovered as separate phases below the eutectic temperatures. The work shows that EFC is a promising alternative to EC for the recovery of Ni and Co sulfates from spent Li-ion batteries.

Keywords Eutectic freeze crystallization · Li-ion battery recycling · $NiSO_4$ · $CoSO_4$ · Hydrometallurgy

Introduction

With the ever-growing need for lithium-ion batteries, particularly from the electric mobility industry, a large amount of lithium-ion batteries are bound to retire in the near future, thereby leading to serious disposal problems and detrimental impacts on environment and energy conservation [1]. The composition of the cathode material

---

Y. Ma (✉) · M. Svärd · K. Forsberg
Department of Chemical Engineering, KTH Royal Institute of Technology, Teknikringen 42, 11428 Stockholm, Sweden
e-mail: yiqianm@kth.se

J. Gardner
Department of Chemistry, KTH Royal Institute of Technology, Teknikringen 30, 11428 Stockholm, Sweden

R. T. Olsson
Department of Fiber and Polymer Technology, KTH Royal Institute of Technology, Teknikringen 56, 11428 Stockholm, Sweden

is complex but rich in valuable metals. The recovery of Ni, Co, Mn, and Li from cathode material of NMC or NCA batteries using a hydrometallurgical process has been widely studied [2–5]. In general, the process can be summarized as pretreatment → leaching → separation/purification → product precipitation/crystallization [4, 6]. Solvent extraction is an effective method to get purified Ni or Co sulfate solution [3, 7]. After solvent extraction, evaporative crystallization (EC) is usually employed to produce Ni and Co sulfates [2]. EC can attain relatively high crystal growth rates with high purity yield, but it is not an energy-efficient process [8]. An alternative to evaporative crystallization is eutectic freeze crystallization (EFC). Figure 1 shows the phase diagram of a generic binary salt-water system. The process of EFC can be represented by path A → B → E or C → D → E as shown in Fig. 1 [9]. This technique is based on cooling an aqueous salt stream down to the eutectic point of the respective ice-salt system where ice and salt crystals are formed simultaneously. The separation of the resulting ice and crystals can then be achieved by gravity, as ice has a lower density but salt crystals have a higher density than the solution [10].

Eutectic freeze crystallization is strongly preferred to evaporative crystallization, as it uses much less energy and has the ability to recover pure water and treat corrosive streams [9]. EFC could be a promising approach for the production of metal salts in the recycling of Li-ion batteries. The recovery of $NiSO_4 \cdot 7H_2O$ and $CoSO_4 \cdot 7H_2O$ from acidic strip solution using EFC was then investigated in this work.

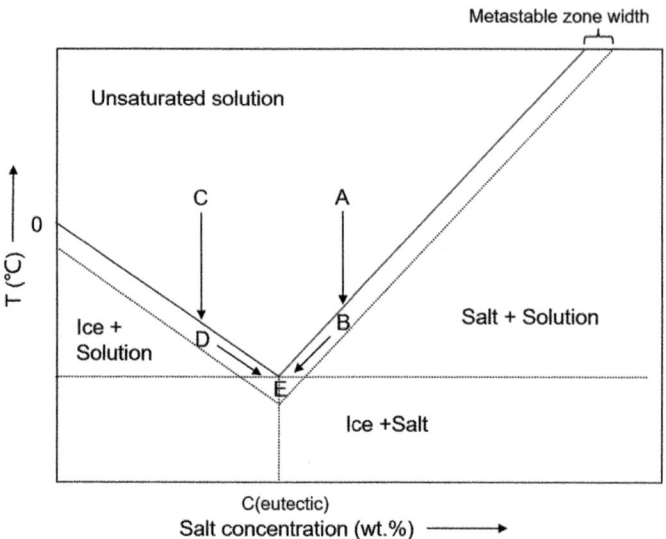

Fig. 1 Phase diagram of a binary salt-water system. (Color figure online)

Materials and Methods

Thermodynamic Modelling

Phase diagrams are used to show the thermodynamic equilibrium behavior of the binary systems in this study. OLI Stream Analyzer [11] was used to model the relationship between temperature and the solubility of metal (Co, Ni) sulfate in H_2O and 0.5 mol/kg H_2SO_4, respectively. In fact, metal sulfate in diluted H_2SO_4 is related to the ternary phase diagram of MSO_4-H_2SO_4-H_2O, but the binary section in the ternary phase diagram is concise and can also show the effect of acid when compared with the binary phase diagram of MSO_4-H2O (M = Ni, Co).

Feed Solution and Analysis

In the hydrometallurgical recycling of NMC or NCA batteries, solvent extraction is frequently used to get purified Ni or Co solution. In the solvent extraction, 1–2 mol/L H_2SO_4 usually acts as a stripping agent, and the strip liquor of Ni or Co contains about 100 g/L Ni^{2+} or Co^{2+}, and 0.5 mol/kg H_2SO_4 [2, 12]. To simulate the strip liquors of Ni and Co, 20 wt% $NiSO_4$ in 0.5 mol/kg H_2SO_4 solution and 20 wt% $CoSO_4$ in 0.5 mol/kg H_2SO_4 solution were prepared by dissolving analytical grade $NiSO_4 \cdot 6H_2O$, $CoSO_4 \cdot 7H_2O$ and 98 wt% H_2SO_4 in distilled water. After batch eutectic freeze crystallization experiments and filtration, metal sulfate hydrates could be obtained. An optical microscope was used to observe the color and shape of the resulting crystals. The chemical phases of the crystals were determined by X-ray diffraction (XRD).

Batch Experiment

The batch EFC experiments were performed in a jacketed glass vessel with a volume of 250 ml equipped with a mechanical stirrer. The sketch of the setup is shown in Fig. 2. The coolant was 30% by volume ethylene glycol in water. The temperature of the coolant was controlled by a chiller (refrigerated/heating circulator, Julabo FP50-HP, 230 V/60 Hz). The temperature was measured with two PT-100 sensors connected to a temperature monitor with an accuracy of 0.1 °C. One sensor measured the temperature in the reactor, while the other sensor measured the temperature of the coolant. 200 g synthetic strip solution was used for each batch experiment. The agitator speed was kept at 300 rpm. After EFC of ice and salt, filtration was conducted to obtain the salt crystals.

Fig. 2 Batch setup

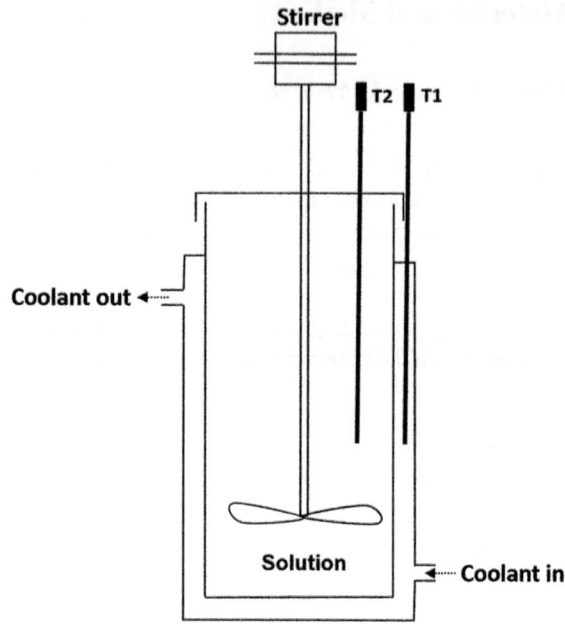

Fig. 3 Binary phase diagrams of NiSO₄-H₂O and NiSO₄–0.5 mol/kg H₂SO₄ for eutectic freeze crystallization. (Color figure online)

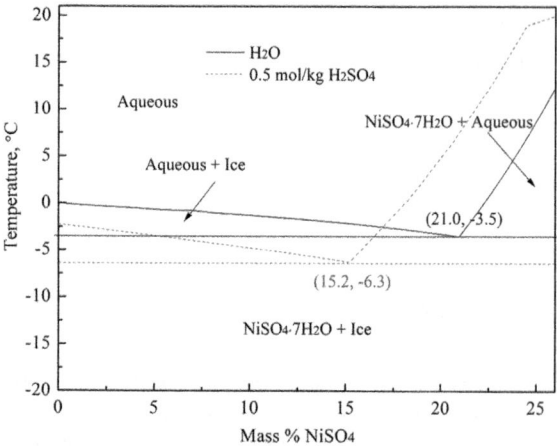

Results and Discussion

Binary Phase Diagrams

The binary phase diagrams of Ni sulfate and Co sulfate dissolved in both pure H₂O and 0.5 mol/kg H₂SO₄ in the temperature range −20 to 40 °C are presented in Figs. 3 and 4, respectively. The solubility data was predicted by the Mixed Solvent

Fig. 4 Binary phase diagrams of CoSO$_4$-H$_2$O and CoSO$_4$–0.5 mol/kg H$_2$SO$_4$ for eutectic freeze crystallization. (Color figure online)

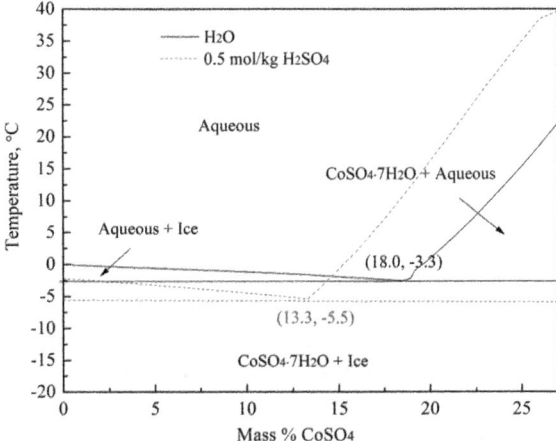

Fig. 5 The cooling curves of batch experiment for 20 wt% NiSO$_4$ in 0.5 mol/kg H$_2$SO$_4$ solution. (Color figure online)

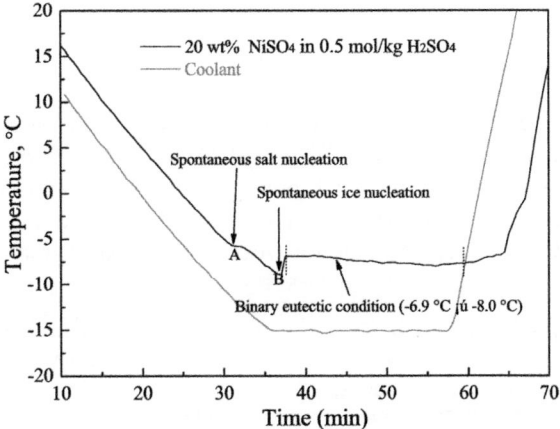

Electrolyte (MSE) model of OLI Stream Analyzer. The diagrams show the regions of stability for NiSO$_4$·7H$_2$O and CoSO$_4$·7H$_2$O as a function of concentration and temperature for EFC. As can be seen, the phase diagrams of nickel sulfate and cobalt sulfate in H$_2$O and 0.5 mol/kg H$_2$SO$_4$ at low temperatures are very similar. At the same temperature, the solubility of both NiSO$_4$ and CoSO$_4$ in dilute sulfuric acid solution is lower than that in water, and the eutectic temperature is also lower in the solution with 0.5 mol/L H$_2$SO$_4$, as illustrated in Figs. 3 and 4. The results show that H$_2$SO$_4$ can decrease the solubility of the metal sulfate salts and the binary eutectic temperature.

Experimental Results

The temperature cooling profile of the 0.5 mol/kg H_2SO_4 nickel sulfate solution as a function of time is presented in Fig. 5. The temperature fluctuations (points A and B) characterize the nucleation of crystals. The system reached salt nucleation at a temperature of –5.8 °C (point A) where the salt crystals were visible to the naked eyes. Upon further cooling, there was an increase in salt crystallization until point B was reached where ice also began to crystallize out. The temperature in the reactor suddenly increased to –6.9 °C due to the release of the crystallization enthalpy. After point B, simultaneous crystallization of ice and $NiSO_4·7H_2O$ proceeded. Under the binary eutectic condition, the temperature decreased slowly from –6.9 to –8.0 °C, due to the increase of H_2SO_4 concentration in the solution (caused by the crystallization of ice and salt), resulting in the decrease of the eutectic temperature of ice and $NiSO_4·7H_2O$.

The EFC experiment for the 0.5 mol/kg H_2SO_4 cobalt sulfate solution resulted in a similar temperature cooling profile (Fig. 6). Spontaneous salt nucleation and ice nucleation occurred at –5.9 and –8.8 °C, respectively. The eutectic temperature of ice and $CoSO_4·7H_2O$ decreased from –6.5 to –7.4 °C in 40 min under the binary eutectic condition.

XRD analysis results confirmed that the crystals obtained from batch experiments were $NiSO_4·7H_2O$ and $CoSO_4·7H_2O$ from batch experiments. Figure 7 shows the pictures and micrographs of $NiSO_4·7H_2O$ and $CoSO_4·7H_2O$ obtained. As can be seen, the $NiSO_4·7H_2O$ crystals have a spiculate or bar shape whereas the $CoSO_4·7H_2O$ crystals are irregular spherical or granular.

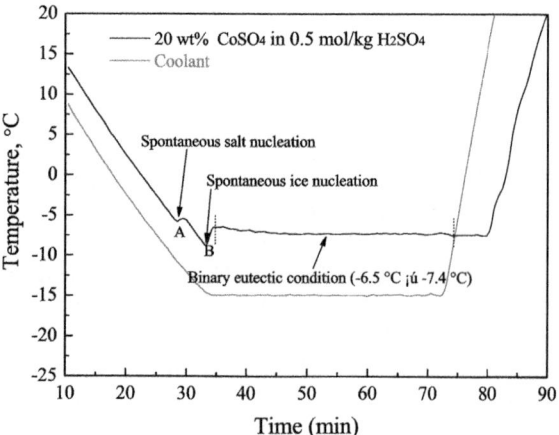

Fig. 6 The cooling curves of batch experiment for 20 wt% $CoSO_4$ in 0.5 mol/kg H_2SO_4 solution. (Color figure online)

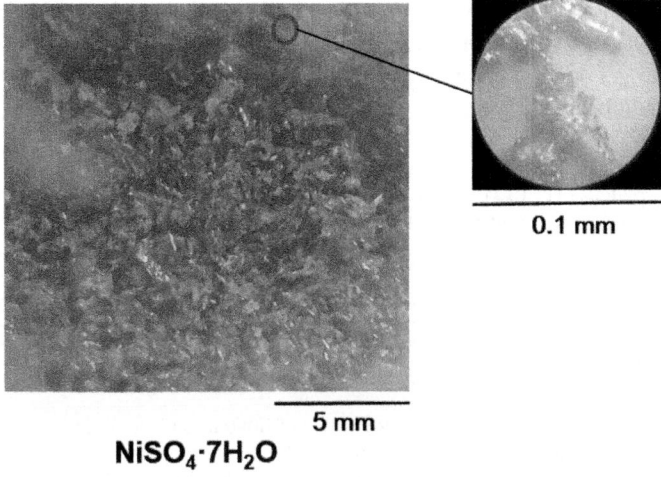

NiSO$_4$·7H$_2$O

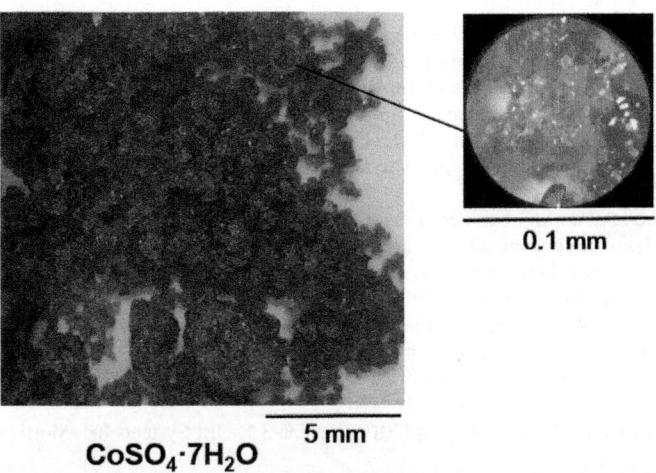

CoSO$_4$·7H$_2$O

Fig. 7 Crystals of NiSO$_4$·7H$_2$O and CoSO$_4$·7H$_2$O from batch experiments. (Color figure online)

Conclusion

This work demonstrates EFC is a promising technology for the recovery of NiSO$_4$·7H$_2$O and CoSO$_4$·7H$_2$O from acidic strip liquors in the recycling of Li-ion batteries. The binary phase diagrams from OLI Stream Analyzer indicate that NiSO$_4$·7H$_2$O and CoSO$_4$·7H$_2$O can crystallize along with ice from both H$_2$O and dilute H$_2$SO$_4$ solution using EFC, and the binary eutectic temperature decreases as the concentration of acid increases. The recovery of NiSO$_4$·7H$_2$O and CoSO$_4$·7H$_2$O from two synthetic acidic strip liquors was then tested to verify the feasibility. The crystals of $NiSO_4 \cdot 7H_2O$ and $CoSO_4 \cdot 7H_2O$ were obtained successfully. Further study is needed to get optimal parameters and test the actual strip liquors in the recycling of Li-ion batteries.

Acknowledgements This study was carried out within the Processes for Efficient Recycling of Lithium-Ion Batteries (PERLI) project with grant number 48228-1. The authors are indebted to Swedish Energy Agency for financial support.

References

1. Huang B, Pan Z, Su X, An L (2018) Recycling of lithium-ion batteries: recent advances and perspectives. J Power Sourc 399:274–286
2. Hu J, Zhang J, Li H, Chen Y, Wang C (2017) A promising approach for the recovery of high value-added metals from spent lithium-ion batteries. J Power Sourc 351:192–199
3. Peng C, Liu F, Lundström M (2019) Selective extraction of lithium (Li) and preparation of battery grade lithium carbonate (Li_2CO_3) from spent Li-ion batteries in nitrate system. J Power Sourc 415:179–188
4. Wang H, Friedrich B (2015) Development of a highly efficient hydrometallurgical recycling process for automotive Li–Ion batteries. J Sustain Metall 1:168–178
5. Chen X, Chen Y, Zhou T, Liu D, Hu H, Fan S (2015) Hydrometallurgical recovery of metal values from sulfuric acid leaching liquor of spent lithium-ion. Waste Manage 38:349–356
6. Zhang J, Hu J, Zhang W, Chen Y, Wang C (2018) Efficient and economical recovery of lithium, cobalt, nickel, manganese from cathode scrap of spent lithium-ion batteries. J Clean Prod 204:437–446
7. Virolainen S, Fini MF, Laitinen A, Sainio T (2017) Solvent extraction fractionation of Li-ion battery leachate containing Li, Ni, and Co. Sep Purif Technol 179:274–282
8. Ulrich J, Jones J (2004) Industrial crystallization: developments in research and technology. Chem Eng Res Des 82:1567–1570
9. Lu H, Wang J, Wang T, Wang N, Bao Y, Hao H (2017) Crystallization techniques in wastewater treatment: an overview of applications. Chemosphere 173:474–484
10. Lu X (2014) Novel applications of eutectic freeze crystallization. PhD Thesis, Technische Universiteit Delft, ISBN: 9789461863416
11. OLI systems Inc Stream Analyser (2010) Version 3.1, OLI Systems Inc, Morris Plains, New Jersey
12. Kang J, Senanayake G, Sohn J, Shin S (2010) Recovery of cobalt sulfate from spent lithium ion batteries by reductive leaching and solvent extraction with Cyanex 272. Hydrometallurgy 100:168–171

Recovery of Valuable Metals from End-of-Life Lithium-Ion Battery Using Electrodialysis

Ka Ho Chan, Monu Malik, and Gisele Azimi

Abstract A novel electrochemical separation process was developed to recover lithium from an end-of-life lithium-ion battery of an electric vehicle using an environmentally friendly and cost-effective process based on electrodialysis. Lithium, nickel, manganese, and cobalt were first extracted from the cathode active material of a spent lithium-ion battery through a hydrometallurgical leaching process using $H_2SO_4+H_2O_2$ leachant under the optimal operating conditions. After leaching, nickel, manganese, and cobalt were recovered as complex anions coupled with ethylenediaminetetraacetic acid chelating agent, whereas lithium was recovered as lithium hydroxide using electrodialysis. The results showed that almost 100% of lithium was separated from nickel, manganese, and cobalt. Future work is underway to improve and optimize the separation process.

Keywords Spent Lithium-Ion Batteries (LIBs) · Cathode active material · Electrodialysis · Separation · Lithium

Introduction

Lithium-ion batteries (LIBs) have been widely used in electronic devices, electric vehicles, and energy storage systems because of their high energy density, high voltage, long storage life, low self-discharge rate, and wide operating temperature range [1]. With the growing demands for LIBs, a serious shortage of lithium (Li) and cobalt (Co), and significant environmental issues, the sustainable recycling of postconsumer LIBs is imperative. Recycling can bring environmental and economic benefits, as it minimizes environmental pollution and provides an alternative route of strategic materials for LIB production.

K. H. Chan · M. Malik · G. Azimi (✉)
Laboratory for Strategic Materials, Department of Chemical Engineering and Applied Chemistry, 200 College Street, Toronto, ON M5S3E5, Canada
e-mail: g.azimi@utoronto.ca

G. Azimi
Department of Materials Science and Engineering, University of Toronto, 184 College Street, Toronto, ON M5S3E4, Canada

Traditional metal recycling processes from spent LIBs comprise pretreatment, metal extraction, and product separation [2]. During the product separation process, metals in the leachate can be separated and recovered through a series of selective precipitation and/or solvent extraction steps. Although these methods are efficient to recover the valuable metals from postconsumer LIBs, they face several drawbacks including complicated recycling routes, high reagent consumption, and high waste emission [3]. Therefore, more sustainable and green processes are desired to recover valuable metals from spent LIBs.

Electrodialysis is a promising alternative separation technique that is commonly used in wastewater treatment, purification of biological solutions, and demineralization of mixed feeds [4]. Electrodialysis has several advantages over solvent extraction and selective precipitation, such as low energy consumption, minimal hazardous waste production, and long membrane lifetime [5]. Electrodialysis is a membrane process in which ions with different charges are separated under the influence of an electrical potential difference. The most important components in the electrodialysis process are ion-exchange membranes, such as anion exchange membranes and cation exchange membranes, which are arranged alternatively between an anode and a cathode. Under the influence of applied potential, anions move towards the anode and cations move towards the cathode, passing through the membranes, which results in the separation of ions.

Most previous studies focused on using electrodialysis for the recovery of Li from primary sources (brines, seawater, and ores), and only a few studies have reported the use of electrodialysis to recover Li from secondary sources (spent LIBs). Currently, second-generation ($LiNi_xCo_yMn_{1-x-y}O_2$) cathode material has become the most dominant cathode type in LIBs over first-generation ($LiCoO_2$) cathode material due to its better performance in electric vehicles [2, 6], and no previous studies have reported the recovery of second-generation cathode material using electrodialysis.

In the current study, an innovative separation method using electrodialysis coupled with ethylenediaminetetraacetic acid (EDTA) was developed to separate Li from an end-of-life LIB of an electric vehicle. The speciation distribution behavior of EDTA with Li, Ni, Mn, and Co at various pH was studied using OLI software. All four metals were first extracted through a hydrometallurgical leaching process using $H_2SO_4+H_2O_2$ leachant under the optimal operating conditions. After leaching, the leachate with a desired amount of EDTA was fed to a lab-scale electrodialysis system and the feasibility of separating Li from a multi-metallic mixture was investigated.

Materials and Methods

Chemicals and Materials

The following reagents were employed in this work: concentrated sulfuric acid (ACS Reagent grade, 95.0–98.0 wt%, VWR), concentrated sodium hydroxide (ACS

Reagent grade, 50 wt%, VWR), hydrogen peroxide (ACS Reagent grade, 30 wt%, VWR), ethylenediaminetetraacetic acid (99.5%, EM Science), and deionized water (0.055 μS, Millipore).

Feed Solution Preparation and Electrodialysis Experiment

To prepare the feed solution, the battery pack from an electric vehicle was first discharged and dismantled, and the cathode active material was removed from the aluminum foils using ultrasonication. The cathode active material was leached in $H_2SO_4+H_2O_2$ leachant under the optimal operating conditions (temperature = 50 °C, H_2SO_4 concentration = 1.5 M, H_2O_2 concentration = 1.0 wt%, and liquid-to-solid ratio = 20 mL g^{-1}) in which ~100% extraction efficiency of Li, Ni, Mn, and Co was achieved [7]. Figure 1 shows the experimental apparatus for the separation of Li from spent LIBs using electrodialysis. The lab-scale electrodialysis stack was purchased from ElectroCell Inc. Platinum/titanium was used as anode and stainless steel 316 was used as cathode. The electrodialysis stack consisted of five compartments, namely feed, lithium, metal-EDTA complex, and electrode rinse compartments, divided by one anion exchange membrane (Neosepta® AMX), one cation

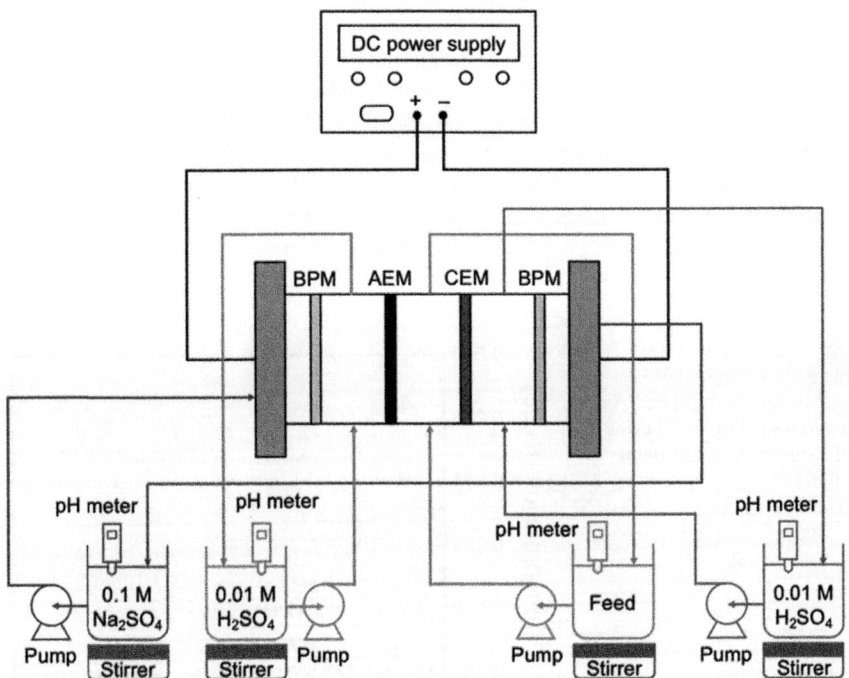

Fig. 1 Schematic diagram of the experimental setup of electrodialysis for metal separation. (Color figure online)

exchange membrane (Neosepta® CMX), and two bipolar membranes (Fumatech). The effective membrane area was 100 cm². The $H_2SO_4+H_2O_2$ leachate mixed with an appropriate amount of EDTA was supplied to the feed compartment, and 0.01 M sulfuric acid was supplied to both Li and metal-EDTA complex recovery compartments. Both anode and cathode compartments were fed by 0.1 M sodium sulfate solution. All solutions were supplied to the electrodialysis stack using pumps with a constant flow rate of 0.75 L min^{-1}. The pH of the solutions in each compartment was monitored by pH meters. A power supply was used to set up a constant voltage across the electrodes. During the experiment, samples were taken at different time intervals from all compartments (except electrode rinse compartment) and then diluted with 5 wt% HNO_3 to determine the metal concentrations by Inductively Coupled Plasma Optical Emission Spectrometry (ICP-OES, PerkinElmer Optima 8000).

Results

Principle of Recovery of Lithium from Spent LIBs Utilizing Electrodialysis Coupled with EDTA

The proposed method for separating Li from Ni, Mn, and Co in the $H_2SO_4+H_2O_2$ leachate by electrodialysis consisted of two steps. In the first step, the stoichiometric quantity of EDTA that is required to complex with Ni, Mn, and Co ions was added to the $H_2SO_4+H_2O_2$ leachate, and Ni, Mn, and Co ions formed complex anions with EDTA. The speciation distribution behavior of EDTA with Li, Ni, Mn, and Co as a function of pH was modeled using OLI software, as shown in Fig. 2. Dominant complex anions of NiY^{2-}, MnY^{2-}, and CoY^{2-} formed when pH was in the range of 6 and 11, whereas Li remained as free ions in this pH range.

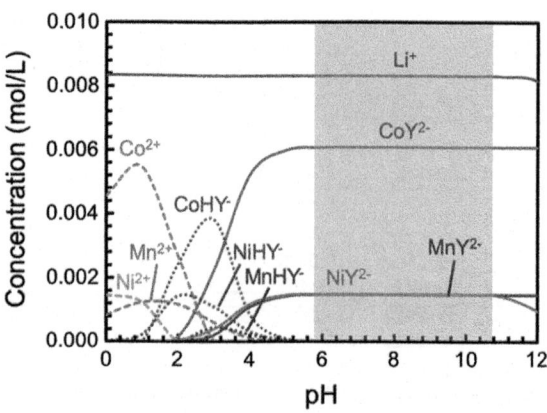

Fig. 2 Speciation distribution diagram of Li, Ni, Mn, and Co with EDTA at various pH modeled using OLI software. (Color figure online)

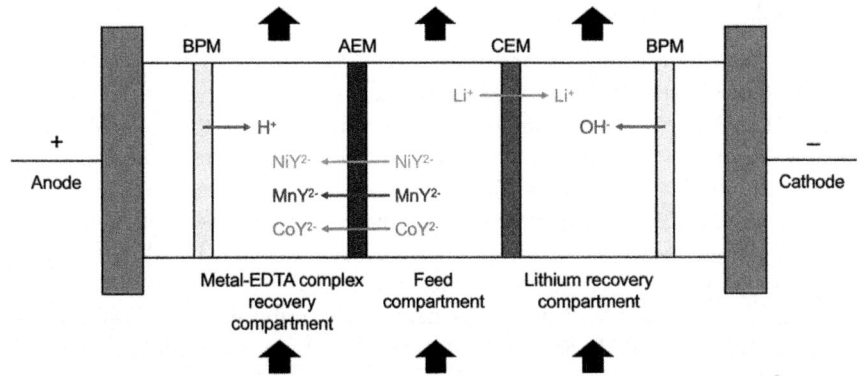

Fig. 3 Principle of Li recovery from spent LIBs utilizing electrodialysis coupled with EDTA. (Color figure online)

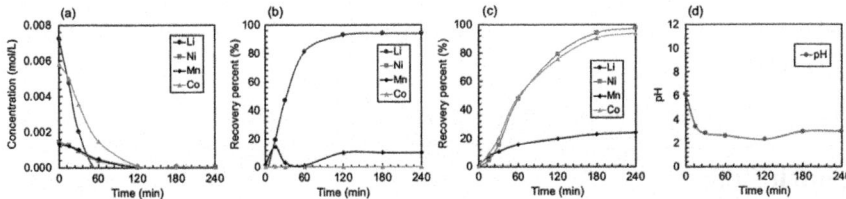

Fig. 4 Results of the preliminary test **a** Concentration changes with time in the feed compartment. **b** Recovery of Li in the Li recovery compartment. **c** Recovery of Ni, Mn, and Co in the metal-EDTA complex recovery compartment. **d** pH change in the feed compartment. (Color figure online)

Figure 3 shows the principle of recovery of Li from spent LIBs utilizing electrodialysis coupled with EDTA. The electrodialysis stack consisted of five compartments divided by one anion exchange membrane (AEM), one cation exchange membrane (CEM), and two bipolar membranes (BPM). When the electric potential was applied through the electrodialysis stack, Li ions passed through the cation exchange membrane to the Li recovery compartment and then combined with hydroxide ions transported from the bipolar membrane to produce lithium hydroxide (LiOH). On the other side, NiY^{2-}, MnY^{2-}, and CoY^{2-} ions passed through the anion exchange membrane to the metal-EDTA complex recovery compartment and then combined with hydrogen ions transported from the bipolar membrane to form a neutral species. Therefore, the separation of a mixture of anions and cations could be achieved by electrodialysis as a result of the charge difference.

Electrodialysis Results

Figure 4 shows the results of one of the preliminary tests. The initial feed solution was adjusted to pH 6 by 2 M NaOH solution, and the voltage applied to the electrodialysis stack was set at 18 V. The concentrations of all four metals in the feed compartment decreased with time, while that of the corresponding metals in the Li and metal-EDTA complex recovery compartments increased. The recovery of Li in the Li recovery compartment was 94.1%, while the recovery of Ni, Co, and Mn in the metal-EDTA complex recovery compartment was 97.2%, 93.8%, and 24.0%, respectively. However, it was observed that pH was decreasing during the process in the feed compartment and some Mn precipitates formed in the Li recovery compartment. The reason behind this result is that once pH dropped in the feed compartment, MnY^{2-} was converted to Mn^{2+}, which could pass through the cation exchange membrane and then combine with hydroxide ions transported from the bipolar membrane to form precipitates (as shown in Fig. 2). Future work is underway to improve and optimize this electrodialysis separation process using statistical analyses combined with experimental and fundamental investigations to elucidate the electrodialysis mechanism.

Conclusions

An electrochemical process for separating Li from the $H_2SO_4+H_2O_2$ leachate utilizing electrodialysis coupled with EDTA was proposed. Electrodialysis is an emerging green process with a promising future in the LIB industry over the solvent extraction and selective precipitation. In this work, when a stoichiometric amount of EDTA was added to the $H_2SO_4+H_2O_2$ leachate at pH 6, only Ni, Mn, and Co were combined with EDTA to form complex anions, while Li remained as free cations. After that, when the electrodialysis stack was operated at a constant voltage of 18 V, almost 100% of Li was separated from the other three metal ions. The challenge with this process was low Mn recovery due to precipitation. Future work is underway to improve the recovery of Mn, to optimize the electrodialysis separation, and to elucidate the electrodialysis mechanism.

References

1. Li L, Zhang X, Li M et al (2018) The recycling of spent lithium-ion batteries: a review of current processes and technologies. Electrochem Energy Rev 1:461–482. https://doi.org/10.1007/s41918-018-0012-1
2. Zheng X, Zhu Z, Lin X et al (2018) A mini-review on metal recycling from spent lithium ion batteries. Engineering 4:361–370. https://doi.org/10.1016/J.ENG.2018.05.018

3. Lv W, Wang Z, Cao H et al (2018) A critical review and analysis on the recycling of spent lithium-ion batteries. ACS Sustain Chem Eng 6:1504–1521. https://doi.org/10.1021/acssuschemeng.7b03811
4. Jiang C, Wang Y, Wang Q et al (2014) Production of lithium hydroxide from lake brines through electro-electrodialysis with bipolar membranes (EEDBM). Ind Eng Chem Res 53:6103–6112. https://doi.org/10.1021/ie404334s
5. Mosadeghsedghi S, Mortazavi S, Sauber ME (2018) The feasibility of separation of rare earth elements by use of electrodialysis. In: Davis BR, Moats MS, Wang S et al (eds) Extraction 2018. Springer International Publishing, Cham, pp 1831–1838
6. Gratz E, Sa Q, Apelian D, Wang Y (2014) A closed loop process for recycling spent lithium ion batteries. J Power Sources 262:255–262. https://doi.org/10.1016/J.JPOWSOUR.2014.03.126
7. Chan KH, Malik M, Anawati J, Azimi G (2020) Recycling of end-of-life lithium-ion battery of electric vehicles. In: Azimi G, Forsberg K, Ouchi T et al (eds) Rare Metal Technology 2020. Springer International Publishing, Cham, pp 23–32

Lithium Adsorption Mechanism for Li$_2$TiO$_3$

Rajashekhar Marthi and York R. Smith

Abstract Layered H$_2$TiO$_3$ has shown to be a promising selective lithium adsorbent to extract lithium from brine solutions. Despite the promising performance of these materials, the lithium adsorption mechanism of layered H$_2$TiO$_3$ is still not properly understood. It is currently accepted that lithium adsorption takes place via Li$^+$-H$^+$ ion-exchange reaction without involving any breakage of chemical bonds. However, in this study we show that Li$^+$-H$^+$ ion exchange involves the breaking of surface O–H bonds present in the HTi$_2$ layers along with the formation of O-Li bonds. Using FTIR and Raman spectroscopy, we also show that the isolated surface hydroxyls are actively involved in lithium ion exchange compared to hydrogen-bonded surface hydroxyl groups, which are present in the interlayer spacings. This newly proposed mechanism also explains the lower observed adsorption capacity from theoretical values.

Keywords Lithium · Surface hydroxyls · Ion exchange

Introduction

Lithium is a critical metal which is used for a number of applications including battery and energy applications, ceramics, and lubrication, to name a few. Lithium demand is expected to reach around 200–1000 kt by 2050 due to an increasing number of electric vehicles [1]. To meet this demand, lithium-enriched brines can be a potential "green resource" for lithium supply in the global market. Ion sieve technology has gained attention in the past decade to concentrate lithium from brines [2, 3]. Ion sieves are lithium-selective adsorbents that have high selectivity towards lithium over other alkali and alkaline earth metal ions present in the brines (e.g., separation factor > 100) [4, 5]. The most commonly used ion sieves for lithium concentration from the brine solutions are λ-MnO$_2$ and layered H$_2$TiO$_3$ [6–9]. Spinel structured λ-MnO$_2$ ion sieve contains (1 × 3) tunnel structures which are large enough to allow lithium

---

R. Marthi · Y. R. Smith (✉)
Department of Materials Science & Engineering, University of Utah, Salt Lake City, UT 84112, USA
e-mail: york.smith@utah.edu

intercalation and narrow enough to prevent the intercalation of larger sized ions [10, 11]. However, Mg^{2+} which has the same ionic size compared to lithium is not able to intercalate due to its large dehydration enthalpy [12]. Ammundsen et al. [13] showed that lithium adsorption in λ-MnO_2 is an ion-exchange reaction which occurs via the breaking of surface OH groups and the formation of O-Li groups. Despite its high selectivity, the λ-MnO_2 ion sieve has poor recyclability due to manganese dissolution during the desorption process in dilute acids [14, 15].

Recently, layered H_2TiO_3 has gained attention as a lithium-selective adsorbent [7]. Layered H_2TiO_3 is obtained by the delithiation of monoclinic β-Li_2TiO_3 in dilute acids. Monoclinic β-Li_2TiO_3 has a layered structure where layers of Li atoms are sandwiched between $LiTi_2$ layers. Delithiation of Li_2TiO_3 gives layered H_2TiO_3 which contains layers of HTi_2 layers stacked along the C axis. Due to stable titanium, and a higher theoretical adsorption capacity (~128 mg/g), H_2TiO_3 is a better option than the λ-MnO_2 ion sieve. However, the actual maximum lithium adsorption capacity of H_2TiO_3 under optimal conditions is less than 50 mg/g [7, 16], and its weaker acidic strength affects its performance in actual salt lake brines [17]. Also, the lithium adsorption mechanism of layered H_2TiO_3 is not well understood. Therefore, in this study we have used thermogravimetric and spectroscopic analytical methods to establish a lithium adsorption mechanism in layered H_2TiO_3. Our results show that lithium adsorption occurs via the breaking of surface hydroxyl bonds and the formation of O-Li bonds. Our proposed mechanism contradicts the currently accepted ion-exchange mechanism which was first proposed by He et al. and suggests that after adsorption, lithium ions are held electrostatically in the interlayer spacings between HTi_2 layers [16]. The study suggested that this does not involve any breaking of covalent bonds. The He study also states that HTi_2 layers are not involved in lithium adsorption process. However, we show that isolated hydroxyl groups present in the HTi_2 layers are more active towards lithium ion exchange compared to the hydrogen-bonded OH groups in the interlayer spacings. Using our newly established mechanism, we are able to explain why the theoretical maximum adsorption capacity is not achieved by H_2TiO_3 ion sieves under optimal conditions.

Experimental Methods

Synthesis and Application of Layered H_2TiO_3 Ion Sieve

A monoclinic Li_2TiO_3 ion sieve precursor was synthesized by performing a solid-state calcination reaction between Li_2CO_3 and TiO_2. A mixture of Li_2CO_3 and TiO_2 (Li/Ti molar ratio $=$ 2) was prepared by grinding the reagents in a mortar using a pestle. The mixture was then transferred into a ceramic crucible and calcined in a box furnace at 700 °C for 4 h to obtain β-Li_2TiO_3. The ion sieve was obtained by magnetically stirring β-Li_2TiO_3 in 0.2 M HCl for 24 h (S/L ratio $=$ 1 g / 200 mL). The ion sieves were separated from the acid using vacuum filtration and were later washed with plenty of deionized water. The filter cake was dried in an oven at 60 °C overnight.

Lithium adsorption studies were performed by magnetically stirring 0.1 g of the layered H_2TiO_3 in 100 mL of 0.1 M LiCl buffered solution (pH = 9.50, adjusted using ammoniacal buffer). The supernatant was separated using a syringe filter and was later analyzed using ICP-MS (Agilent 3500 CE). The lithium-enriched adsorbent ($Li_xH_{2-x}TiO_3$) was washed with plenty of deionized water to remove excess buffer solution and later dried in an oven at 60 °C overnight. The measured value of lithium adsorption capacity was ~ 40 mg/g.

Characterization

X-ray diffraction was performed using Cu K_α irradiation to study the crystalline structure present in ion sieves (MiniFlex Rigaku, step size = 0.01°, scan rate = 1°/min). The relative amount of surface hydroxyl groups present in the ion sieves before and after lithium adsorption were quantified by performing thermogravimetry analysis (TGA) in nitrogen atmosphere from 25 to 600 °C using a DSC-TGA 650 (TA instruments). Corresponding differential scanning calorimetric (DSC) data were used to identify the different stages of weight loss in the ion sieves. The role of surface hydroxyl groups in lithium adsorption studies was studied by performing FTIR spectroscopy (Thermo Scientific Nicolet iN10mx). Reflectance measurements were taken from 700 to 4000 cm^{-1} with an exposure time of 5 s. The deconvolution of the peaks was performed using a combination of Gaussian and Lorentzian peak fitting methods (Origin Lab software). The breaking and formation of O-Li bonds during delithiation and adsorption were studied using Raman spectroscopy (Thermo Scientific DXR Raman Microscope, 532 nm laser, exposure time 4 s, high-resolution grating up to 1846 cm^{-1}).

Results

X-ray Diffraction

The β-Li_2TiO_3 ion sieve precursor is obtained as a calcination product of Li_2CO_3 and TiO_2. The X-ray diffractogram in Fig. 1 shows that the diffraction peaks belong to β-Li_2TiO_3 (ICDD card no. 33–0831), which occupies a monoclinic crystal structure. Upon delithiation in dilute acid, a layered H_2TiO_3 ion sieve is obtained. The peak corresponding to (002) plane shifts towards higher 2θ values indicating that after delithiation, there is a shrinkage in the unit cell volume. This shrinkage in the cell volume is due to the replacement of H^+ ions with Li^+ ions [5]. It can be observed that after delithiation, there is a significant change in the diffraction patterns of H_2TiO_3

Fig. 1 X-ray diffraction of Li$_2$TiO$_3$, H$_2$TiO$_3$, and Li$_x$H$_{2-x}$TiO$_3$. It can be seen that after delithiation, certain diffraction peaks present in Li$_2$TiO$_3$ disappear due to weak Li-O bonds. These disappeared peaks do not reappear even after lithium adsorption. (Color figure online)

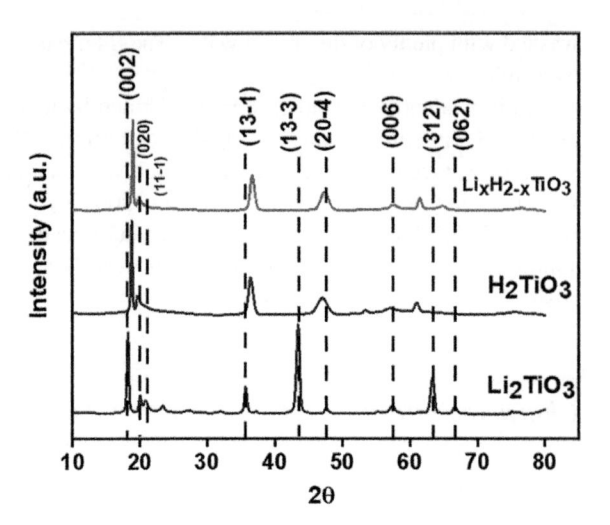

compared to the β-Li$_2$TiO$_3$ precursor material. The diffraction peaks corresponding to (020), (13–3), (006), (312), and (062) planes disappear. Tarakina et al. [18] and Yu et al. [19] have attributed the disappearance of these peaks to a high concentration of stacking faults along the C axis. Moreover, Zhang et al. [20] showed using first principle calculations that the disappeared peaks correspond to the planes having low O-Li bond energy. The peaks corresponding to (002), (13–1), and (–204) planes do not disappear even after the delithiation of Li$_2$TiO$_3$ indicating that H$_2$TiO$_3$ maintains the layered arrangement. Therefore, the delithiation of β-Li$_2$TiO$_3$ ion sieve precursor leads to the formation of layered H$_2$TiO$_3$ with a high concentration of stacking faults. It is interesting to note that after lithium adsorption, the disappeared peaks do not reappear in Li$_x$H$_{2-x}$TiO$_3$. This indicates that lithium adsorption on H$_2$TiO$_3$ does not involve any structural changes. In brief, the delithiation of Li$_2$TiO$_3$ results in structural changes which arise due to defect chemistry, whereas lithium adsorption on H$_2$TiO$_3$ does not lead to any new structural changes.

Thermogravimetric Analysis

Figure 2 shows the weight loss and heat flow profile of layered H$_2$TiO$_3$ before and after lithium adsorption. Based on the heat flow profile, the weight losses can be classified into two regions. Region I corresponds to the weight loss due to physically adsorbed water molecules [21]. It can be observed that in layered H$_2$TiO$_3$, physically adsorbed water molecules are completely removed at 230 °C, whereas physically adsorbed water molecules are completely removed in Li$_x$H$_{2-x}$TiO$_3$ at 276 °C. Due to a stronger interaction between physically adsorbed water molecules and Li$_x$H$_{2-x}$TiO$_3$, there is a shift in Region I towards higher temperatures. Region II corresponds to the weight loss due to the removal of surface hydroxyl groups [22]. It can be observed

Fig. 2 TGA–DSC profile of H_2TiO_3 and $Li_xH_{2-x}TiO_3$. The weight loss due to dehydroxylation in H_2TiO_3 is greater than $Li_xH_{2-x}TiO_3$ (2.96 vs. 1.29 wt%) indicating that surface hydroxyl groups are involved in lithium adsorption. (Color figure online)

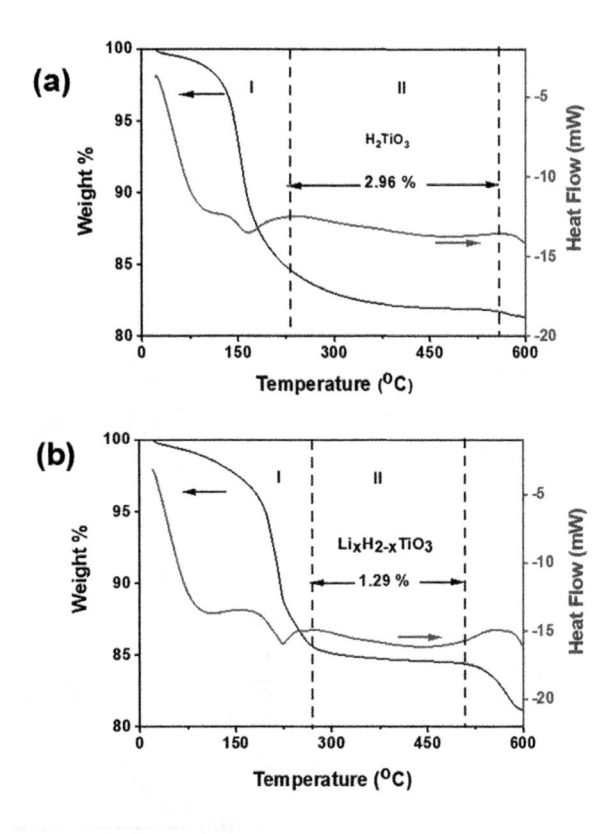

from Fig. 2 that in layered H_2TiO_3, the weight loss due to dehydroxylation is 2.96 wt%, whereas in $Li_xH_{2-x}TiO_3$ the weight loss due to dehydroxylation is only 1.29 wt%. This indicates that after lithium adsorption, the relative amount of hydroxyl groups decreases indicating that surface hydroxyl groups are involved in the Li^+-H^+ ion-exchange reaction.

Spectroscopy Results

The stretching vibrations of hydroxyl groups were analyzed using FTIR. Figure 3 shows that layered H_2TiO_3 contains a broad peak between 2000 and 4000 cm^{-1} with a narrow shoulder peak at around 3500 cm^{-1}. The deconvolution of the spectrum shows that three types of hydroxyl groups are present in layered H_2TiO_3. Peak '1' corresponds to hydroxyl groups which interact with each other via hydrogen bonds and peak '2' corresponds to the stretching vibrations of OH groups from physically adsorbed water [23]. Peak '3' is associated with isolated surface hydroxyl groups which do not interact with each other [23]. However, after lithium adsorption, the intensity of ratio of peak '3' to peak '1' decreases from 0.53 to 0.35 indicating

Fig. 3 FTIR analysis shows the absorbance of surface hydroxyl groups before and after lithium adsorption. Peak '1' represents hydrogen-bonded OH groups, peak '2' represents stretching vibrations of OH groups from physically adsorbed water molecules, and peak '3' represents the stretching vibrations of isolated OH groups. After lithium adsorption, the ratio of absorbance of peak '3' to peak '1' significantly decreases indicating that isolated OH groups are actively involved in the lithium adsorption compared to hydrogen-bonded OH groups. (Color figure online)

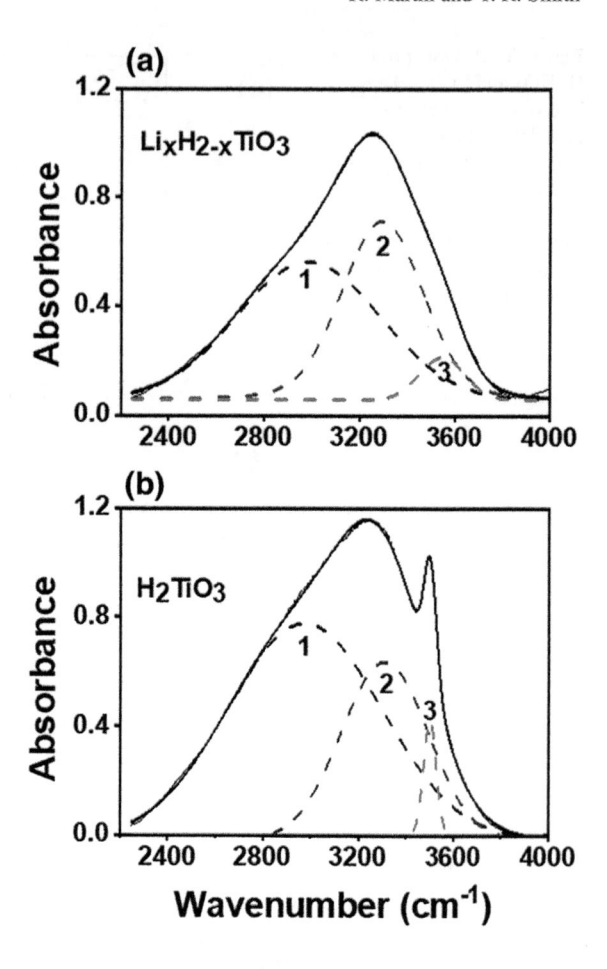

that isolated surface hydroxyl groups are more actively involved in lithium adsorption than hydrogen-bonded hydroxyl groups. During the delithiation of β-Li$_2$TiO$_3$, lithium atoms from Li layers are eluted first, followed by lithium elution from LiTi$_2$ layers. Denisova et al. [23] showed that during delithiation, isolated surface OH groups are formed on HTi$_2$ layers and hydrogen-bonded OH groups are formed in the interlayer spacings. Therefore, FTIR spectroscopy studies show that lithium adsorption occurs in HTi$_2$ layers, which contradicts the currently accepted mechanism which states that HTi$_2$ layers are not involved in lithium adsorption.

Figure 4 shows the Raman spectra of Li$_2$TiO$_3$, H$_2$TiO$_3$ before and after lithium adsorption. The bands corresponding to 250–420 cm^{-1} are from O-Li stretching and bending vibrations (represented as dashed lines) [24]. It can be observed that after the delithiation of Li$_2$TiO$_3$, the bands corresponding to O-Li stretching and bending vibrations disappear. After lithium adsorption, these bands reappear indicating that O-Li bond breakage takes place during the delithiation, and O-Li bonds are formed after lithium adsorption. It should also be noted that the bands at wavenumbers

Fig. 4 Raman spectra show the stretching and bending vibration of O-Li bonds. During the delithiation of Li$_2$TiO$_3$, the O-Li bands disappear in H$_2$TiO$_3$ indicating breaking of O-Li bonds during delithiation. The O-Li bands reappear after lithium adsorption indicating that O-Li bond formation takes place. (Color figure online)

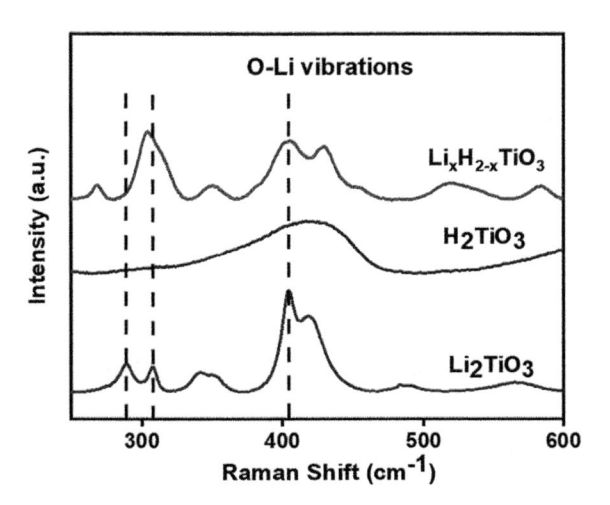

greater than 420 cm^{-1} correspond to the Ti–O stretching or bending vibrations. It is also interesting to note that due to decreases in the intensity of O-Li vibrations after delithiation, the doublet band at 400–420 cm^{-1} in Li$_2$TiO$_3$ transforms into a singlet band in H$_2$TiO$_3$. After lithium adsorption, this doublet band reappears due to the formation of O-Li bonds. It is also interesting to note that the O-Li bands in Li$_x$H$_{2-x}$TiO$_3$ are shifted towards lower wavenumbers compared to the Li$_2$TiO$_3$ ion sieve precursor.

In an ideal stoichiometric Li$_2$TiO$_3$, 75% of lithium atoms are in Li layers and 25% of lithium is in LiTi$_2$ layers. Therefore, after delithiation, 75% of hydrogen-bonded OH groups and 25% of isolated OH groups are present in the H$_2$TiO$_3$. Based on the spectroscopic data, isolated OH groups are more actively involved in lithium adsorption compared to hydrogen-bonded OH groups. Since most of the hydroxyl groups are hydrogen bonded, the lithium adsorption capacity and its rate of lithium adsorption is mainly limited by the number of isolated OH groups. The isolated hydroxyl groups are essentially bridged hydroxyl groups formed by Ti-O-Ti bonds which are present in the HTi$_2$ layers. The Li$^+$-H$^+$ ion exchange can be represented as

$$\text{TiO} - \text{H (s)} + \text{Li}^+(\text{aq}) = \text{TiO} - \text{Li (s)} + \text{H}^+(\text{aq}) \quad (1)$$

The role of HTi$_2$ in lithium adsorption was also discussed by Wang et al., where they showed that higher lattice oxygen content in HTi$_2$ layers favors lithium adsorption capacity [25–27]. Therefore, by controlling the concentration of surface hydroxyl groups via point or stacking fault defects, lithium adsorption properties can be improved.

Conclusion

Current lithium adsorption mechanisms state that lithium adsorption occurs via ion exchange which involves electrostatic interactions with no chemical bond breaking and formation. However, in this work, we showed that lithium adsorption on H_2TiO_3 involves the breaking of surface hydroxyl groups and the formation of O-Li bonds. We also showed using spectroscopic studies that isolated surface hydroxyl groups are more actively involved in lithium adsorption than hydrogen-bonded hydroxyl groups. We also explained the reason for lower adsorption capacity than its theoretical maximum. This work also shows that lithium adsorption properties can be improved by controlling the concentration of isolated surface hydroxyl groups.

References

1. Speirs J, Contestabile M, Houari Y, Gross R (2014) The future of lithium availability for electric vehicle batteries. Renew Sustain Energy Rev 35:183–193. https://doi.org/10.1016/j.rser.2014.04.018
2. Xu X, Chen Y, Wan P, Gasem K, Wang K, He T, Adidharma H, Fan M (2016) Extraction of lithium with functionalized lithium ion-sieves. Prog Mater Sci 84:276–313. https://doi.org/10.1016/j.pmatsci.2016.09.004
3. Safari S, Lottermoser BG, Alessi DS (2020) Metal oxide sorbents for the sustainable recovery of lithium from unconventional resources. Appl Mater Today 19:100638. https://doi.org/10.1016/j.apmt.2020.100638
4. Luo X, Zhang K, Luo J, Luo S, Crittenden J (2016) Capturing Lithium from wastewater using a fixed bed packed with 3-D MnO_2 ion cages. Environ Sci Technol 50:13002–13012. https://doi.org/10.1021/acs.est.6b02247
5. Marthi R, Smith YR (2019) Selective recovery of lithium from the Great Salt Lake using lithium manganese oxide-diatomaceous earth composite. Hydrometallurgy 186:115–125. https://doi.org/10.1016/j.hydromet.2019.03.011
6. Lawagon CP, Nisola GM, Mun J, Tron A, Torrejos REC, Seo JG, Kim H, Chung WJ (2015) Adsorptive Li+ mining from liquid resources by H_2TiO_3: Equilibrium, kinetics, thermodynamics, and mechanisms. J. Ind Eng Chem 35:347–356. https://doi.org/10.1016/j.jiec.2016.01.015
7. Chitrakar R, Makita Y, Ooi K, Sonoda A (2014) Lithium recovery from salt lake brine by H_2TiO_3. Dalt Trans 43:8933–8939. https://doi.org/10.1039/c4dt00467a
8. Zhang QH, Li SP, Sun SY, Yin XS, Yu JG (2009) Lithium selective adsorption on 1-D MnO_2 nanostructure ion-sieve. Adv Powder Technol 20:432–437. https://doi.org/10.1016/j.apt.2009.02.008
9. Hong HJ, Park IS, Ryu T, Ryu J, Kim BG, Chung KS (2013) Granulation of $Li_{1.33}Mn_{1.67}O_4$ (LMO) through the use of cross-linked chitosan for the effective recovery of Li+ from seawater. Chem Eng J 234:16–22. doi:https://doi.org/10.1016/j.cej.2013.08.060
10. Devaraj S, Munichandraiah N (2008) Effect of crystallographic structure of MnO_2 on its electrochemical capacitance properties. J Phys Chem C 112:4406–4417. https://doi.org/10.1021/jp7108785
11. H, Feng KO, Kanoh Q (1999) Manganese oxide porous crystal. J Mater 9:319–333. doi:https://doi.org/10.1039/a805369c
12. Marcus Y (1991) Thermodyn Solvat Ions 87:2995–2999

13. Ammundsen B, Jones DJ, Rozière J, Berg H, Tellgren R, Thomas JO (1998) Ion exchange in manganese dioxide spinel: proton, deuteron, and lithium sites determined from neutron powder diffraction data. Chem Mater 10:1680–1687. https://doi.org/10.1021/cm9800478
14. Liu DF, Sun SY, Yu JG (2019) $Li_4Mn_5O_{12}$ desorption process with acetic acid and Mn dissolution mechanism. J Chem Eng Japan 52:274–279. https://doi.org/10.1252/jcej.18we209
15. Ji ZY, Zhao MY, Zhao YY, Liu J, Peng JL, Yuan JS (2017) Lithium extraction process on spinel-type $LiMn_2O_4$ and characterization based on the hydrolysis of sodium persulfate. Solid State Ionics 301:116–124. https://doi.org/10.1016/j.ssi.2017.01.018
16. He G, Zhang L, Zhou D, Zou Y, Wang F (2015) The optimal condition for H_2TiO_3–lithium adsorbent preparation and Li+ adsorption confirmed by an orthogonal test design. Ionics (Kiel). 21:2219–2226. https://doi.org/10.1007/s11581-015-1393-3
17. Marthi R, Smith YR (2020) Application and limitations of a H_2TiO_3–diatomaceous earth composite synthesized from titania slag as a selective lithium adsorbent. Sep Purif Technol, 117580. doi:https://doi.org/10.1016/j.seppur.2020.117580
18. Tarakina NV, Neder RB, Denisova TA, Maksimova LG, Baklanova YV, Tyutyunnik AP, Zubkov VG (2010) Defect crystal structure of new $TiO(OH)_2$ hydroxide and related lithium salt Li2TiO3. Dalt Trans 39:8168–8176. https://doi.org/10.1039/c0dt00354a
19. Yu CL, Wang F, Cao SY, Gao DP, Hui HB, Guo YY, Wang DY (2015) The structure of H_2TiO_3—a short discussion on "lithium recovery from salt lake brine by H_2TiO_3." Dalt Trans 44:15721–15724. https://doi.org/10.1039/c4dt03689a
20. Zhang L, Zhou J, He G, Zhou D, Tang D, Wang F (2018) Extraction difficulty of lithium ions from various crystal planes of lithium titanate. J Wuhan Univ Technol Mater Sci Ed 33:1086–1091. https://doi.org/10.1007/s11595-018-1939-0
21. Zhuravlev LT (2000) The surface chemistry of amorphous silica. Zhuravlev model, Colloids Surfaces A Physicochem. Eng Asp 173:1–38. https://doi.org/10.1016/S0927-7757(00)00556-2
22. Mueller R, Kammler HK, Wegner K, Pratsinis SE (2003) OH surface density of SiO_2 and TiO_2 by thermogravimetric analysis. Langmuir 19:160–165. https://doi.org/10.1021/la025785w
23. Denisova TA, Maksimova LG, Polyakov EV, Zhuravlev NA, Kovyazina SA, Leonidova ON, Khabibulin DF, Yur'eva EI (2006) Metatitanic acid: synthesis and properties. Russ J Inorg Chem 51:691–699. doi:https://doi.org/10.1134/S0036023606050019
24. Nakazawa T, Naito A, Aruga T, Grismanovs V, Chimi Y, Iwase A, Jitsukawa S (2007) High energy heavy ion induced structural disorder in Li_2TiO_3. J Nucl Mater 367–370 B:1398–1403. doi:https://doi.org/10.1016/j.jnucmat.2007.04.003
25. Wang S, Zheng S, Wang Z, Cui W, Zhang H, Yang L, Zhang Y, Li P (2018) Superior lithium adsorption and required magnetic separation behavior of iron-doped lithium ion-sieves. Chem Eng J 332:160–168. https://doi.org/10.1016/j.cej.2017.09.055
26. Wang S, Li P, Cui W, Zhang H, Wang H, Zheng S, Zhang Y (2016) Hydrothermal synthesis of lithium-enriched β-Li_2TiO_3 with an ion-sieve application: excellent lithium adsorption. RSC Adv 6:102608–102616. https://doi.org/10.1039/c6ra18018c
27. Wang S, Zhang M, Zhang Y, Zhang Y, Qiao S, Zheng S (2019) Hydrometallurgy High adsorption performance of the Mo-doped titanium oxide sieve for lithium ions. Hydrometallurgy 187:30–37. https://doi.org/10.1016/j.hydromet.2019.05.004

Study on the Production of Lithium by Aluminothermic Reduction Method

Huimin Lu and Neale R. Neelameggham

Abstract At present, the global production of 45,000 tons of lithium is obtained by molten salt electrolysis using lithium chloride as a raw material. Due to the release of chlorine gas in the production process, the environment is seriously polluted and the energy consumption is high. In this paper, lithium carbonate is used as a raw material, and aluminum powder is used as a reducing agent to extract metal lithium in a vacuum reduction furnace by aluminothermic reduction. The lithium recovery rate is 85% and the purity of metal lithium reaches 99.5%. In the laboratory, a continuous vacuum lithium reduction furnace was developed. No harmful gases were released during the entire production process. It was environmentally friendly. The production of lithium by aluminothermic reduction is a very promising method. The continuous lithium vacuum reduction furnace is the key equipment of this process, which lays the foundation for future industrial application.

Keywords Aluminothermic reduction · Vacuum reduction furnace · Aluminum metal powder · Lithium carbonate · Lithium

Introduction

At present, consumer electronic products represented by electric vehicles, notebook computers, and mobile phones have strong demand for metal lithium batteries, which is the main driving force for the growth of metal lithium demand [1, 2]. With the advancement and development of science and technology, people's requirements for environmental protection are getting higher and higher, which promotes the development of electric vehicles, electric bicycles, and new energy vehicles. Pb-acid batteries are heavy, inconvenient, and easy to cause traffic accidents. Therefore,

---

H. Lu (✉)
Beijing Oufei Jintai Technology Co., Ltd, No. 12, Xingyuan Street, Miyun Economic Development Zone, Beijing 101500, China
e-mail: lhm0862002@aliyun.com

N. R. Neelameggham
IND LLC, 9859 Dream Circle, South Jordan, UT 84095, USA
e-mail: neelameggham@gmail.com

© The Minerals, Metals & Minerals Society 2021
G. Azimi et al. (eds.), *Rare Metal Technology 2021*, The Minerals, Metals & Materials Series, https://doi.org/10.1007/978-3-030-65489-4_4

the replacement of lead-acid batteries by metal lithium batteries will further accelerate in the future, especially the popularization of new energy vehicles, which will greatly stimulate the demand for lithium batteries.

In addition to being an energy material, lithium has a large number of other important applications. When lithium is added to molten metal or alloy, it will react with impurities such as hydrogen, oxygen, sulphur, nitrogen, and other impurities in the metal or alloy to form compounds with low density and low melting point, which can not only remove these gases, but also change the metal denser and can eliminate bubbles and other defects in the metal, thereby improving the grain structure of the metal and improving the mechanical properties of the metal. In terms of alloys, lithium is easily fused with any metal except iron. As a component of light alloys, ultra-light alloys, wear-resistant alloys, and other non-ferrous alloys, lithium can greatly improve the properties of the alloy. For example, lithium-magnesium alloy is a high-strength lightweight alloy. It not only has good thermal conductivity, electrical conductivity, and ductility, but also has the characteristics of corrosion resistance, abrasion resistance, good impact resistance, and resistance to high-speed particle penetration. It is known as "Tomorrow's Aerospace Alloy," widely used in aerospace, national defence, and military industries. As the world's requirements for lightweight structural materials, weight reduction and energy saving, environmental protection, and sustainable development continue to increase, lithium-magnesium alloys will be applied to areas where lightweight structural materials are more needed. The alloys formed by adding lithium to beryllium, zinc, copper, silver, cadmium, and boron are not only tougher or harder, but also increase tensile strength and elasticity. The lithium content in these alloys varies from a few thousandths to a few percent [3].

In the nuclear industry, after being bombarded by neutrons, lithium isotopes will split into tritium and helium. After the "deuterium–tritium" fusion reaction, in addition to the formation of a helium nucleus, there is an excess medium containing a lot of energy. Therefore, it is only necessary to maintain the concentration of lithium nuclei in the nuclear fusion reaction system to achieve "deuterium–tritium" nuclear fusion and maintain its continued reaction. Therefore, lithium is also known as a "high energy metal." In addition, due to its large heat capacity, wide liquidus temperature range, high thermal conductivity, low viscosity, and low density, metallic lithium is used as a coolant in nuclear fusion or nuclear fission reactors [4].

There are two main types of lithium in nature. One is in the form of lithium containing ores such as spodumene, lepidolite, and permeate feldspar, and the other is in the form of lithium ions in salt lake brines, underground brine, and seawater. At present, the mainstream process of industrial production of metal lithium is molten salt electrolysis and vacuum reduction. The molten salt electrolysis method uses lithium chloride (LiCl) as the raw material and potassium chloride as the electrolyte to play the role of stability, cooling, and conductivity. The electrolysis temperature is about 450–500 °C. The disadvantage of this method is that the lithium in the spodumene is first extracted into lithium carbonate and then converted into lithium chloride. The production cost is high. For every ton of lithium produced, 5 tons of chlorine will be generated, which will seriously pollute the air. The vacuum reduction method for lithium smelting also uses lithium carbonate or lithium hydroxide monohydrate as raw materials, but the existing vacuum reduction method for lithium smelting equipment has the problems of the incapability of continuous production and low labour efficiency [5–7].

In view of the above shortcomings, the purpose of this paper is to provide a continuous vacuum lithium reduction furnace and a method for lithium production. In this paper, lithium carbonate is used as a raw material, and aluminum powder is used as a reducing agent to extract metal lithium in a continuous vacuum reduction furnace by aluminothermic reduction.

Materials and Experimental Equipment

Raw Materials

CaO: industrial grade (Beijing Chemical Reagent Factory), purity >98.0%; Al powder: Industrial grade (Beijing Chemical Reagent Factory), purity >99.0%; Argon: >99.999% (Miyun Gas Manufacturing Co. Ltd.); Li_2CO_3: industrial grade, purity >99.0%.

Laboratory Experiments Equipment

Laboratory experiments equipment for producing lithium were conducted in a 120 kW continuous reduction of lithium test furnace, as pictured in Fig. 1. Process conditions were as follows: the reaction is performed in semi-continuous feeding charge 30 kg each within the lithium test furnace that operates around 1200°C and a vacuum of 10–20 Pa over a two- to four-hour period.

Schematic Diagram of Laboratory Equipment

1—reduction chamber, 2—induction heating device, 3—heating power supply, 4—electrical control system, 5—feeding chamber, 6—moving device, 7—isolation valve 4, 8—crucible lifting device, 9—isolation valve 3, 10—discharge chamber, 11—graphite crucible, 12—hydraulic system, 13—water cooling system, 14—vacuum unit, 15—isolation valve 2, 16—isolation valve 1, 17—crystallizer, 18—isolation valve 5, 19—isolation valve 6, 20—isolation valve 7, 21—metal lithium tank.

Fig. 1 120 kW continuous reduction of lithium test furnace. (Color figure online)

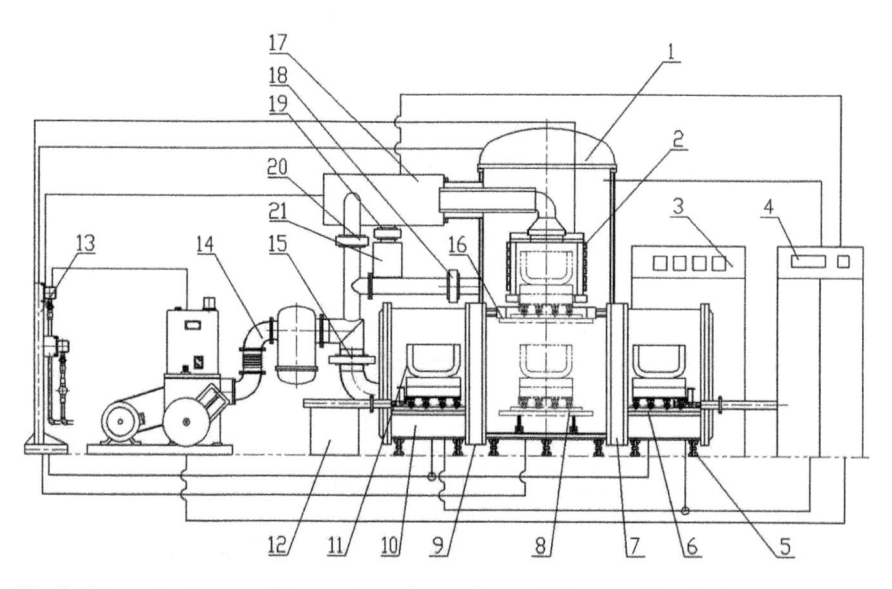

Fig. 2 Schematic diagram of the structure of a continuous lithium smelting device

Metal Lithium Purity and Impurity Analysis

Take the reduced product lithium in a dry argon-filled glove box. After cutting off the surface oxide layer, quickly weigh 3–6 g and place it in a polytetrafluoroethylene beaker. Then take the sample beaker out of the glove box and put it in a desiccator with water at the bottom to fully oxidize for several days. Wet it with secondary water first, then dissolve it with concentrated hydrochloric acid, transfer it into a volumetric flask, use atomic absorption spectrophotometer to analyze lithium content, and use inductively coupled plasma atomic emission spectrometer to determine impurity content (sodium, potassium, magnesium, calcium, aluminum, and iron).

Analysis of Metal Lithium Recovery Rate

The materials before and after reduction are dissolved in hydrochloric acid, and the lithium content is determined by atomic absorption spectroscopy to obtain the lithium recovery rate.

Results and Discussion

Preparation Principle of Lithium Metal

In this paper, lithium oxide (Li_2O) is used as a raw material to extract lithium metal. Lithium oxide is obtained by thermal decomposition of lithium carbonate. The equilibrium pressures of carbon dioxide for lithium carbonate at 810, 890, and 1270 °C are 2, 4.3, and 101 kPa, respectively. In order to prevent the melting of lithium carbonate with a melting point of 735 °C during thermal decomposition and make it difficult to discharge carbon dioxide, the lithium carbonate and the flux inhibitor (CaO) are usually briquetted in a certain proportion, and then the agglomerates are thermally decomposed under vacuum. The product is a mixture of lithium oxide and calcium oxide, which facilitates the vacuum thermal reduction process. Using aluminum powder as a reducing agent, the reaction formulae extracting lithium metal are as follows:

$$21Li_2O + 12CaO + 14Al = 42Li + 7Al_2O_3 \cdot 12CaO \qquad (1)$$

$$3Li_2O + 3CaO + 2Al = 6Li + Al_2O_3 \cdot 3CaO \qquad (2)$$

Comprehensive Experiment for Producing Lithium

Start of the first furnace: use commercially available lithium carbonate as raw material, mix with calcium oxide as additives, and aluminum powder as reducing agent. The purity of lithium carbonate is above 99%. The purity of calcium oxide is greater than 98%. Mixing ratio (mass ratio): Lithium carbonate: calcium oxide = 2:1, mix uniformly, first roast, temperature 750°C, time 100 min. After sintering, grind to 60 meshes, add aluminum powder with a particle size of 60 meshes, and add 8% of the total mass of the calcined lithium carbonate + calcium oxide mixture. 2% CaF_2 is used to obtain mixed raw materials and pressed into pellets.

Open the door of the feed chamber (5) manually, put the pellets made of lithium-containing materials, additives, and reducing agent into the graphite crucible (11), close the door of the feed chamber (5); through the electrical control system (4), open the isolation valve 4 (7), move the graphite crucible (11) to the crucible lifting device (8) through the moving device (6); open the isolation valve 1 (16), and use the crucible lifting device (8) to move the graphite crucible (11) that is raised into the induction heating device (2); the vacuum unit (14) is turned on through the electrical control system (4), the isolation valve 2 (15) and the isolation valve 3 (9) are opened, and the discharge chamber (10) and the feed chamber (5) are evacuated; open isolation valve 5 (18), isolation valve 6 (19), and isolation valve 7 (20) to the reduction chamber (1), crystallizer (17), and metal lithium tank (21) vacuum; start the water cooling system (13) to the vacuum unit (14), reduction chamber (1), crystallizer (17), and induction heating device (2) water cooling; start the heating power supply (3) for induction heating the device (2) is heated, the vacuum degree and temperature are displayed in the electrical control system (4), and the reduction reaction is performed. The induction heating device (2) heats up to 550 °C at a heating rate of 14 °C/min, constant temperature for 50 min, and vacuum degree maintains 1 Pa; continue to heat up to 1150 °C at a heating rate of 14 °C/min, and control the vacuum of the vacuum reactor to 1050 Pa until no more lithium vapor is generated. Keep the temperature constant for about 3 h, and the lithium will be collected in the crystal in liquid form. In the container (17), the crystallization chamber (17) is connected to the metal lithium tank (21), and a graphite filter is also provided. The metal lithium tank (20) contains liquid paraffin, and the metal lithium tank (21) is periodically removed from the crystallization chamber (17). The metal lithium tank (21) contains liquid paraffin. The metal lithium tank (21) is preferably removed from the crystallizer (17) regularly, connected to a glove box filled with high purity argon, and then the valve of the lithium metal tank is opened to obtain lithium ingot. The purity of the lithium ingot is better than 99.5%; the chemical analysis results of the impurities of lithium are listed in Table 1; the recovery rate is 85%, and the current efficiency reaches 90%; after the reduction reaction is completed, the graphite crucible (11) is lowered by the crucible lifting device (8), the isolation valve 1 (16) is closed, the isolation valve 4 (7) is closed, and the mobile device (6) is moved into the discharge chamber (10)); close the isolation valve 3 (9), open the door of the discharge chamber (10) to break the vacuum; the hydraulic system (12) is connected with the crucible lifting device (8) and controls the lifting of the crucible lifting device (8); the water cooling system

Table 1 The chemical analysis results of the impurities of lithium (mass%)

Na	K	Mg	Ca	Al	Fe
0.018	0.008	0.002	0.002	0.003	0.002

(13) is respectively connected with the vacuum unit (14), the reduction chamber (1), the induction heating device (2), and the crystallization chamber (17).

The second furnace starts: close the isolation valve 4 (7), manually open the feeding chamber (5), fill the graphite crucible (11) with materials, evacuate the feeding chamber (5) to the required vacuum degree of the process, and open the isolation valve 4(7), move the graphite crucible (11) to the crucible lifting device (8) through the moving device (6), open the isolation valve 1 (16), and move the graphite crucible (11) through the crucible lifting device (8) to ascend into the induction heating device (2). At this time, the vacuum degree and the temperature of the reduction chamber (1) meet the process requirements of the reduction reaction. The reduction reaction starts, and the lithium will be collected in the crystallizer (17) in liquid form, and flows into the metal lithium tank (21). After the reaction is completed, the graphite crucible (11) is lowered through the crucible lifting device (8), the isolation valve 1 (16) is closed, the isolation valve 3 (9) is opened, and the mobile device (6) is moved into the discharge chamber (10); close the isolation valve 3 (9), open the door of the discharge chamber (10) to break the vacuum, take out the graphite crucible (11), clean up the residue of the graphite crucible (11), close the door of the discharge chamber (10), face the discharge chamber (10) vacuum to the required vacuum degree of the process, and cycle in turn.

The lithium extraction tailings obtained from the discharging chamber (10) include $12CaO \cdot 7Al_2O_3(s)$ and $3CaO \cdot Al_2O_3(s)$ (slag). The chemical composition is calcium oxide 53wt %, alumina 42wt%, the balance is other oxides.

Conclusions

1. The process of extracting lithium from the aluminum powder heat-reducing lithium oxide is reasonable. The Al powder and the calcium oxide can extract the metallic lithium from the lithium oxide at 0.13-133 Pa of vacuum and from 1100 to 1200 °C.
2. Laboratory scale of comprehensive experimental studies has shown that the extraction of lithium metal is feasible, lithium recovery rate reaches 85%, and the purity of the lithium ingot is better than 99.5%. This is a cleaning process with no chlorine gas emissions.
3. Laboratory continuous reduction furnace improves lithium production efficiency and reduces costs, it is worthy for further study, and it has a bright future. Continuous process with aluminum powder for the extraction of metal lithium is simple, of clean production, and low cost.

References

1. Liu W, Zhang J, Wang Q, Xie X, Lou Y, Xia B (2014) Ionics 20:1553–1560
2. Wei D, Wang H (2017) J Am Ceram Soc 100:1073–1079
3. Ding MS, Li Q, Li X, Xu W, Xu K (2017) J Phys Chem C 121:11178–11183
4. Kostrivas A, Lippold JC (1999) Int Mater Rev 44:217–237
5. Smeets AAJ, Fray DJ (1991) Extraction of lithium by vacuum thermal reduction with aluminum and silicon [J]. Trans Inst Min Metal sect C 100:42–47
6. Jena B (1989) Role of magnesium in the aluminothermic reduction of spodumene [J]. Prod Electrolysis Light Metal, 257–262
7. Inoue Makoto, Tanno Osamu, Kojima Yo (1990) Refinement of magnesium by vacuum sublimation [J]. J Japan Ins Light Metal 40(12):879–883

Effect of Synthesis Method on the Electrochemical Performance of $LiNi_xMnCo_{1-x-y}O_2$ (NMC) Cathode for Li-Ion Batteries: A Review

Monu Malik, Ka Ho Chan, and Gisele Azimi

Abstract With high specific capacity, high nominal voltage, low self-discharge, and low cost, layered $LiNi_xMnCo_{1-x-y}O_2$ (NMC) cathode has gained interest for second-generation lithium-ion batteries. Because the performance of Li-ion batteries highly depends on the composition, crystallography, morphology, and other parameters decided during synthesis, interest in developing high-performance Li-ion batteries has motivated researchers to develop novel synthesis methods to control these parameters. Although significant progress has been made in several synthesis methods such as sol–gel, hydrothermal, solid-state reaction, and emulsion drying at lab scale, an industrially viable synthesis approach is needed for the cost-effective production of highly efficient NMC cathode materials. This paper summarises the common synthesis methods investigated at laboratory and industrial scales for new and regenerated NMC cathode from end-of-life Li-ion batteries and their effect on the battery performance. It is found that co-precipitation is the most commonly used method to produce NMC cathode, while spray pyrolysis is commonly used at the industrial scale.

Keywords Co-precipitation · Lithium-ion · $LiNi_xMnCo_{1-x-y}O_2$ cathode · Solid-state · Spray pyrolysis

Introduction

Rechargeable lithium-ion (Li-ion) batteries have received significant attention over the last few decades after their commercialization by Sony in the early 1990s because of their high volumetric and gravimetric capacity, high efficiency, and long cycle life [1]. With the advancement in technology and rapidly increasing demand for

---

M. Malik · K. H. Chan · G. Azimi (✉)
Department of Chemical Engineering and Applied Chemistry, University of Toronto, 200 College Street, Toronto, ON M5S35, Canada
e-mail: g.azimi@utoronto.ca

G. Azimi
Department of Materials Science and Engineering, University of Toronto, 184 College Street, Toronto, ON M5S34, Canada

electric vehicles, more advanced, high capacity, high-performance Li-ion batteries are required. Since most of the commercial cells use graphitic anode and transition metal oxide cathode, several methods have been developed to prepare these different electroactive materials (especially cathodes) with the goal of controlling the particle morphology, composition, physical properties, and surface functionality. The performance of these batteries significantly depends on the various parameters during the synthesis of cathode material. There are several cathode materials used for the Li-ion batteries such as lithium iron phosphate ($LiFePO_4$, LFP), lithium cobalt oxide ($LiCoO_2$, LCO), lithium manganese oxide ($LiMn_2O_4$, LMO), lithium nickel cobalt aluminum oxide ($LiNi_xCo_yAl_{1-x-y}O_2$, NCA), and lithium nickel manganese cobalt oxide ($LiNi_xMn_yCo_{1-x-y}O_2$, NMC) [2].

Because of the synergic effect of the nickel (Ni), cobalt (Co), and manganese (Mn), NMC is one of the most successful combinations of these metals as a new-generation energy-storage material providing very good performance in terms of specific power, life span, cost, safety, and excels in specific energy. Ni provides high specific energy but poor stability while Mn helps achieve low internal resistance by forming a spinel structure but offers low specific energy [3]. Combining these two materials in the NMC cathode compensates for their drawbacks. However, efforts are being made to reduce the cobalt content from these cathodes due to the higher cost and limited supply of cobalt. There are several methods to synthesize the NMC cathode such as combustion, sol–gel, hydrothermal/solvothermal, and emulsion drying, the most common being co-precipitation, followed by spray pyrolysis and solid-state reaction methods [4]. Each method has its own advantages and disadvantages and can be selected based on the application and production scale.

The co-precipitation method can be further divided into three categories: hydroxide co-precipitation, carbonate co-precipitation, and oxalate co-precipitation, where hydroxide co-precipitation is one of the most economical methods to produce high tap density particles. However, it requires an inert atmosphere, otherwise, it can lead to different structures and impurities. On the other hand, spray pyrolysis provides several advantages such as uniform atomic-level mixing, effective particle morphology control, short residence time, and excellent reproducibility, but suffers from the formation of hollow secondary particles. The solid-state reaction method is one of the most simple and cost-effective methods to produce an NMC cathode, but requires significant mechanical mixing and long holding time. These methods have been thoroughly investigated by various researchers, which helps to understand the impact of these synthesis techniques on the properties of the cathode material such as tap density, particle size and shape, crystallinity, and ultimately electrochemical performance.

Although some review papers reported the methods of synthesizing cathode materials for Li-ion batteries, none of them have focused on NMC cathode alone, which is considered as second-generation cathode for Li-ion batteries. This paper provides a review of the major synthesis methods used to prepare NMC cathode material and provides a comprehensive and systematic literature review covering several original research articles. Each method is briefly discussed while highlighting the effect of various synthesis parameters on the properties and electrochemical performance of NMC cathode material. The results obtained from the literature and optimized conditions for the co-precipitation method are also discussed.

Co-precipitation Method

Although the co-precipitation method has a long history of application in water treatment, metal mining, and the pharmaceutical industry, its application for Li-ion battery materials is more recent. The method is used for several types of cathodes for Li-ion batteries, and it has gained popularity for the synthesis of NMC cathode in the last 5–10 years. This method is often used for batch production of the cathode materials, and the process is usually carried out in a continuous reactor system shown in Fig. 1. The method is also commonly used to synthesize regenerated cathode from end-of-life Li-ion batteries. The process of co-precipitation has several parameters that can significantly affect the morphology and composition of the particles including temperature, atmosphere, stirring speed, pH, and chelating agent concentration. A variation in these co-precipitation conditions and chemistry of the particle can result in different particle morphologies including sphere, cube, needle, cubical rod, ellipsoid, hollow sphere, star, dumbbell, plate, and others. Figure 2 present a few examples of these unique morphologies obtained from the literature. The most common shape of the NMC secondary particle prepared from co-precipitation is spherical.

The process of synthesizing a typical NMC cathode material is shown in Fig. 1a, where the salt of different transition metals in the stoichiometric amount is mixed inside a continuous stirred tank reactor (CSTR) with water to prepare a homogenous solution. The prepared solution is kept at a predefined temperature using an

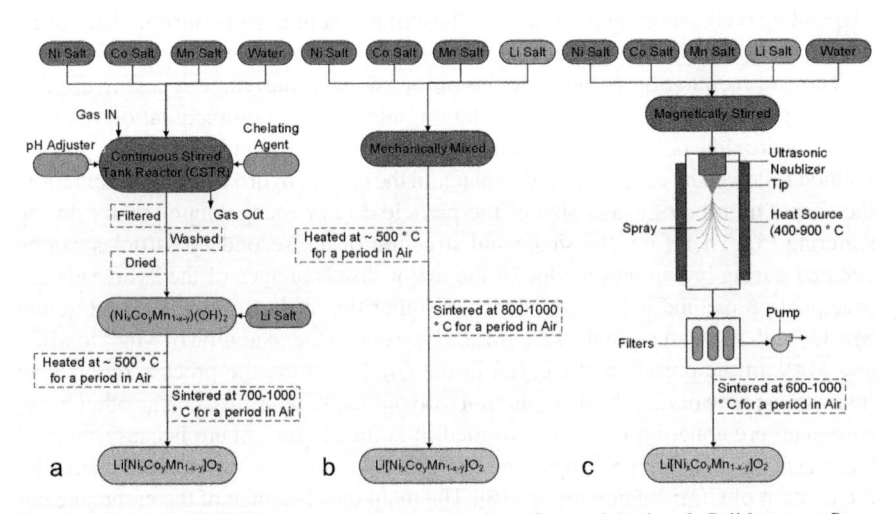

Fig. 1 Schematic of NMC battery synthesis methods: **a** Co-precipitation, **b** Solid state, **c** Spray pyrolysis. (Color figure online)

Fig. 2 Cathode particles of different morphologies obtained from co-precipitation reactions: **a** sphere, **b** cube, **c** needle, **d** rod, **e** ellipsoid, and **f** star. (Modified from [8–11])

external heat source while stirring continuously, and pH is usually adjusted using sodium hydroxide (NaOH) solution fed directly to the CSTR to form precipitates. A chelating agent such as ammonium hydroxide (NH$_4$OH) is used to control the reaction parameters. A gas such as nitrogen or argon can be flown through the CSTR depending on the type of the co-precipitation method to maintain an inert atmosphere to avoid the oxidation of metal ions such as Mn and Ni. The obtained (Ni$_x$Mn$_y$Co$_{1-x-y}$)(OH)$_2$ precipitates are filtered, washed with deionized water, and dried before mixing with a Li salt. The mixture of co-precipitates and Li salt is heated to around 500 °C (depending on the melting point of Li salt used) for melting and uniform distribution of Li and sintered at 700–1000 °C in air to obtain the final cathode material.

The co-precipitation process can be further divided into three types: hydroxide co-precipitation, carbonate co-precipitation, and oxalate co-precipitation. Among these, hydroxide is the most popular, effective, and economical co-precipitation method followed by carbonate and oxalate. In the case of hydroxide co-precipitation, the shape, morphology, and size of the particle do not change significantly during sintering [5]. Therefore, the shape and structure of the secondary particles can be decided during precipitation. One of the major disadvantages of the hydroxide co-precipitation method is the coexistence of impurities such as MnOOH, MnO$_2$, and Mn$_3$O$_4$ with manganese hydroxide (Mn(OH)$_2$) due to the oxidation of Mn^{2+} to Mn^{3+} and Mn^{4+} in the presence of oxygen in the air. Therefore, the process requires an inert atmosphere during the precipitation to avoid such impurities. On the other hand, carbonate precipitation can be performed in an open atmosphere because most of the transition metal cations including Mn remain in the divalent oxidation state by the fixation of CO$_3^{2-}$ anion groups [6]. The main disadvantage of the carbonate co-precipitates is increased porosity of secondary particles that become more susceptible to fracture during calendaring, and the compositional difference between the desired transition metal ratio and feed material due to the difference in precipitates solubility [7]. The process of oxalate co-precipitation which uses oxalic acid is similar to the other two methods but it is significantly cheaper and more environmentally friendly, making it effective for industrial applications.

Effect of Synthesis Parameters

As discussed earlier, co-precipitation can be used to synthesize particles with different morphologies such as rods, cubes, and plates, but the spherical shape is the most preferred one because spheres tend to pack relatively densely leading to the desirable high tap density and electrode loading. This section is focused on the effect of pH, chelating agent concentration, and stirring speed on particle morphology of the hydroxide co-precipitates.

The change in pH has a significant effect on hydroxide precipitation both in terms of chemistry and the complexes of different species that form in the solution. In the case of hydroxide precipitation, it has been reported in the literature that the increase in the pH leads to a decrease in the size of secondary particle NMC hydroxide due to higher supersaturation and faster precipitation speed [1, 12]. In a previous study, authors varied pH to 11, 11.5, and 12 with a constant value of chelating agent concentration (0.24 M) and stirring speed (1000 rpm). As shown in Fig. 3a–c, the particle size significantly reduced from around 10 μm to 2–3 μm, and tap density reduced from 2.28 to 1.51 g cm^{-3} when pH increased from 11 to 12.

Another study showed that the use of a chelating agent such as NH_4OH during precipitation helps increase the tap density of material without sacrificing the specific capacity [12]. The authors varied the concentration of the NH_4OH in the range of 0.12–0.36 M and reported the most regular spherical particle and the narrowest size distribution with the highest concentration within their investigation ranges, as shown in Fig. 3d–f. The tap density also increased from 2.04 to 2.23 g cm^{-3} with an increase in chelating agent concentration from 0.12 to 0.36 M.

The stirring speed during the precipitation also plays an important role in determining the morphology of the secondary particle (Fig. 3g–i). In this figure, stirring speed was varied from 400 to 1000 rpm, while using optimized condition for NH_4OH chelating agent (0.36 M) and pH (11). At 400 rpm, bulky secondary particles that formed with random size distribution were observed. As the stirring speed increased from 400 to 1000 rpm, more spherical and uniform secondary particles composed of more densely packed primary particles formed. Moreover, the tap density increased from 2.12 to 2.39 g cm^{-3}.

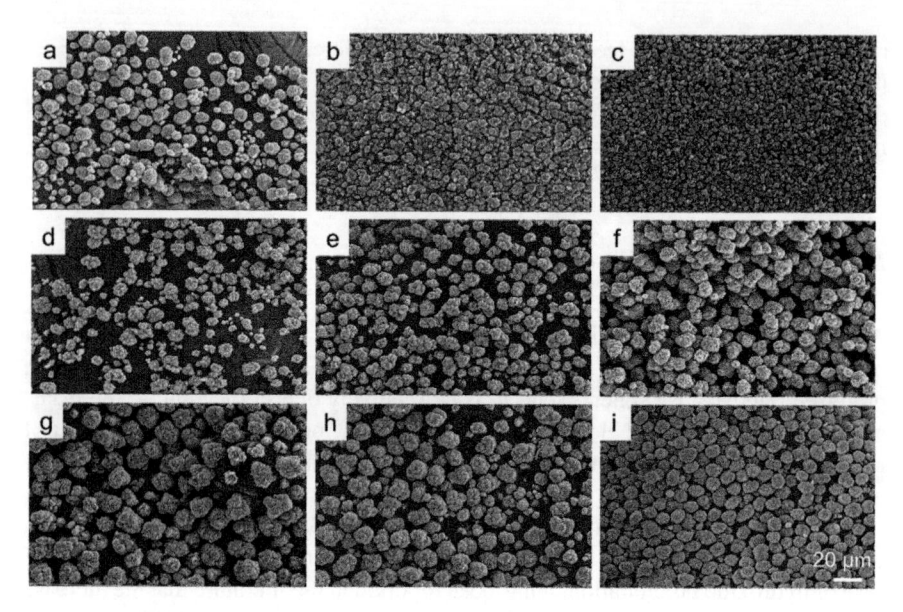

Fig. 3 SEM images of $(Ni_{1/3}Co_{1/3}Mn_{1/3})(OH)_2$ precipitated at different synthesis conditions: at pH **a** 11, **b** 11.5, and **c** 12, with 0.36 M NH_4OH and stirring speed of 1000 rpm; at NH_4OH concentration of **d** 0.12, **e** 0.24, and **f** 0.36 M, with stirring speed of 1000 rpm and pH 11; and at stirring speed of **g** 400, **h** 600, and **i** 1000 rpm with 0.36 M NH_4OH and pH 11

Effect of Sintering Conditions on Electrochemical Performance

After co-precipitation, the hydroxides are dried and usually mixed with lithium hydroxide (LiOH) and normally heated to the melting point of LiOH for uniform distribution of Li. The obtained mixture is grounded and sintered again at 700–1000 °C in air or pure oxygen for the crystals to align together and form a layered structure. Therefore, sintering conditions can also impact the morphology of the final cathode material and its electrochemical performance in a full cell. Several researchers investigated the effect of sintering conditions on the electrochemical performance of NMC cathode some of which are presented in Table 1 [12–15]. Lee et al. [12] varied the sintering temperature in the range of 900–1000 °C and reported the optimum electrochemical performance with a discharge capacity of 168 mAh g^{-1} using cathode sintered in air at 1000 °C. However, Kim et al. [15] obtained the same discharge capacity with cathodes sintered at a lower temperature (700 °C) but for a longer period. This shows that sintering for longer periods compensates for lower temperatures. Most of these studies performed the sintering process in air and reported similar size secondary particles (10–20 μm) and similar discharge capacity (~168 mAh g^{-1}).

Table 1 Cathode sintering conditions, properties, test conditions, and electrochemical performance of $LiNi_xMnCo_{1-x-y}O_2$ (NMC) cathode with different synthesis methods

Reference	Cathode Type	Reactor Wall Temperature (°C)	Sintering Temperature (°C)	Sintering Time (h)	Atmosphere	Primary Particle Size (nm)	Secondary Particle Size (μm)	Voltage Range (V)	Current Density (mA g^{-1})	Discharge Capacity (mAh g^{-1})	Capacity Retention (%)/Cycles
Co-Precipitation											
Lee et al. [12]	$LiNi_{0.33}Co_{0.33}Mn_{0.33}O_2$		1000	10	Air	500–1000	10	2.8–4.4	20	168	
Kim et al. [15]	$LiNi_{0.8}Co_{0.1}Mn_{0.1}O_2$		750	20	Air	N/A	~10–15	2.8–4.3	20	168	94 (40)
Lee et al. [14]	$LiNi_{0.4}Co_{0.3}Mn_{0.3}O_2$		900	15	Air	N/A	~15–20	3.0–4.3	20	165	98 (50)
Woo et al. [13]	$LiNi_{0.8}Co_{0.1}Mn_{0.1}O_{1.96}F_{0.04}$		770	12	Oxygen	100–800	N/A	2.8–4.3	20	190	97 (50)
Solid-State Reaction Method											
Tan and Liu [16]	$LiNi_{1/3}Co_{1/3}Mn_{1/3}O_2$		900	12	Air	200	N/A	2.5–4.5	1000	137	94 (30)
Zhou et al. [20]	$Li_{1.13}Ni_{0.2}Co_{0.2}Mn_{0.47}O_2$		800	8	Air	85–250	N/A	2.0–4.8	80	170	96 (50)
He et al. [21]	$LiNi_{1/3}Co_{1/3}Mn_{1/3}O_2$		950	20	Air	200–500	N/A	3.0–4.3	16	159	81 (20)
Li et al. [17]	$LiNi_{0.5}Mn_{0.2}Co_{0.3}O_2$		950	25	Oxygen	N/A	1000–2000	3.0–4.6	40	176	86 (25)
Spray Pyrolysis											
Lengyel et al. [18]	$Li_{1.2}Mn_{0.54}Ni_{0.13}Co_{0.13}O_2$	575	900	2	Air	230	1.8–2.0	2.0–4.8	67	212	
Park et al. [22]	$LiNi_{1/3}Co_{1/3}Mn_{1/3}O_2$	500	900		Air			2.8–4.4	20	163	96 (50)
Ju and Kang [23]	$LiNi_{1/3}Co_{1/3}Mn_{1/3}O_2$	900	800	3	Air		1.1	2.8–4.5		154	84 (30)
Lengyel et al. [19]	$Li_{1.26}Mn_{0.6}Ni_{0.07}Co_{0.07}O_2$	450	900	2	Air	175	1.7		28	311	95 (50)

Solid-State Reaction Method

Solid-state synthesis is one of the oldest and simplest methods in the literature to prepare battery active materials and is commonly used for the preparation of nanometer-sized NMC cathode material in the industry due to low requirements of equipment and easy process control. The synthesis process is simple in which all transition metal oxides and Li salt are mixed and sintered at around 800–1000 °C similar to co-precipitation shown in Fig. 1b. The main advantage of this method is that the correct stoichiometry of Li and transition metals can be obtained easily, but it requires strong mechanical mixing and long hold time at high temperature. In this case, metal constituents have to move large distances through solid phases to find their desired atomic positions, and elevated temperature assists in accelerating the diffusion process and long hold time helps reach the structure equilibrium. Therefore, sintering time and temperature play a critical role in the morphology and phase purity of the cathode material. Tan and Liu [16] synthesized the $LiNi_{1/3}Co_{1/3}Mn_{1/3}O_2$ cathode using α-MnO_2 nanorods as the source of Mn and investigated the effect of metal source and sintering temperature (800–950 °C) on cation mixing and electrochemical performance. The optimum performance of the cell was reported with cathode sintered at 900 °C, which gave a discharge capacity of 136.9 mAh g^{-1} at a high current density of 1000 mA g^{-1}. Li et al. [17] reported the highest discharge capacity of 176 mAh g^{-1} with cathode sintered in an oxygen environment at 950 °C, but at a much lower current density of 40 mA g^{-1}, as presented in Table 1.

Spray Pyrolysis

Spray pyrolysis is a versatile method that is normally used to produce commercial transition metal oxide materials at a large scale, but it is not commonly used in academic research due to increased equipment cost and complexity. This method has several advantages such as short residence time, high production rate and purity, narrow particle size distribution, and excellent reproducibility [18]. Spray pyrolysis is a continuous synthesizing technique, where the solution of transition metal and Li salt in water is atomized/aerosolized to form micron-sized droplets using a nebulizer as shown in Fig. 1c. These droplets are carried to a reactor at elevated temperature, where each droplet serves as a microreactor and undergoes a series of physical and chemical processes to form the cathode particle. These particles are collected at the outlet by filters and depending on the reactor conditions, they can be directly used as a cathode or further sintered at an elevated temperature similar to other methods. The morphology of the cathode material from this method depends on the type of nebulizer, reactor temperature, solution concentration, and residence time. Table 1 shows selected properties of cathode material synthesized using the spray pyrolysis method and their electrochemical performance. Most previous studies reported the optimum reactor temperature of around 500 °C with an average specific discharge capacity of above 150 mAh g^{-1}. One of the common issues with the spray pyrolysis method is the formation of the hollow shell-like particles, which significantly reduces the tap density and electrode volumetric capacity. Lengyel et al. [19] partially resolved the low tap density issue using the flame-assisted spray technology-slurry spray pyrolysis method and achieved 1.05 g cm^{-3}, however, this value is still lower than that obtained by other synthesis methods.

Conclusions

In the present paper, the most common methods to synthesize NMC cathode material were summarised, and the effect of various parameters in these methods on particle morphology and electrochemical performance is discussed. Among the available methods, co-precipitation is most popular due to low cost, high production rate, ease of controlling the process parameter, and achieving high tap density. Although precipitation reaction is straightforward, careful control of reaction parameters is highly essential to obtain the desired properties. Hydroxide is the most common synthesis method among the three co-precipitation techniques due to the ease of controlling the morphology of cathode particles, but it requires an inert atmosphere which increases the production cost. The solid-state reaction method is the simplest method to produce NMC cathode material because the correct stoichiometry can be achieved easily. However, the process is considered energy-intensive because of the need for extensive mechanical mixing and long hold time during sintering at elevated temperatures. Spray pyrolysis is a versatile method to produce NMC cathode material at a large scale with high reproducibility. Spray pyrolysis is a continuous synthesizing technique and has several advantages such as short residence time, high purity, and narrow particle size distribution, but it often leads to the formation of the hollow shell-like particles, which significantly reduce the tap density. Overall, co-precipitation is most suited for batch production of NMC cathode material, while spray pyrolysis is the best method for continuous production.

References

1. Dong H, Koenig GM (2020) A review on synthesis and engineering of crystal precursors produced via coprecipitation for multicomponent lithium-ion battery cathode materials. CrystEngComm
2. Bensalah N, Dawood H (2016) Review on synthesis, characterizations, and electrochemical properties of cathode materials for lithium ion batteries
3. Jiao L, Liu Z, Sun Z, Wu T, Gao Y, Li H, et al. (2018) An advanced lithium ion battery based on a high quality graphitic graphene anode and a Li [Ni0. 6Co0. 2Mn0. 2] O2 cathode. Electrochimica Acta 259:48–55
4. Voronov V, Shvetsov A, Gubin S, Cheglakov A, Kornilov DY, Karaseva A, et al. (2016) Effect of the preparation method of the cathode material LiNi 0.33 Mn 0.33 Co 0.33 O 2 on the electrochemical characteristics of a lithium ion cell. Russian J Inorg Chem 61:1153–1159

5. Zhang S (2007) Characterization of high tap density Li [Ni1/3Co1/3Mn1/3] O2 cathode material synthesized via hydroxide co-precipitation. Electrochim Acta 52:7337–7342
6. Wang D, Belharouak I, Ortega LH, Zhang X, Xu R, Zhou D et al (2015) Synthesis of high capacity cathodes for lithium-ion batteries by morphology-tailored hydroxide co-precipitation. J Power Sour 274:451–457
7. Dong H, Koenig GM Jr (2017) Compositional control of precipitate precursors for lithium-ion battery active materials: role of solution equilibrium and precipitation rate. J Mater Chem A 5:13785–13798
8. Wang G, Yi L, Yu R, Wang X, Wang Y, Liu Z et al. (2017) Li1. 2Ni0. 13Co0. 13Mn0. 54O2 with controllable morphology and size for high performance lithium-ion batteries. ACS Appl Mater Inter 9:25358–25368
9. Cheng F, Zhao J, Song W, Li C, Ma H, Chen J et al (2006) Facile controlled synthesis of MnO2 nanostructures of novel shapes and their application in batteries. Inorg Chem 45:2038–2044
10. Zhang S, Deng C, Fu B, Yang S, Ma L (2010) Synthetic optimization of spherical Li [Ni1/3Mn1/3Co1/3] O2 prepared by a carbonate co-precipitation method. Powder Technol 198:373–380
11. Lin H, Hu J, Rong H, Zhang Y, Mai S, Xing L et al (2014) Porous LiMn 2 O 4 cubes architectured with single-crystalline nanoparticles and exhibiting excellent cyclic stability and rate capability as the cathode of a lithium ion battery. J Materi Chem A 2:9272–9279
12. Lee MH, Kang YJ, Myung ST, Sun YK (2004) Synthetic optimization of Li [Ni1/3Co1/3Mn1/3] O2 via co-precipitation. Electrochim Acta 50:939–948
13. Woo SU, Park BC, Yoon CS, Myung ST, Prakash J, Sun YK (2007) Improvement of electrochemical performances of Li [Ni0. 8Co0. 1Mn0. 1] O2 cathode materials by fluorine substitution. J Electrochem Soc 154:A649
14. Lee KS, Myung ST, Amine K, Yashiro H, Sun YK (2007) Structural and electrochemical properties of layered Li [Ni1− 2x Co x Mn x] O2 (x= 0.1–0.3) Positive Electrode Materials for Li-Ion Batteries. J Electrochem Soc 154:A971
15. Kim MH, Shin HS, Shin D, Sun YK (2006) Synthesis and electrochemical properties of Li [Ni0. 8Co0. 1Mn0. 1] O2 and Li [Ni0. 8Co0. 2] O2 via co-precipitation. J Power Sour 159:1328–1333
16. Tan L, Liu H (2010) High rate charge–discharge properties of LiNi1/3Co1/3Mn1/3O2 synthesized via a low temperature solid-state method. Solid State Ionics 181:1530–1533
17. Li D, Sasaki Y, Kageyama M, Kobayakawa K, Sato Y (2005) Structure, morphology and electrochemical properties of LiNi0. 5Mn0. 5− xCoxO2 prepared by solid state reaction. J Power Sour 148:85–89
18. Lengyel M, Atlas G, Elhassid D, Luo PY, Zhang X, Belharouak I et al. (2014) Effects of synthesis conditions on the physical and electrochemical properties of Li1. 2Mn0. 54Ni0. 13Co0. 13O2 prepared by spray pyrolysis. J Power Sour 262:286–296
19. Lengyel M, Zhang X, Atlas G, Bretscher HL, Belharouak I, Axelbaum RL (2014) Composition optimization of layered lithium nickel manganese cobalt oxide materials synthesized via ultrasonic spray pyrolysis. J Electrochem Soc 161:A1338
20. Zhou LZ, Xu QJ, Liu MS, Jin X (2013) Novel solid-state preparation and electrochemical properties of Li1. 13 [Ni0. 2Co0. 2Mn0. 47] O2 material with a high capacity by acetate precursor for Li-ion batteries. Solid State Ionics 249:134–138
21. He YS, Ma ZF, Liao XZ, Jiang Y (2007) Synthesis and characterization of submicron-sized LiNi1/3Co1/3Mn1/3O2 by a simple self-propagating solid-state metathesis method. J Power Sourc 163:1053–1058
22. Park S, Yoon CS, Kang S, Kim HS, Moon SI, Sun YK (2004) Synthesis and structural characterization of layered Li [Ni1/3Co1/3Mn1/3] O2 cathode materials by ultrasonic spray pyrolysis method. Electrochim Acta 49:557–563
23. Ju SH, Kang YC (2009) The characteristics of Ni–Co–Mn–O precursor and Li (Ni1/3Co1/3Mn1/3) O2 cathode powders prepared by spray pyrolysis. Ceram Int 35:1205–1210

Recovery of Cobalt as Cobalt Sulfate from Discarded Lithium-Ion Batteries (LIBs) of Mobile Phones

Pankaj Kumar Choubey, Archana Kumari, Manis Kumar Jha, and Devendra Deo Pathak

Abstract Cobalt, an exceptional cathode material present in lithium-ion batteries (LIBs), is an essential element for the production of energy storage devices. But, the lifespan of rechargeable batteries is decreasing day-by-day, which become obsolete after reaching their end of life. Therefore, an enormous amount of discarded LIBs are generated. Keeping in mind the above, a novel approach has been made to selectively recover cobalt from sulfate leach liquor of discarded LIBs containing 1.4 g/L Cu, 1.1 g/L Ni, 11.9 g/L Co, 6.9 g/L Mn, and 1.2 g/L Li. Initially, Cu and Ni were extracted by solvent extraction techniques using 10% LIX 84-IC. Almost complete precipitation of cobalt occurred from leach liquor at pH ~3 using ammonium sulfide solutions. Cobalt from the precipitated product was further dissolved in H_2SO_4 in presence of H_2O_2 at elevated temperature. The leach liquor obtained was evaporated to get the cobalt sulfate with a purity of more than 98%.

Keywords Spent LIBs · Leach liquor · Precipitation · Cobalt · Cobalt sulfate

Introduction

Lithium-ion batteries (LIBs) are extensively used in portable electronic devices as well as in electric vehicles to store energy. The unique properties of LIBs such as their high energy storage capacity and high specific energy have increased their demand to fulfill various technological applications [1, 2]. But with the passage of time, the energy storing capacity of LIBs decreases, thus resulting in the huge generation of waste LIBs. In 2017, ~12 billion units of obsolete LIBs were generated, which is expected to reach ~25 billion units by the end of 2020 [1]. Presently ~10% spent LIBs are reported to be recycled by the formal sector while the rest of the

---

P. K. Choubey · A. Kumari · M. K. Jha (✉)
Metal Extraction and Recycling Division, CSIR-National Metallurgical Laboratory, Jamshedpur 831007, India
e-mail: mkjha@nmlindia.org

P. K. Choubey · D. D. Pathak
Department of Chemistry, Indian Institute of Technology (ISM), Dhanbad 826004, India

batteries remains untreated or recycled in an illegal manner. This illegal treatment of scrap batteries results in the loss of valuable metals [2]. These discarded batteries contain a significant amount of cobalt (Co) and manganese (Mn) along with other valuable metals like lithium (Li), nickel (Ni), etc. Hence, it is necessary to develop sustainable recycling routes for the recovery of Co from these discarded LIBs due to a lack of abundant natural resource and to mitigate their demand–supply gap. Several researchers studied different techniques but generally, pyro- and hydro-metallurgical routes were employed for the recovery of Co from discarded LIBs [3–5]. Hanisch et al. (2015) treated the spent LIBs at high temperature (600 °C) to get the roasted product and further leached them in 2 M H_2SO_4 at 60 °C in the presence of 10% hydrogen peroxide to leach out Co [3]. 4 mol/L HCl is reported to be used to enhance the Co leaching efficiency up to 99% at 80 °C in 60 min keeping the pulp density at 20 g/L. But a quite long reaction time (180 min) was required to get the same leaching efficiency of Co in 3 mol/L H_2SO_4 at 90 °C in the presence of 0.25 g/L $Na_2S_2O_3$ (reductant) [4, 5]. Apart from this, organic acids like citric acid, malic acid, and ascorbic acid have also been used to leach out Co from spent LIBs. Almost 98% cobalt was leached in 1.0 mol/L oxalic acid at 80 °C in the presence of hydrogen peroxide. The advantage of this leaching system lies in the selective precipitation of Co as oxalate salt (CoC_2O_4), while other metals Li, Mn, Ni, and Al remained in the leach liquor [6, 7]. Ascorbic acid also acts as a self-reductant and hence eliminates the requirement of any other reducing agent during Co leaching [8].

Subsequently, precipitation studies were carried out to recover the metals from leach liquor of LIBs using a number of precipitating agents [4, 9]. Initially, Mn was precipitated with potassium dichromate ($KMnO_4$) at pH 2. After the separation of manganese, Ni and Co were removed as nickel-dimethylglyoxime (Ni-DMG) and cobalt hydroxide complex, respectively [4]. Finally, lithium was precipitated as lithium carbonate using sodium carbonate as a precipitant at 90 °C in a range of pH 11 to 12 [9].

In this paper, a novel process flow-sheet has been reported to selectively recover Co as hydrated cobalt sulfate ($CoSO_4 \cdot 7H_2O$) from leach liquor of spent LIBs using precipitation followed by sulfate leaching and evaporation. Various process parameters were optimized for the recovery of pure Co in sulfate form.

Experimental

Materials

Spent LIBs supplied by the local market were used for experimental purposes. The spent LIBs are mainly composed of cathode, anode, separator, and electrolyte. The cathodic materials contain lithium, cobalt, manganese, and nickel in the form of their oxides such as lithium cobalt oxide ($LiCoO_2$), lithium cobalt manganese nickel and oxide ($LiCoMnNiO_2$). Initially, sodium chloride solution was used to discharge

Table 1 Typical composition of spent LIBs (wt %)

Li	Co	Mn	Ni	Cu	Graphite	Others
2.1	13.4	7.1	2.3	2.5	55.1	~20

Table 2 Chemical composition of leach liquor (g/L)

Li	Co	Mn	Ni	Cu
1.2	11.9	6.9	1.1	1.4

the LIBs prior to pre-treatment. Thereafter, LIBs were crushed using the scutter crusher, and separated into different components such as black powder, metallic concentrate, and plastic materials. The obtained black powder was dissolved in sulfuric acid at optimized leaching conditions [10]. The detailed composition of spent LIBs and leach liquor used in this study are presented in Tables 1 and 2, respectively. Ammonium sulfide (Analytical grade, supplied by Loba Chemie Pvt. Ltd., Mumbai, India), sulfuric acid, hydrogen peroxide, and sodium hydroxide supplied by E. Merck, Mumbai India were used during the experiments (E. Merck, Mumbai, India, analytical Grade).

Methodology

Precipitation Procedure

Bench-scale studies were carried out in a beaker (capacity 200 mL) containing 100 mL leach liquor with the addition of different amounts of ammonium sulfide (2–10% v/v) under a constant stirring facility to uniformly mix the solution. During precipitation, the pH of the solution was kept constant ~3.0 throughout the experiments by adding ammonium sulfide and sulfuric acid solution. Solution temperature was also kept constant (30 °C) with the help of a temperature-controlling sensor. The slurry was filtered and metal content in the filtrate was analyzed using Inductively Coupled Plasma-Optical Emission Spectroscopy (ICP-OES) (VISTA MPX, CCD Simultaneous, Make: Australia) and Atomic Absorption Spectrometer (AAS) (Model no. AAS 200, Make Perkin Elmer, USA).

Leaching Procedure

The leaching experiments were carried out in a temperature-controlled leaching reactor fitted with a reflux condenser to avoid loss of solution during the experiments. Different concentrations of sulfuric acid were used as a leaching agent at pre-selected temperature (30–60 °C) with the help of temperature-sensitive hot plate having a magnetic stirring facility to maintain the stable temperature. On reaching the set temperature, the desired amount of sample was charged into the leaching reactor under a stirring speed of 400 rpm, which was found sufficient to ignore the effect of mass transfer on leaching. Sampling was carried out at different time intervals during experiments to observe the effect on leaching of metals. Once the leaching was completed, the solution was filtered and the residue left was dried in a vacuum oven and kept for further analysis.

Results and Discussion

Leach liquor of spent LIBs containing 1.4 g/L Cu, 1.1 g/L Ni, 11.9 g/L Co, 6.9 g/L Mn, and 1.2 g/L Li was used for the recovery of cobalt. At first, Cu and Ni were removed by a solvent extraction process using 10% LIX 84-IC as an extractant. Thereafter, the precipitation process was used to selectively precipitate the cobalt from the raffinate obtained after the separation of Cu and Ni. The detailed discussion is presented below.

Precipitation of Cobalt from Leach Liquor

In order to selectively precipitate cobalt from the leach liquor, precipitation studies were carried out using different concentrations of ammonium sulfide varying from 2 to 10% (v/v), while maintaining pH ~3. The result shows that (Fig. 1) the precipitation of cobalt increases with increase in ammonium sulfide concentration due to an increase in the content of sulfur ion which facilitates the formation of insoluble cobalt sulfide species resulting in the enhancement of cobalt precipitation. About 30.1%

Fig. 1 Effect of ammonium sulfide concentration on Co precipitation. [Solution: Leach liquor of spent LIBs; Precipitant: ammonium sulfide; pH: 3; Time: 30 min]. (Color figure online)

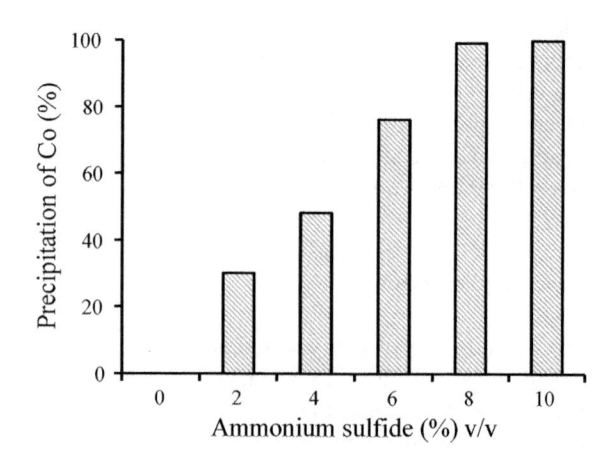

Fig. 2 Effect of sulfuric acid concentration on Co leaching. [Solid: Precipitated cobalt sulfide; Acid: H_2SO_4; Additive: 10% H_2O_2; Time: 30 min]. (Color figure online)

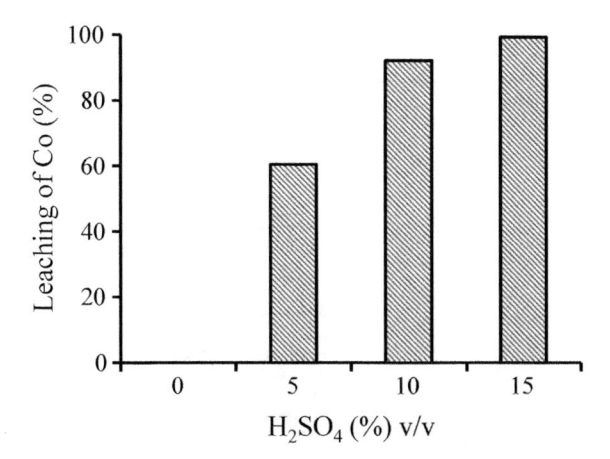

Co was precipitated with 2% ammonium sulfide (v/v) while 99.2% Co precipitated with 8% ammonium sulfide (v/v). Lithium and manganese remained in the filtrate. Hence, 8% ammonium sulfide (v/v) was chosen as the optimum concentration for the precipitation of cobalt from leach liquor of spent LIBs. Precipitated cobalt sulfide was further dissolved in sulfuric acid solution to recover the cobalt in the form of cobalt sulfate.

Leaching of Cobalt from the Obtained Precipitate/Cobalt Sulfide

In order to get Co as hydrated cobalt sulfate, the precipitate of cobalt sulfide was further leached in sulfuric acid in a range varying from 5 to 15% (v/v) in the presence of 10% (v/v) hydrogen peroxide at elevated temperature. Figure 2 shows that the leaching of cobalt was found to increase with the increase in the acidic strength of the solution, which facilitated the dissociation of the sulfide molecule of cobalt. As a result, cobalt leaching efficiency was increased. Further, the increase in the acid concentration above 15% had no significant effect on the leaching of cobalt. Hence, 15% H_2SO_4 was considered as optimum acid concentration for cobalt leaching from precipitated cobalt sulfide.

In addition, reductant concentration (H_2O_2) was also varied from 5 to 15% at elevated temperature to leach out cobalt from the precipitated product (cobalt sulfide). It was found that 75.2% Co dissolved in 15% H_2SO_4 at 60 °C in 30 min when 5% H_2O_2 was added to the leachant, while cobalt leaching was enhanced up to 99.2% when 10% H_2O_2 was added to the solution under the same experimental conditions. The leaching of cobalt was found to increase with the addition of hydrogen peroxide due to the generation of sufficient dissolved oxygen or nascent oxygen, which might be responsible for the enhancement of cobalt leaching efficiency. Further, the role of

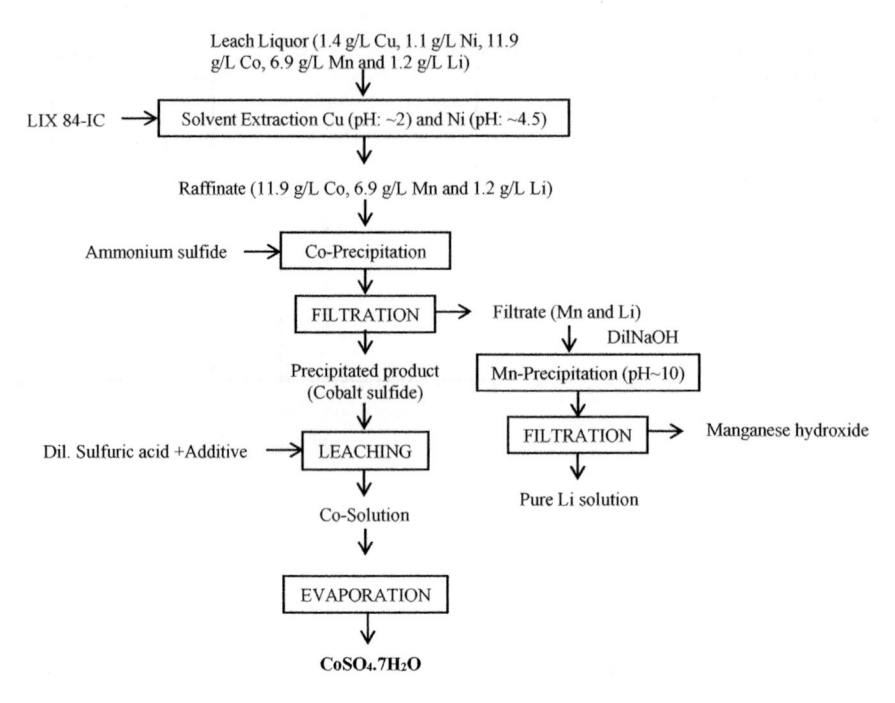

Fig. 3 Developed process flow-sheet for the recovery of cobalt from leach liquor of spent LIBs

temperature was also studied and found that leaching of cobalt gradually increased with an increase in solution temperature up to 60 °C due to an increase in the rate of reaction. But above 60 °C, the temperature had no significant effect on the enhancement of cobalt leaching efficiency. Therefore, 60 °C temperature was considered as the optimum temperature for cobalt leaching from precipitated cobalt sulfide. Finally, a process flow-sheet has also been developed for the separation of copper, nickel, cobalt, manganese, and lithium from leach liquor of spent LIBs as presented in Fig. 3.

Conclusion

Based on the lab-scale precipitation and leaching studies, the following conclusions have been drawn for the recovery of cobalt from leach liquor of spent LIBs.

1. 99.2% cobalt was precipitated from leach liquor of spent LIBs in 30 min at room temperature using 8% (v/v) ammonium sulfide as a precipitating agent, while lithium and manganese remained in the solution.
2. It was found that 60.4% cobalt dissolved from the precipitated product with 5% H_2SO_4 at 60 °C in 30 min in the presence of 10% H_2O_2 (v/v).
3. Complete leaching of cobalt (99%) occurred using 15% H_2SO_4 at 60 °C in 30 min with the addition of 10% H_2O_2 (v/v).

4. Finally, leach liquor of the precipitated product was evaporated to get cobalt sulfate of purity more than 98% as shown in Fig. 3.

Acknowledgements The authors are thankful to the Director, CSIR-National Metallurgical laboratory, Jamshedpur for giving permission to publish this paper.

References

1. Church C, Wuennenberg L (2019) Sustainability and second life: the case of lithium and cobalt recycling. International institute for sustainable development (Report), pp 1–56
2. Choubey PK, Chung KW, Kim MS, Lee JC, Srivastava RR (2017) Advance review on the exploitation of the prominent energy storage element lithium. Part II: From sea water and spent lithium ion batteries (LIBs). Miner Eng 110:104–121
3. Hanisch C, Loellhoeffel T, Diekmann J, Markley KJ, Haselrieder W, Kwade A (2015) Recycling of lithium ion batteries: a novel method to separate coating and foil of electrodes. J Clean Prod 108:301–311
4. Wang RC, Lin YC, Wu SH (2009) A novel recovery process of metal values from the cathode active materials of the lithium-ion secondary batteries. Hydrometallurgy 99:194–201
5. Wang J, Chen M, Chen H, Luo T, Xu Z (2012) Leaching study of spent Li-ion batteries. Procedia Environ Sci 16:443–450
6. Sun L, Qiu K (2012) Organic oxalate as leachant and precipitant for the recovery of valuable metals from spent lithium-ion batteries. Waste Manag 32:1575–1582
7. Zeng X, Li J, Shen B (2015) Novel approach to recover cobalt and lithium from spent lithium-ion battery using oxalic acid. J Hazar Mater 295:112–118
8. Li L, Lu J, Ren Y, Zhang XX, Chen RJ, Wu F, Amine K (2012) Ascorbic-acid-assisted recovery of cobalt and lithium from spent Li-ion batteries. J Power Sources 218:21–27
9. Nayl AA, Elkhashab RA, Badaw SM, El-Khateeb MA (2015) Acid leaching of mixed spent Li-ion batteries. Arab J Chem 1–8
10. Jha MK, Kumari A, Jha AK, Umar KV, Hait J, Pandey BD (2013) Recovery of lithium and cobalt from waste lithium ion batteries of mobile phone. Waste Manag 33(9):1890–1897

Part II
Li, Co, Au, Ag, PGMs, Te, Na, W, and In

Environmental Aspects of the Electrochemical Recovery of Tellurium by Electrochemical Deposition-Redox Replacement (EDRR)

P. Halli, M. Rinne, B. P. Wilson, K. Yliniemi, and M. Lundström

Abstract The current study investigates the energy consumption and the corresponding global warming potential (GWP) of tellurium recovery from multimetal solution by the use of a tailored electrochemical recovery approach based on electrodeposition-redox replacement (EDRR). A three-electrode cell was used to recover Te from synthetically prepared pregnant leach solution similar to the PLS of leached Doré slag (30% aqua regia, $[Cu] = 3.9$ g/L, $[Bi] = 4.6$ g/L, $[Fe] = 1.4$ g/L, and $[Te] = 100–500$ ppm). The enrichment of Te on the electrode (with 100 EDRR cycles) had a calculated global warming potential of 3.7 CO_2-eqv from a solution with 500 ppm Te based on a Finnish energy mix. In comparison, a decrease of Te concentration to 100 ppm increased the corresponding environmental impact to 16.9 CO_2-eqv. Overall, GWP was shown to be highly dependent on the geographical area, i.e. the dominating energy production methods.

Keywords Tellurium · Metals circular economy · Energy efficiency · Life cycle assessment (LCA)

Introduction

Tellurium is a rare metalloid element, which has an abundance in the Earth's crust of approximately 1 ppb [1] *cf.* gold of ~5 ppb [2]. Although the application of tellurium is at present rather limited—semiconductors [3], thermoelectrics [4], and CdTe solar panels [5], which have potentially higher efficiency compared to the traditional solar panels [6]—the heightened demand for renewable energy systems [7] will require access to increased levels of tellurium. However, as the content of

---

P. Halli · M. Rinne · B. P. Wilson · M. Lundström (✉)
School of Chemical Engineering, Department of Chemical and Metallurgical Engineering, Research Group of Hydrometallurgy and Corrosion, Aalto University, Vuorimiehentie 2 K, P.O. Box 16200, 00076 Aalto, Finland
e-mail: mari.lundstrom@aalto.fi

K. Yliniemi
School of Chemical Engineering, Department of Chemistry and Materials Science, Aalto University, Kemistintie 1, P.O. Box 16100, 00076 Aalto, Finland

tellurium in Earth's crust is relatively low, the production of applications employing tellurium is also relatively low but the demand is increasing. Currently, the primary source of tellurium comes from the treatment of copper anode slime via a variety of different pyro- and hydrometallurgical unit processes [8, 9]. Nevertheless, in order to promote a more ambitious circular economy of metals, different methods for the recovery of minor/trace concentrations of tellurium (and other rare metals) which are currently lost are essential.

The electrodeposition redox-replacement (EDRR) technique has recently been investigated as an alternative route for the recovery of metals like gold, silver, platinum, and tellurium in the field of hydrometallurgy [10–14]. The method is based on a multistep approach that involves the electrodeposition of a less noble—typically base metal—material on an electrode surface with short pulses (typically a few μs to a few s) followed by a predefined time under open-circuit conditions (OCP) to allow the spontaneous redox replacement of the electrodeposited layers by the nobler target element. The parameters applied during EDRR include deposition potential (E_1), cut-off potential (E_2) or deposition current (i), deposition time (t_1), cut-off time (t_2), and the number of EDRR cycles (n). More details about the overall procedure can be found elsewhere [10–14]. This tailored electrochemical method for the recovery of minor metals or trace amounts of metals from complex multimetal solutions offers a way to increase the sustainability of metal processing.

Life cycle assessment (LCA) is a standardized, systematic method (ISO 14040) to evaluate the environmental impacts of a product or a process through the calculation of a number of impact category indicators. One of the main indicators used is global warming potential (GWP), which relates to atmospheric heat captured due to the associated greenhouse gases emitted and allows different processes or methods to be compared based on their ecological effects. This study demonstrates the environmental impact, specifically GWP, related to energy consumption by a novel direct Te recovery method, EDRR.

Experimental

The base solution utilized consisted of 30% aqua regia [10] (1:3 molar ratio of nitric and hydrochloric acids, VWR Chemicals, Belgium) diluted with Milli-Q ion-exchanged water (Merck Millipore, USA). Due to the aqua regia medium, the high concentration metals—Bi, Cu, and Fe—were sourced from nitrite and chloride salts (bismuth(III) nitrate pentahydrate, Alfa Aesar, Germany; copper(II) chloride dehydrate, VWR Chemicals, Belgium; and iron(III) nitrate nonhydrate, Alfa Aesar, Germany) in order to avoid any potential sulfate contamination. Tellurium content of the solution was varied between 100 and 500 ppm through the use of a Te AAS standard (Sigma-Aldrich, USA) and all chemicals were of technical grade. The electrochemical setup was a typical three-electrode cell with saturated calomel electrode (SCE, B521, SI Analytics, Germany) as a reference electrode (RE), 0.1 mm thick (A = 24 cm^2) platinum plate as a counter electrode (CE), and 0.1 mm thick

($A = 0.4$ cm^2) platinum plate as a working electrode (WE). Both Pt plates were from Kultakeskus (Finland) with a purity of 99.5%. A potentiostat (IviumStat 24-bit CompactStat, Ivium Technologies, the Netherlands) was employed for the electrochemical measurements. In order to determine the mass of the deposit, the working electrode was weighed before and after the electrochemical experiments by Mettler Toledo (XSE205, DualRange, USA), weightings done in triplets. SEM–EDS (scanning electron microscope, Mira^3 Tescan GM, Czech Republic; energy dispersion spectroscopy, Thermo Scientific 50 mm^2 UltraDry, USA) was used to estimate the purity of the deposit with ten spot analyses performed for each sample, and the average values are reported.

During the EDRR experiments, a previously determined deposition potential (E_1) of −500 mV versus SCE and cut-off potential (E_2) of +150 mV versus SCE was used [10]. Deposition time (t_1) was 2 s, the cut-off time (t_2) 1000 s, and the total number of applied EDRR cycles (n) was 100. The specific energy (E_s) consumption [kWh/kg] of the EDRR method for recovering tellurium was calculated by using Eq. (1) [13]:

$$E_s = \frac{E_{cell} \cdot \int_0^{t_1} i(t)dt}{m_{Te}} \quad (1)$$

where E_{cell} is the cell voltage [V], $i(t)$ is the function of current with time, i.e. electrical current is integrated over the time spent for the ED step, resulting in consumed electrical charge [As], t_1 is the time spent for the ED step [s], and m_{Te} is the mass of recovered tellurium [g].

Results and Discussion

The measured tellurium content in the deposits on the electrode surface is presented in Table 1. The enrichment Te on the electrode surface was almost the same (~76–80 wt% by SEM–EDS), regardless of the tellurium content on the original solution. Other metals that were found to predominate in the surface metal deposit analysis included Cu (from 2.4 to 7.8 wt%) and Bi (from 13.0 to 16.3 wt%). Tellurium within the deposits is suggested to be either in the pure metallic form or as an alloy with

Table 1 Purity of Te in the metal deposits (analysed by SEM–EDS) on the electrode surface after conducting the EDRR experiments with $E_1 = -500$ mV and $E_2 = +150$ mV versus SCE, $t_1 = 2$ s, $t_2 = 1000$ s, and $n = 100$

Te content in the solution (ppm)	Purity of Te on the electrode (wt%)
500	76
250	79
125	80
100	80

Cu and Bi (e.g. CuTe and Bi_2Te_3). The purity of the recovered tellurium was also found to be in good agreement with previously published results [10]. Moreover, in solutions with 500, 250, 125, and 100 ppm tellurium, the corresponding tellurium recoveries on the electrode with 100 cycles were 0.34%, 0.43%, 0.35%, and 0.36%, respectively. Although the recovery of tellurium is rather low, it is worth noting that the total processing time was also short. For example, when $n = 100$, the total EDRR time spent with a 500 ppm tellurium solution was only ~1 h. Furthermore, it is assumed that an increase in the number of cycles (n) will lead to enhanced levels of tellurium recovery.

The calculated specific energy consumption for tellurium recovery—defined by the deposit mass multiplied by the Te content (SEM–EDS)—varied from ~14 to ~63 kWh/kg with tellurium content of 500 ppm originally in solution to 100 ppm in solution, Fig. 1. Moreover, one of the limiting factors for EDRR is mass and/or charge transfer [10–14] and the ion movement further away from the electrode surface requires in the order of a magnitude more time *cf.* the ions that are closer to the electrode surface. Therefore, also the energy consumptions are increased when decreasing the tellurium content due to a lower amount of ions in the solution, leading to a situation, where there is a lower amount of ions closer to the electrode surface. However, the lower content of tellurium has not a direct effect on the recovered purity (Table 1) or on the recovery rates, but more energy is consumed during the process. The specific energy consumption for tellurium recovery was found to be approximately the same with the two highest concentrations employed (250–500 ppm), however, it was found to drastically increase as the level of Te in the solution decreased, Fig. 1.

In order to get more understanding about the environmental impact of the selected tellurium recovery method, the global warming potential (GWP) of EDRR was calculated by using GaBi software [15] (ecoinvent 3.5, [16]). GWP indicates the measured

Fig. 1 Calculated specific energy consumptions of the EDRR method as a function of tellurium in the solution. (Color figure online)

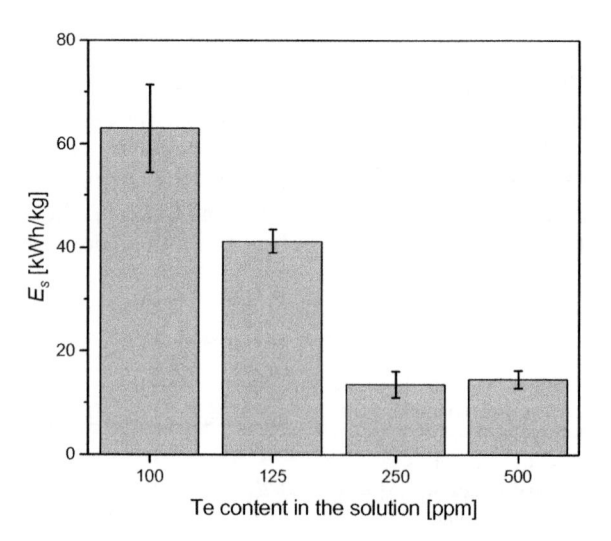

Fig. 2 GWP values for tellurium recovery by EDRR from 100 and 500 ppm solutions in the largest tellurium producing countries, in the European continent and Finland based on the specific energy consumption based on the ecoinvent 3.5 database. (Color figure online)

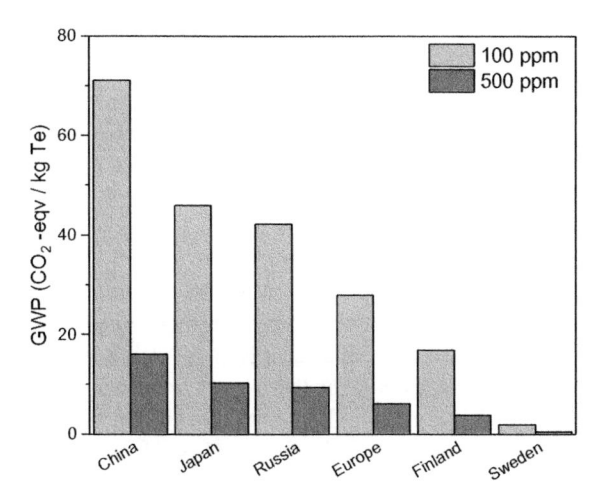

heat trapped in the atmosphere by various greenhouse gases when compared to CO_2 over a 100-year time window. GWP is scaled for one unit (kg) of CO_2, therefore, e.g. methane has 21 CO_2-eqv and N_2O 310 CO_2-eqv. The selected countries for GWP determination were the top four tellurium producing countries (China, Japan, Sweden, and Russia [17]), along with Finland and the European average without Switzerland. The database values for the electricity production mixes were from the year 2018/2019, and the GWP values were calculated for the electricity consumption of tellurium recovery from 100 (~63 kWh/kg) and 500 ppm (~14 kWh/kg) solutions.

The results suggest that the GWP values for the recovery of tellurium from 100 ppm solutions was considerably larger than from 500 ppm solutions, Fig. 2. The average energy production mix (ecoinvent 3.5, [15]) suggests that the impacts of producing tellurium are expected to be larger in China, Japan, and Russia compared to Europe. The main factor for the increased GWP values results from the database that includes high-level GWP fossil fuel-driven energy sources, whereas nuclear power is listed as GWP energy sources that produce no direct emissions. Nevertheless, the database *does not* take into account the other aspects related to power sources, for example, the radioactive waste produced by nuclear power or the huge demand for critical metals needed required for renewable energy applications like solar panels [18, 19].

In Russia, the share of nuclear power utilized is relatively large, whereas the Fukushima incident in 2011 has largely decreased nuclear power use in Japan [20] and consequently, increased the share of imported fossil fuels in Japan. Although Finland's electricity sector is far less CO_2—intensive than the European average, Finland (16.9 and 3.7 kg CO_2-eqv/kg) was still reliant on fossil fuels *cf.* Sweden (1.8 and 0.4 kg CO_2-eqv/kg), Fig. 2. Although both Nordic countries primarily utilize nuclear and hydropower as their sources of low-emission electricity, the extent is much larger in Sweden, which explains the discrepancy [21].

This study suggests that the concentration of tellurium in solution has a marked effect on the energy consumption required for Te recovery. Nevertheless, the GWP value related to tellurium recovery is not necessarily dependent on the process, process optimization, and process conditions, but rather that the origin of energy, *i.e.* the energy production method used is also a key factor. For example, despite the higher energy consumption required for EDRR with a 100 ppm tellurium solution, the related GWP value for tellurium recovery is calculated to be lower in Sweden when compared to tellurium recovery from a 500 ppm solution in Finland. Therefore, the local electricity production methods used for tellurium recovery by EDRR will have a significant impact on the environment and overall global warming potential. Therefore, the individual energy mix, which is averaged across a whole country, is not always accurate as there are often local variations. Nevertheless, the results presented in Fig. 2 highlight both the opportunities and challenges related to energy production methods, as the usage of fossil fuels and their direct greenhouse gas emissions ultimately have a dominating impact on the sustainability of the industrial scale application of "electricity-based technologies" like EDRR for tellurium recovery.

Conclusions

The specific energy consumption of the EDRR process in terms of recovering tellurium was determined to increase with decreasing Te solution content. Correspondingly, the global warming potentials (GWP) of the EDRR method were also found to be increased with reduced tellurium concentrations. With 250–500 ppm tellurium in solution, the emissions were only ~16% of those calculated for the 100 ppm solution of tellurium. Moreover, the geographical location, and especially the national or local energy production mix, was found to have a major impact effect on the level of GWP expected.

Acknowledgements This work has been financed and supported by the "GoldTail" (Grant 319691, PH, MR, and BW) and "NoWASTE" (Grant 297962, KY and ML) projects funded by Academy of Finland. The research also made use of the Academy of Finland funded "RawMatTERS Finland Infrastructure" (RAMI) based at Aalto University.

References

1. Ibers J (2009) Tellurium in a twist. Nat Chem 1(6):508. https://doi.org/10.1038/nchem.350
2. Foster RP (1993) Gold metallogeny and exploration. Springer, 426 pp. ISBN: 978-0-412-56960-9
3. Tang G, Qian Q, Wen X, Zhou G, Chen X, Sun M, Chen D, Yang Z (2015) Phosphate glass-clad tellurium semiconductor core optical fibers. J Alloy Compd 633:1–4. https://doi.org/10.1016/j.jallcom.2015.02.007

4. Aspiala M, Taskinen P (2016) Thermodynamic study of the Ag–Sb–Te system with an advanced EMF method. J Chem Thermodyn 93:261–266. https://doi.org/10.1016/j.jct.2015.08.025
5. Maani T, Celik I, Heben MJ, Ellingson RJ, Apul D (2020) Environmental impacts of recycling crystalline silicon (c-SI) and cadmium telluride (CDTE) solar panels. Sci Total Environ 735:138827. https://doi.org/10.1016/j.scitotenv.2020.138827
6. Imamzai M, Aghaei M, Thayoob YHM, Forouzanfar M (2012) A review on comparison between traditional silicon solar cells and thin-film CdTe solar cells. In: Proceedings national graduate conference, 5 pp
7. Sherwani AF, Usmani JA, Varun (2010) Life cycle assessment of solar PV based electricity generation systems: a review. Renew Sustain Energy Rev 14(1): 540–544. https://doi.org/10.1016/j.rser.2009.08.003
8. Biswas J, Jana RK, Kumar V, Dasgupta P, Bandyopadhyay M, Sanyal SK (1998) Hydrometallurgical processing of anode slime for recovery of valuable metals. Environ Waste Manag 216–224. ISSN: 0971-9407
9. Robles-Vega A, Sanchez-Corrales VM, Castillon-Barraza F (2009) An improved hydrometallurgical route for tellurium production. Miner Metall Process 26(3):169–173. https://doi.org/10.1007/BF03402231
10. Halli P, Wilson BP, Hailemariam T, Latostenmaa P, Yliniemi K, Lundström M (2020) Electrochemical recovery of tellurium from metallurgical industrial waste. J Appl Electrochem 50. https://doi.org/10.1007/s10800-019-01363-6.
11. Halli P, Heikkinen JJ, Elomaa H, Wilson BP, Jokinen V, Yliniemi K, Franssila S, Lundström M (2018) Platinum recovery from industrial process solutions by electrodeposition-redox replacement. ACS Sustain Chem Eng 6(11):14631–14640. https://doi.org/10.1021/acssuschemeng.8b03224
12. Korolev I, Spathariotis S, Yliniemi K, Wilson BP, Abbott AP, Lundström M (2020) Mechanism of selective gold extraction from multi-metal chloride solutions by electrodeposition-redox replacement. Green Chem. https://doi.org/10.1039/d0gc00985g
13. Wang Z, Halli P, Hannula P, Liu F, Wilson BP, Yliniemi K, Lundström M (2019) Recovery of silver from dilute effluents via electrodeposition and redox replacement. J Electrochem Soc 166(8):E266–E274. https://doi.org/10.1149/2.0031910jes
14. Hannula P-M, Pletincx S, Janas D, Yliniemi K, Hubin A, Lundström M (2019) Controlling the deposition of silver and bimetallic silver/copper particles onto a carbon nanotube film by electrodeposition-redox replacement. Surf Coat Technol 374:305–316. https://doi.org/10.1016/j.surfcoat.2019.05.085
15. Sphera, GaBi-software, GaBi solutions
16. Ecoinvent, database ecoinvent 3.5. www.ecoinvent.org. Accessed 12 June 2020
17. U.S. Geological Survey (2020) Mineral commodity summaries 2020, U.S. Geological Survey, 200 pp. https://doi.org/10.3133/mcs2020.
18. IAEE (2018) Renewable energy materials supply implications. IAEE Energy Forum. https://www.iaee.org/documents/2018EnergyForum1qtr.pdf. Accessed 12 June 2020
19. Anctill A, Fthenakis V (2012) Critical metals in strategic photovoltaic technologies: abundance versus recyclability. Progr Photovolt 21(6):1253–1259. https://doi.org/10.1002/pip.2308
20. U.S. Energy Information Administration, Country Analysis Brief: Japan (2017), 21 pp. Internet: https://www.eia.gov/international/analysis/country/JPN. Accessed 12 June 2020
21. ENTSO-E (2018) Electricity in Europe 2017

Sodium Metal from Sulfate

Jed Checketts and Neale R. Neelameggham

Abstract A new method of making sodium metal from sodium sulfate is discussed. Anhydrous sodium sulfate as may be made from sodium sulfate waste solutions is reduced with aluminum metal. The reactor design to make sodium metal along with aluminum and sulfur oxides minimizing wastes is explored. Thermochemical tools are used in this development. Experiments carried out in this regard and how they fit the thermochemistry are evaluated in this paper.

Keywords Sodium metal · Sodium sulfate · Aluminothermic · Lantern reactor

Introduction

During the 1990s, Powerball Technologies, Salt Lake City, Utah, developed a method to provide hydrogen on demand, where metallic sodium was encapsulated in polyethylene. The encapsulated sodium metal is cut by a mechanism in water kept inside a pressure vessel, and the hydrogen generated by sodium reaction with water builds up pressure and is made available to external devices, such as fuel cells or burners on demand [1, 2]. Powerball Technologies explored methods of regenerating sodium metal from the sodium hydroxide made during the hydrogen production [3, 4]. Known techniques were electrolysis, carbothermic reduction, and methanothermic reductions. Market conditions were not suitable towards the turn of the twenty-first century and Powerball projects were temporarily suspended. However, bench-scale possibilities of making sodium metal were continued to be analyzed.

The possible availability of waste sodium sulfate from processes such as boric acid production lead to exploring the use of sodium sulfate as a raw material for sodium production. In addition, the availability of secondary and tertiary metallic aluminum posed an interesting potential for developing a non-carbon reduction

---

J. Checketts
Powerball Technologies, Gardena, CA, USA

R. Neelameggham (✉)
IND LLC, South Jordan, Utah, USA
e-mail: neelmeggham@outlook.com

approach. Thermochemical analysis was carried out followed by evaluating reactor designs as discussed in this paper. A preliminary paper was presented during the 2019 TMS annual meeting in San Antonio on general aluminothermic techniques for sodium reduction [5].

Thermochemistry

Anhydrous sodium sulfate can be reacted with aluminum in a stepwise fashion as shown below:

$$\text{Heat of Reaction}$$
$$\Delta H°, \text{ kcal}$$

$$3Na_2SO_4 + 8Al = 3Na_2S + 4Al_2O_3 \quad -874.3 \quad (1)$$

$$3Na_2S + 2Al = 6Na + Al_2S_3 \quad +147.8 \quad (2)$$

Adding,

$$3Na_2SO_4 + 10Al = 6Na + Al_2O_3 + Al_2S_3 \quad -726.5 \quad (3)$$

Other reactions of importance in this process are

$$\text{Heat of Reaction}$$
$$\Delta H°, \text{ kcal}$$

$$2Al + 3S = Al_2S_3 \quad -121.6 \quad (4)$$

$$Al_2S_3 + 1.5O_2 = Al_2O_3 \quad -277.5 \quad (5)$$

The reactions (1) and (3) are exothermic, but reaction (2) is endothermic at the standard 298.15 K state; reaction (2) becomes feasible in making sodium vapor beyond 1250 K. The melting points are Al 650 °C, Na_2SO_4 886 °C, Al_2S_3 1100 °C, and Na_2S 1126 °C. It can be noted that reaction (3) can be adjusted to produce at high temperatures, sodium vapor leaving behind alumina and aluminum sulfide. The aluminum sulfide content can be further oxidized to produce more alumina showing the way of producing a valuable by-product which is clean, and the sulfur oxides can be handled to produce sulfuric acids. It was felt that there may be a need to utilize reaction (4) to an extent to provide the activation energy needed in addition to utilizing electrical resistor heating of the charge to overcome phase change heating to further facilitate initiating the reaction.

It is to be noted that other catalytic lower temperature reactions are possible with lime additions, but they will not be able to produce alumina by-products. This study started evaluating reactor designs suitable for doing that using mainly reaction (3) with an initiator amount of reaction (4).

Reactor Design

For the bench-scale reaction, an electrical resistance heater contained within a proprietary Lantern-style reactor was developed. The intent is to initiate the exothermic portion of the reactions as quickly as possible to minimize and avoid unnecessary side reactions. It was also necessary to minimize materials of construction which may react with aluminum and or sodium sulfide-sulfate combinations. We are developing a power resistance heater reactor which mainly initiates the reaction by providing activation energy, and the reaction temperature would then moderate power input. This we term as the Lantern Reactor for the bench-scale use.

In the Lantern reactor, carbon resistors were used. Two sets of configurations were evaluated. The first one utilized 4 each of 4 in. long by 0.25 in. diameter carbon rods making a resistor length of 16 in. The resistivity of the carbon is in the range 1100–1400 $\mu\Omega$ in. as measured in our experiments, which includes the internal connectivity resistances of the carbon rods. This produced a power of about 300 W, which was found to be insufficient to trigger the reaction with a charge of about 100 gm. The design was modified to give a resistor length of 8 in. but with an electrical area equivalent to 2×0.25 in. diameter rods, which gave close to 850 W and was found sufficient to trigger the reaction in 5 min or less. Table 1 shows a set of measurements of the resistance heater in the reactor.

Table 1 Lantern reactor resistances

	Voltage (V)	Current (A)	Resistance (Ω)
4 each 4″ long electrode in series			
Set 1	11.6	27	0.430
Set 1	11.4	31	0.368
Set 2	11.6	24.7	0.470
Set 2	11.5	23.7	0.485
2 each pair of 4″ long electrodes in series			
Set 3	9.8	93	0.105
Set 3	8.72	93.1	0.094

The contents in the reactor were initially encapsulated with aluminum foil and the lantern reactor setup was contained inside a graphite crucible. The heat melted the aluminum foil. The aluminum foil was replaced with steel shim stock wrap, which also reacted, possibly by reaction with melting aluminum, forming the low melting inter-metallic Fe_3Al. In both cases, the immediate outer container, graphite crucible, held up the product mass. The graphite crucible containing the lantern reactor was kept in a pressure vessel with means to get the leads for electrical connections to the sides and vacuum connection to a second vessel followed by a vacuum pump. Figures 1, 2, 3 and 4 show the different views of the Lantern reactor.

At the time of writing this manuscript, the bench-scale setup had only evolved to the extent that the reactants were completely utilized in producing mixed products. Some modifications to the reactor assembly may have to be made in having sodium vapors isolated going into the 2nd vessel and condense there. The product mix is planned to be analyzed chemically. The Separation of Lantern product solids into components is in progress and will be reported as available.

The variations in voltage are from the time effect which changes the temperature of the resistors, which was not measured.

Fig. 1 Lantern reactor and graphite box container (*Courtesy* Powerball Technology) (Color figure online)

Fig. 2 Lantern Reactor after cooling following 3 min resistance heating triggering a reaction with aluminothermic sodium sulfate reactant (*Courtesy* Powerball Technology) (Color figure online)

We might choose to concentrate on reaction 2—while adopting carbothermic conversion of Na_2SO_4 if the separation of solid alumina from solid sodium sulfide proves cost-ineffective.

Conclusion

It has been shown that it is feasible to develop a Power Reactor such as the Lantern reactor to provide activation energy to reactions which are exothermic in nature but still need to be triggered to start.

For the long-term thinking of the evolution of the process, it may be relevant to note that companies such as Rio Tinto are working to utilize renewable energy sources. For instance, it has been reported that Rio Tinto is building a solar plant consisting of 100,000 panels, covering an area of 105 ha. At this stage, we are evolving our reactor design. Once this is done, we will start optimizing for scale-up and cost-effectiveness by having two or more marketable products in creating a sustainable process and minimizing waste.

Fig. 3 Lantern Reactor opened showing products (*Courtesy* Powerball Technology) (Color figure online)

In addition to producing electricity using solar panels, solar heat energy directed at our proposed sulfide to sodium process (in particular, replacing aluminum with carbon in the sulfate to sulfide step as done in conventional processes) may utilize the acreage covered in silicon solar cells even more effectively by concentrating solar energy to effect the chemical change by providing activation energy. Solar energy can be used in the first step of upgrading hydrated sodium sulfate to anhydrous sodium sulfate. At least we could make potential readers of our process aware that it is not our intention to base the chemistry far into the future on a large scale based solely on scrap aluminum shavings.

Fig. 4 Lantern reactor with products. Products are powdery substances easily scraped out into containers (*Courtesy* Powerball Technology) (Color figure online)

References

1. Cheketts JH (1998) Hydrogen generation pelletized fuel. U.S. Patent 5,728,464, March 17, 1998
2. Cheketts JH (1998) Hydrogen generation system and pelletized fuel. U.S. Patent 5,817,157, October 6, 1998
3. Cheketts JH, Hatfield KE, Neelameggham NR (2001) System for extracting sodium metal from sodium hydroxide with methane as a reductant. U.S. Patent 6,221,310, April 24, 2001
4. Checketts JH (2001) System for extracting sodium metal from sodium hydroxide and a reductant of natural gas. U.S. Patent 6,235,235, May 22, 2001
5. Checketts JH, Neelameggham NR (2019) Manufacturing of hydrogen on demand using aluminum can scrap with near zero waste. In: Chesonis C (eds) Light metals 2019. The minerals, metals and materials series. Springer, Cham, pp 1385–1387. https://doi.org/10.1007/978-3-030-05864-7_172

Preparation of High Grade Ammonium Metatungstate (AMT) as Precursor for Industrial Tungsten Catalyst

Alafara A. Baba, Sadisu Girigisu, Mustapha A. Raji, Abdullah S. Ibrahim, Kuranga I. Ayinla, Christianah O. Adeyemi, Aishat Y. Abdulkareem, Mohammed J. Abdul, and Abdul G. F. Alabi

Abstract The increasing demand for pure tungsten and its compounds due to their high tensile strength makes them a versatile material in catalyst, heavy alloy, cemented carbide, among others. This exceptional property makes it to be of high interest by industrialists for use in the engineering and manufacturing industries. Thus, preparation of ammonium metatungstate (AMT) from a Nigerian wolframite ore by hydrometallurgical technique was examined in sulphuric and phosphoric acid. During leaching, parameters such as leachant concentration, reaction temperature, and particle size on ore dissolution were examined. At optimal conditions (2.0M H_2SO_4 + 0.15M H_3PO_4, 75 °C, −63 μm), 93.7% of the ore dissolved within 120 min. The calculated activation energy of 6.93 kJ/mol supported the proposed diffusion mechanism. The leachate at optimal conditions was treated to obtain a pure tungstate solution. The purified solution was beneficiated to prepare a high grade AMT (($NH_4)_6[H2W_{12}O_{40}]\cdot 4H_2O$: 96-901-3322, *m.p.*: 98.7 °C, *density*: 2.16 g/cm^3), which serves as an intermediate for some defined industries.

Keywords Wolframite · Leaching · H_2SO_4 · H_3PO_4 · Ammonium metatungstate (AMT) · Industrial applications

---

A. A. Baba (✉) · M. A. Raji · A. S. Ibrahim · K. I. Ayinla
Department of Industrial Chemistry, University of Ilorin, P.M.B. 1515, Ilorin 240003, Nigeria
e-mail: alafara@unilorin.edu.ng; baalafara@yahoo.com

S. Girigisu (✉) · C. O. Adeyemi
Department of Science Laboratory Technology, Federal Polytechnic Offa, P.M.B. 420, Offa, Kwara State, Nigeria
e-mail: sadisu.girigisu@fedpoffaonline.edu.ng

A. Y. Abdulkareem
National Mathematical Centre, P.M.B 118, Sheda Abuja, Kwali, Nigeria

M. J. Abdul
Department of Mechanical Engineering, Federal Polytechnic Offa, P.M.B. 420, Offa, Kwara State, Nigeria

A. G. F. Alabi
Department of Material Science and Engineering, Kwara State University, P.M.B 1530, Malete, Nigeria

Introduction

Tungsten is of critical importance to the modern society, not only because of its scarcity and strategic importance [1] but also due to its demand earlier proposed to increase at an annual rate between 4.6 and 6.4% from 2013 to 2018 [2]. Important tungsten compounds such as tungsten oxides, metal tungstates among others can be used as catalysts, for medical diagnosis and pigments [3]. Most importantly, ammonium metatungstate (AMT) is mainly used as a raw material for the production of tungsten catalysts for a variety of reactions including oxidation, hydroxylation, hydrogenation, and polymerization. Here, suitable carriers are often impregnated with alkali-free solutions of tungsten and processed to remove the water and ammonia, which can then be refined as catalysts in petrochemical industries.

One of the most important tungsten minerals that can be used for the above purposes are the monotungstates such as scheelite (calcium tungstate, $CaWO_4$), stolzite (lead tungstate, $PbWO_4$), and wolframite which is a combination of ferrous tungstate or ferberite ($FeWO_4$) and manganous tungstate or hurbenite (MnO_4). From any of the aforementioned minerals, tungstate solution is firstly obtained through leaching and then undergoes solvent extraction/beneficiation to obtain ammonium tungstate solution followed by crystallizing AMT via evaporation [4]. For example, in the classical acid-leaching procedure, scheelite concentrates are decomposed to produce tungstic acid (H_2WO_4) using hydrochloric acid solution, this results into the formation of a dense solid layer of tungstic acid that covers the surface of the scheelite particles, which results into reducing the rate of dissolution process. In order to avoid the formation of this tungstic acid in the leaching process, an oxidant (H_3PO_4) is often introduced into the solution until the characteristic yellow paste tungstic acid resulted in complexation with H_3PO_4 acting as a chelating agent. In this work, the possibility of producing a high grade AMT was examined through treatment of an indigenous wolframite ore using H_2SO_4/H_3PO_4 for the dissolution process. The leached liquor at optimal conditions was beneficiated to obtain high grade AMT via precipitation and crystallization methods.

Materials and Methods

Material

Wolframite sample used for this study was sourced from the North Eastern part of Anka community (N12″ 13″ 16″ E005″ 49″ 35″), Anka Local Government Area of Zamfara State, Nigeria. The sample was crushed and grinded into different particle sizes using standard Tyler screen sieve of different sizes of the range: $-90+75\,\mu m$, $-112+90\,\mu m$, and $-300+112\,\mu m$. The finest particle with the largest specific surface area determined to be ($-90+75\,\mu m$) was used throughout the dissolution process, unless otherwise stated. The raw ore together with the residue obtained at optimal conditions was accordingly characterized using X-ray diffractometry (XRD), X-ray flouresence (XRF), and scanning electron microscope (EDS) techniques respectively.

Leaching Test

Two g of the wolframite ore was treated with equal volume of sulphuric and phosphoric acid in the glass reaction flask at varying reaction temperature range between 27 and 75 °C and at various leaching times of 5–120 min. At the end of each leaching run, the solution was filtered, washed and the residue was dried at room temperature (27 ± 2 °C). The content of tungsten in the filtrate was analyzed by thiocynate spectrometric method using a spectrometer (Becman Couter Du 730) at 420 nm [5]. The optimal acid concentration which gave the highest dissolution was used for further optimization of other parameters. The rate of dissolution of the ore was examined using the Avrami and Shrinking core model for better understanding of the dissolution mechanism as described by Zhu et al. [6]. The residue obtained at optimal condition (75 °C in 2.0 mol/L H_2SO_4 + 0.15 mol/L H_3PO_4) was accordingly characterized using XRD, XRF, and SEM–EDX techniques.

Solvent Extraction/Beneficiation Studies

Solvent extraction being a proven technique for the extraction and separation of metals from dilute and complex leach solutions has been used for tungsten recovery in this study. The resultant leachate used for solvent extraction was analyzed using spectrophotometer techniques [7]. The efficiency of the Aliquat 336 extractant used for the extraction of tungsten from the leached liquor obtained was studied by obtaining the percent extracted in each case. The results obtained at maximum extractant concentration of 0.25 M Aliquat 336 during tungsten extraction was stripped from loaded organic phase by 0.1 M ammonium hydroxide. The process yielded 93.7% pure tungsten which was treated to produce ammonium metatungstate solution, further beneficiated through evaporation and crystallization to obtain a high grade crystal ammonium metatungstate of industrial value.

Results and Discussion

Characterization Studies

The mineralogical analysis of the raw ore by XRF shows that wolframite mineral under study contained 15.71 wt% Si, 1.38 wt% K, 1.33 wt% Fe, 0.52 wt% Zn,

1.43 wt% Pb, and 51.45 wt% W as major elements. The mineral purity of the raw wolframite ore analysed by XRD ascertains the ore to contain admixtures of *Ferberite* (Fe$_{2.00}$W$_{2.00}$O$_{8.00}$: 96-900-8125), *Stolzite* (Pb$_{4.00}$W$_{4.00}$O$_{16.00}$: 96-00-9812), *Quartz* (Si$_{3.00}$O$_{6.00}$: 96-500-0036), *Slvite* (Fe$_{2.00}$W$_{2.00}$O$_{8.00}$: 96-900-3119), and *Wulfenite* (Pb$_{4.00}$Mo$_{4.00}$O$_{0.00}$: 96-900-8125). The SEM–EDX results further support the elemental characterization by XRF and XRD, respectively.

Leaching Studies

The leaching experiment shows that the wolframite ore dissolution in H$_2$SO$_4$/H$_3$PO$_4$ solution increases rapidly with increasing acid concentration, temperature with decreasing particle diameter at moderate stirring as follows:

Effect of Leachant Concentration

The effects of the initial phosphoric acid concentration with tetraoxosulphate (VI) acid on the conversion of wolframite ore were studied. The conversion of wolframite increases with an increase in the initial phosphoric acid concentration (0.01–0.35 M). After 90 min the conversion rate of wolframite ore is well above 90% with 0.15 MH$_3$PO$_4$ and 2.0 M H$_2$SO$_4$ solution, indicating that increasing the initial H$_3$PO$_4$ concentration is an effective method to raise the decomposition of wolframite. The addition of phosphoric acid aids the removal of the yellow tungstic acid formation with sulphuric acid solution and thus, improves the dissolution rate and increasing the wolframite ore dissolution.

Effect of Reaction Temperature

The effect of reaction temperature on the leaching rate was examined at 27–75 °C for the -63 μm particle size using 2.0 M H$_2$SO$_4$ + 0.15 M H$_3$PO$_4$ solution at reaction time of 120 min. At a set of experimental conditions, about 55.6%, 60.5%, 62.85%, 71.1%, and 93.7% dissolution were achieved with reaction temperatures of 27 °C, 40 °C, 55 °C, 65 °C, and 75 °C, respectively.

Effect of Particle Size

The effect of particle size on the rate of wolframite dissolution was studied over the ranges of three particle sizes of -63, -75, and -150 μm, respectively. It was established that an increase in surface area per unit of mass gives rise to a higher reaction rate [8]. Thus, the rate of wolframite dissolution of smallest particle size was higher at optimal condition (75 °C in 2.0 mol/L H_2SO_4 + 0.15 mol/L H_3PO_4 at reaction time of 120 min).

Solvent Extraction/Beneficiation Studies

Prior to obtaining pure tungsten solution, iron and other gangues in the leach liquor were selectively precipitated at pH 3 to achieve improved extraction and beneficiation efficiencies. About 93.7% purified tungsten solution was successfully extracted at pH 3 as AMT. The produced AMT was further purified by evaporation and crystallization to obtain AMT with 95% purity. The crystallinity of the ammonium metatungstate in the prepared sample was evaluated on the basis of the prominent XRD peaks detected in the 2θ ranges of 12.970, 16.714, 18.479, 19.600, 19.70, 21.035, 22,203, 26.999, and 27.745. The observed narrow slit at 2θ of 12.970 gave crystal lattice (h, k, l = 0, 2, 0) which confirms the presence of ammonium metatungstate [9, 10], with experimental chemical formula of the compound identified as $(NH_4)_6[H2W_{12}O_{40}]0.4H_2O$. The operational hydrometallurgical scheme for purifying a Nigerian wolframite ore to obtain high grade AMT is summarized in Fig. 1.

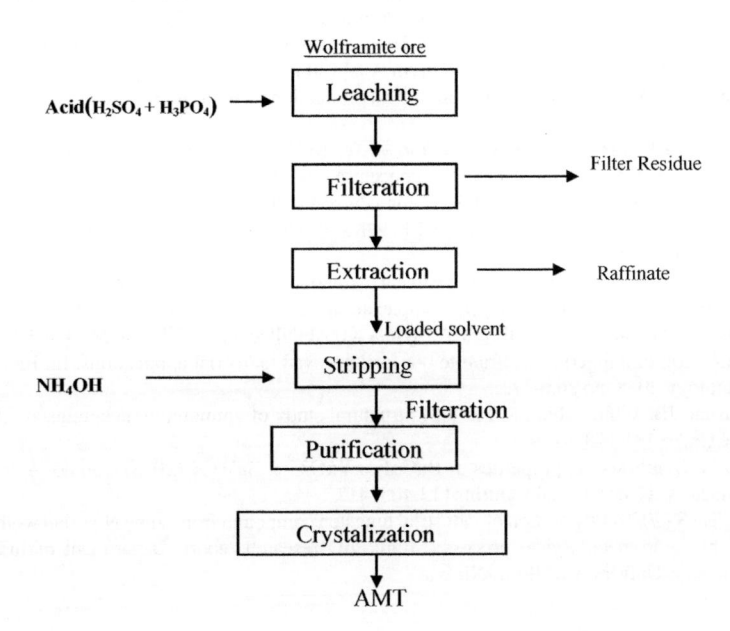

Fig. 1 A Flow chart for the production of ammonium metatungstate (AMT) from an indigenous wolframite ore [11]

Conclusion

The extent of wolframite ore dissolution in this study was highly enhanced by optimizing the following parameters (H_2SO_4/ H_3PO_4) concentration, reaction temperature, and particle size at the optimal conditions of 75 °C in 2.0 mol/L H_2SO_4 + 0.15 mol/L H_3PO_4 and reaction time of 120 min. The obtained experimental data agreed well with the Avrami model, with one diffusional model as the rate-controlling step in the dissolution process. This affirms the possibility of producing high grade and industrially applicable AMT by hydrometallurgical route from an indigenous wolframite ore. The AMT product as characterized, if well tapped, could contribute to the country's economic growth and development in some defined catalytic industries; while saving the country's current hard-earned foreign exchange through petroleum explorations devalued by COVID-19 pandemic.

References

1. European Commission (2020) Communication from the Commission to the European Parliament, the Council, the European Economic and Social Committee and the Committee of the Regions Tackling the challenges in commodity markets and on raw material
2. Dvořáček J, Sousedíková R, Vrátný T, Jureková Z (2017) Global tungsten demand and supply forecast. Arch Min Sci 62(1):3–12. https://doi.org/10.1515/amsc-2017-000
3. Blovská V, Bělina P, Šulcová P (2013) Synthesis of tungstate pigments of the formula $MNd_2W_2O_{10}$ (M = Ni, Zn, Mn). J Therm Anal Calorim 113:83–89
4. Gaur RPS (2006) Modern hydrometallurgical production methods for tungsten. JOM 58(9):45–49
5. Ting-ting LI, Yan-bai S, Si-kai Z, Yao-yu Y, Rui L, Shu-ling G, Cong H, De-zhou W (2019) Leaching kinetics of scheelite concentrate with sodium hydroxide inthe presence of phosphateTrans. Nonferrous Met Soc China 29:634–640
6. Zhu X, Liu X, Zhao Z (2019) Leaching kinetics of scheelite with sodium phytate. Hydrometallurgy 186:83–90
7. Guan WJ, Zhang GQ, Gao CJ (2012) Solvent extraction separation of molybdenum and tungsten from ammonium solution by H_2O_2-complexation. Hydrometallurgy 128:84–90
8. Baba AA, Muhammed MO, Raji MA, Ayinla KI, Abdulkareem AY, Lawal M, Bale RB (2018) Purification of a nigerian wolframite ore for improved industrial applications. In: Rare metal technology 2018, pp 265–272
9. Christian JB, Whittingham MS (2008) Structural study of ammonium metatungstate. J Solid State Chem 181:1782–1791
10. Łącz A, Synthesis PP, properties of BaCe1−xYxO3−δ–BaWO4 (2013) Composite protonic conductors. J Therm Anal Calorim 113:405–412
11. Girigisu S (2020) Preparation of industrial tungsten compound from some Nigerian wolframite ores by hydrometallurgical process. MPhil/PhD research report, Department of Industrial Chemistry, University of Ilorin-Nigeria

Industrial-Scale Indium Recovery from Various e-Waste Resources Through Simulation and Integration of Developed Processes

Basudev Swain, Jae Ryang Park, Kyung Soo Park, Chan-Gi Lee, Hyun Seon Hong, and Jae-chun Lee

Abstract Various e-wastes like waste LCD, LED, and LCD etching industry wastewater are important secondary resources for indium, which is a critical metal. In this research, the industrial-scale indium recovery from e-waste resources like waste LCD, LED, and LCD industry etching wastewater is being emphasized through simulation and integration of the developed processes. A demonstration plant for indium recovery on one ton/day of ITO etching wastewater has been developed with almost complete (99%) recovery of indium. For the indium recovery, integration of the processes can be managed by following two approaches unique to this system, (i) utilization of ITO etching industry wastewaters for the leaching of waste LCD, (ii) integration of leaching processes developed for waste LCD and LED with that of the treatment process for ITO wastewaters. Through the proposed approach, the semi-conductor manufacturing industry and ITO industry can address various pressing issues like (i) waste disposal, (ii) indium recovery, (iii) circular economy.

Keywords Indium · ITO etching · Wastewater · Recycling

Introduction

Global indium (In) consumption for the years 2014, 2015, 2016, and 2017 were 1565, 1643, 1640, and 1683 tons, and during 2007–2017 the consumption increased by 58% from 1060 tons [1]. Grand View Research, Inc. has estimated the demand of In to

B. Swain (✉) · J. R. Park · K. S. Park · C.-G. Lee
Materials Science and Chemical Engineering Center, Institute for Advanced Engineering (IAE), Yongin-Si 17180, Republic of Korea
e-mail: Swain@iae.re.kr; basudevswain@outlook.com

H. S. Hong
Department of Environment & Energy Engineering, Sungshin University, Seoul 142-732, Republic of Korea

J. Lee (✉)
Mineral Resources Research Division, Institute of Geoscience and MineralResources (KIGAM), 124, Gwahak-ro, Yuseong-gu, Daejeon 34132, Republic of Korea
e-mail: jclee@kigam.re.kr

grow at a compound annual growth rate (CAGR) of around 6.4% from 2017 to 2025 [2]. United States Geological Survey (USGS) reported that the world total refinery production of In was 755, 680, 720, 741, 760 tons for the years 2015, 2016, 2017, 2018, and 2019, respectively [3–5]. This indicates that the In metal is at the critical bottleneck from a supply–demand perspective. In the latest report EU commission has classified the In as a critical raw material, the criticality of the In for the EU is determined based on supply risk and economic importance [6]. The US Department of Energy (DOE) [7] has also reported In as a critical material for the energy and emerging technologies [8].

Indium-tin-oxide (ITO) is a transparent, conducting, and easy to deposit material, widely used in field emission displays, electroluminescent displays, electrochromic displays, touch panels, windshields, photovoltaics, optoelectronics, heat reflective coatings, gas sensors, energy-efficient windows, and solar cells [9, 10]. The ITO accounted for 97% of the transparent conducting oxide markets that are mainly used in the liquid crystal displays (LCDs) devices like TV, computer, laptop, smartphone, and notebooks. The demand and usage pattern indicates the massive end of life (EOL) In bearing waste generation, which is a potential resource to address the supply risk of this critical metal. At the EOL, these are added to the e-waste stream. In the e-waste stream, the waste LCD accounts for the second uppermost entity after the discarded computer. In the context of ITO, a significant volume of ITO etching wastewaters is being generated during the manufacturing of LCDs. These e-wastes as well as the etching industry wastewaters (etching solution that is used for LCD manufacturing, after several cycles of use reach the end of life) are the significant resources for In as ITO contains 90% of In_2O_3 [11, 12]. Currently, industrial-scale In recovery processes for e-waste and ITO etching wastewater are not well developed. To successfully address the issues like supply–demand disparity, e-waste management, and ITO industry wastewaters treatment, the In value recovery from such secondary resources is the only attractive solution. Processing of these resources can significantly address certain aspects of urban mining and resource recycling while dwelling upon the circular economy of In.

Waste specific industrial-scale process development is often cost-inefficient and time-consuming. Specific to the existing industrial process, simulation and integration can be a sustainable approach to address the various issues connected with the recycling of e-waste in general and In bearing wastes in particular.

In the current research, the industrial-scale In recovery from various e-wastes like waste LCD, LED, and LCD industry itching wastewater is being emphasized through simulation and integration of the developed processes. As such a demonstration plant to treat one ton/day of ITO etching wastewaters has been developed with almost complete (99%) recovery of In. In our reported previous studies, processes have also been developed for indium recovery from waste LCD and LED. Therefore, it is considered worthwhile to investigate the extraction and recovery of In from all these resources by process simulation. In this investigation, In recovery from the ITO etching wastewaters was conducted at three different scales. Firstly, the lab-scale process was developed and optimized, followed by a bench-scale investigation that was carried out based on the lab-scale investigation, and the complete cross-current process was simulated. Finally, the process has been demonstrated in a 1 t/day plant-scale.

Experimental

Indium rich etching wastewater from the ITO etching industry was supplied by TSM Co. Ltd., Korea. The term 'ITO etching wastewater' will now be used for the In-rich wastewater of this industry. Extractants like Bis-(2,4,4,-trimethyl-pentyl) phosphinic acid (Cyanex-272), and Bis-(2-Ethylhexyl) phosphoric acid (D2EPHA) were supplied by Cytec Korea Inc., and Daihachi (Japan), respectively. Kerosene was used as diluent for solvent extraction. Reagents like NaOH, HCl, and NH_4OH were of analytical grade supplied by Daejung chemical and metal Co., Ltd., Korea.

Before solvent extraction (SX), the requisite pH of the ITO etching wastewater was adjusted by the addition of NaOH or HCl solution. Various process parameters like extractant concentration, organic phase/aqueous phase volume ratio (O/A), Mc-Cabe Thiele isotherm, and stripping parameters were optimized varying one parameter at a time in a lab-scale experiment. This is followed by bench-scale simulation experiments carried out in a 2×10^{-3} m^3 scale. Necessary modifications were verified in the bench-scale operation also. Finally, the plant-scale batch operation was demonstrated in a 50×10^{-3} m^3 scale. The plant-scale operation is capable of handling 2 m^3 etching wastewater per operation. Wet chemical reduction experiments were carried out on lab-scale only.

The concentration of metals in the ITO etching wastewater, raffinate, and stripped solution for each lab-scale, bench-scale, and plant-scale was determined using ICP-AES (OPTIMA 4300DV, Perkin-Elmer, USA) after suitable dilution using 5 vol% of HCl. The maximum deviations accepted were about ±3% in ICP-AES analysis. The metal concentrations in the organic phases were calculated through the mass balance.

Results and Discussion

Lab-Scale Process Development

ITO etching wastewaters received from TSM Co. Ltd, Korea were analyzed as received without any treatment. The ITO etching wastewater was acidic (pH 0.7), the medium was chloride and consisted of Al (1.353 kg/m^3), Cu (6.113 kg/m^3), In (1.806 kg/m^3), Mo (0.621 kg/m^3), and Sn (0.052 kg/m^3). In lab-scale process development, pH isotherms of all constituent metals using Cyanex 272 and D2EPHA were analyzed. In the pH isotherm (not given), Cyanex 272 indicated selective Sn and Mo extraction, whereas D2EPHA showed no selectivity for metal extraction from such liquor. Followed by the effect of each process parameter like Cyanex 272 concentration, D2EPHA concentration, and phase volume ratio (A/O), McCabe–Thiele isotherm for extraction and stripping was investigated, and optimized parameters have been reported elsewhere [11]. For precise presentation, the lab-scale process is summarized in Fig. 1. The complete process consists of Mo and Sn scrubbing followed by their recovery, In purification followed by In sponge recovery, Cu recovery by wet chemical reduction, and treatment of wastewater. Figure 1 also shows that from the ITO etching wastewater, Mo and Sn were extracted completely by SX using 73 kg/m^3 of Cyanex 272 as an extractant in two stages. Then from the loaded Cyanex 272, Mo and Sn were scrubbed sequentially using 4 kmol/m^3 of NH$_4$OH and 2 kmol/m^3 of NaOH. From Mo and Sn free (ITO etching wastewater), In was simultaneously purified by SX using 161 kg/m^3 of D2EPHA as an extractant in two stages and enriched by phase ratio management using A/O ratio at 8.5 for both the stages. From the loaded D2EPHA, the In was stripped by 4 kmol/m^3 of HCl and was recovered through wet chemical reduction using Al powder at 80 °C.

Finally, Cu was recovered (Cu°) from the raffinate of In extraction stage by a simple wet chemical reduction using Fe powder. From the filtrate (wastewater) of this stage, heavy metals were treated with NaOH. The complete process given in Fig. 1, indicated total metal value recovery, and also the efficient handling of the wastewater stream (effluent treatment). Following the lab-scale investigation, the process was validated through a bench-scale investigation and discussed in the next section.

Bench-Scale Simulation of the Process

Efficacy and reproducibility of the hydrometallurgical process developed on the lab-scale for metal value recovery from ITO etching wastewater were investigated in a bench-scale simulation process. Each process parameter was scaled up and verified through continuous cross-current simulation. The stoichiometry involved for Sn and Mo scrubbing and In purification were analyzed through log–log dependencies of distribution versus pH and extractant concentration. Through stoichiometry analysis, ideal loading capacity and real loading capacity of each extractant were analyzed. Suitable models were fitted for scrubbing, extraction, and loading behavior. Using the proposed model, the possibility for quantitative extraction of metals, desired extractant concentration, and the number of stages required were investigated and reported elsewhere [13]. Considering the scope of work outlined earlier and space, only the bench-scale counter-current simulation is summarized in Fig. 2.

In the bench-scale investigation, almost similar results were obtained as on the lab-scale and are schematically presented in Fig. 2. The bench-scale operation consists of 3 steps for complete process development, i.e., Sn and Mo extractive-scrubbing by the same concentration of Cyanex 272 and In purification also by the same concentration of D2EPHA as in lab-scale operation. During purification, In enrichment was also

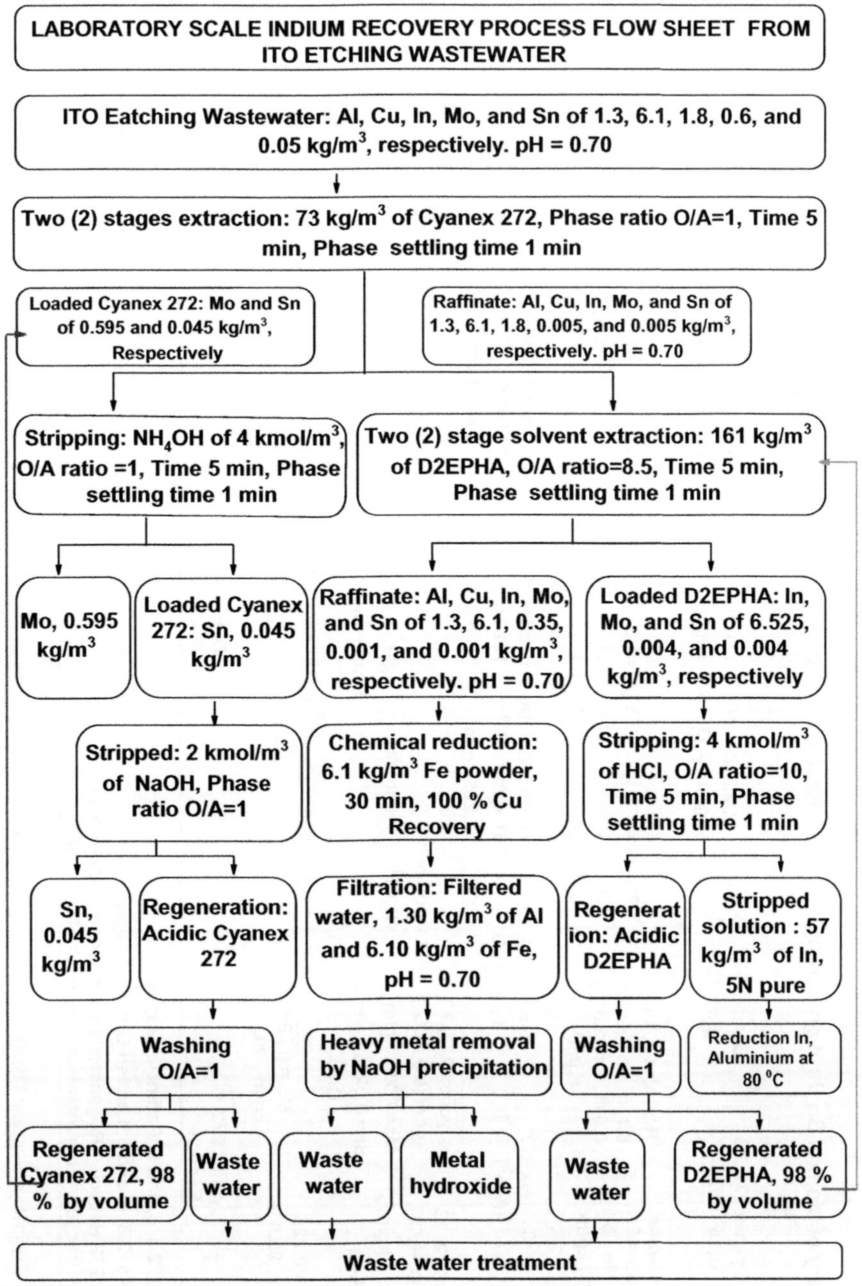

LABORATORY SCALE INDIUM RECOVERY PROCESS FLOW SHEET FROM ITO ETCHING WASTEWATER

ITO Eatching Wastewater: Al, Cu, In, Mo, and Sn of 1.3, 6.1, 1.8, 0.6, and 0.05 kg/m^3, respectively. pH = 0.70

↓

Two (2) stages extraction: 73 kg/m^3 of Cyanex 272, Phase ratio O/A=1, Time 5 min, Phase settling time 1 min

- Loaded Cyanex 272: Mo and Sn of 0.595 and 0.045 kg/m^3, Respectively
- Raffinate: Al, Cu, In, Mo, and Sn of 1.3, 6.1, 1.8, 0.005, and 0.005 kg/m^3, respectively. pH = 0.70

Stripping: NH$_4$OH of 4 kmol/m^3, O/A ratio =1, Time 5 min, Phase settling time 1 min

Two (2) stage solvent extraction: 161 kg/m^3 of D2EPHA, O/A ratio=8.5, Time 5 min, Phase settling time 1 min

- Mo, 0.595 kg/m^3
- Loaded Cyanex 272: Sn, 0.045 kg/m^3
- Raffinate: Al, Cu, In, Mo, and Sn of 1.3, 6.1, 0.35, 0.001, and 0.001 kg/m^3, respectively. pH = 0.70
- Loaded D2EPHA: In, Mo, and Sn of 6.525, 0.004, and 0.004 kg/m^3, respectively

Stripped: 2 kmol/m^3 of NaOH, Phase ratio O/A=1

Chemical reduction: 6.1 kg/m^3 Fe powder, 30 min, 100 % Cu Recovery

Stripping: 4 kmol/m^3 of HCl, O/A ratio=10, Time 5 min, Phase settling time 1 min

- Sn, 0.045 kg/m^3
- Regeneration: Acidic Cyanex 272
- Filtration: Filtered water, 1.30 kg/m^3 of Al and 6.10 kg/m^3 of Fe, pH = 0.70
- Regeneration: Acidic D2EPHA
- Stripped solution: 57 kg/m^3 of In, 5N pure

Washing O/A=1 | Heavy metal removal by NaOH precipitation | Washing O/A=1 | Reduction In, Aluminium at 80 °C

- Regenerated Cyanex 272, 98 % by volume
- Waste water
- Waste water
- Metal hydroxide
- Waste water
- Regenerated D2EPHA, 98 % by volume

Waste water treatment

Fig. 1 Complete hydrometallurgical lab-scale process for metal value recovery from ITO etching wastewater. (Color figure online)

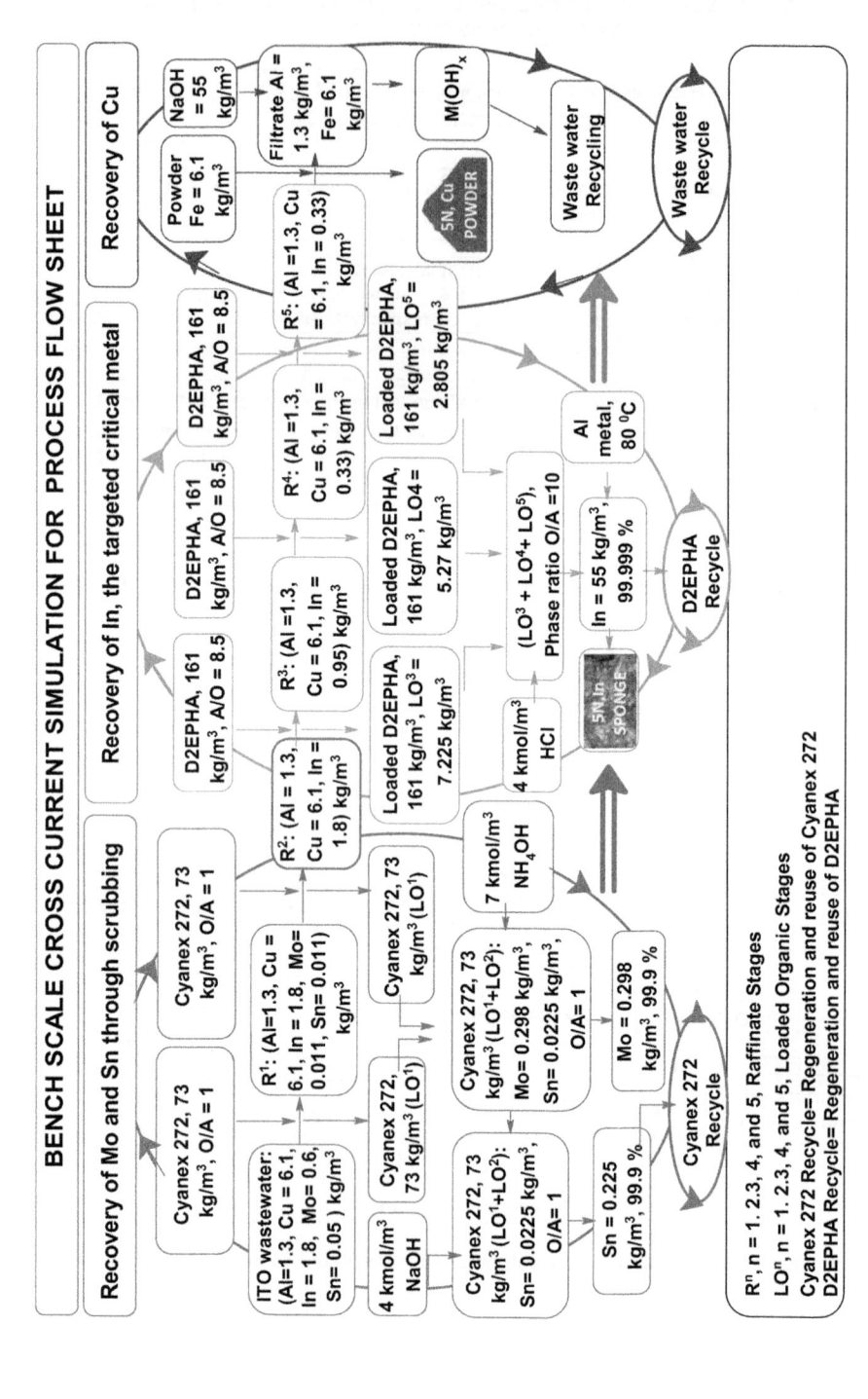

Fig. 2 Complete hydrometallurgical bench-scale process for metal value recovery from ITO etching wastewater. (Color figure online)

achieved at the phase ratio (A/O) of 8.5. In the bench-scale simulation, three-stages of extraction with D2EPHA achieved the quantitative recovery of In. The process simulation indicates the complete reproducibility of the lab-scale process in scaled-up bench-scale data and provides confidence for industrial-scale demonstration.

Pilot-Plant Demonstration

Having relied upon the bench-scale process reflected in Fig. 2, a demonstration plant of one ton per day capacity was developed and complete cross-current operation was demonstrated. The demonstration plant is a batch operated unit, and can handle 0.2 m^3 of ITO etching solution in each batch. Figure 3 shows various views of the demo-plant. Figure 4 demonstrates industrial-scale simulation which has been projected above in Fig. 3. The figure clearly indicates that in a batch 0.05 m^3 of ITO etching wastewater was considered for demonstration. As explained earlier two-stage extractive-scrubbing and three-stage In recovery were carried out in the demonstration plant. During In purification process, it was again enriched by 8.5 times through managing the etching wastewater versus D2EPHA volume ratio approximately at that level (8.5). Through the enrichment and purification process In was thus enriched adequately as designed (8.5 times), and 0.0018 m^3 of enriched pure $InCl_3$ solution was subsequently generated through incorporating scrubbing using 4 $kmol/m^3$ of HCl. From the enriched solution, In sponge was recovered by wet chemical reduction using Al powder at 80 °C.

The developed process only addresses the In recovery from ITO etching wastewater, and does not address In recovery from the EOL substances such as LCD and LED. As the reported lab-scale development to demonstration plant operation optimization process is a time- and cost-intensive process, process development to recover value from each In bearing secondary resources is quite tedious and cost-sensitive matter. The processing time and cost challenges can suitably be addressed through simulation and integration to developed ITO etching wastewater processes. Accordingly, the In recovery process described above can be integrated for value recovery from different secondary resources like In bearing waste LCD or LED, via two different routes explained below.

Integration of Process Through Simulation

Industrial-scale metal recovery process by hydrometallurgical routes either from primary or secondary resources consists of three rudimentary steps, extraction of metal values by leaching (acid or alkali), purification by selective metals separation (precipitation, ion exchange, solvent extraction), and metal value recovery by electrowinning or cementation/precipitation. In the context of In recovery, sequential acid leaching, solvent extraction, and recovery are common hydrometallurgical

Fig. 3 Demostration plant view **a** complete 3-stage batch scale operated plant, **b** typical agitation and phase separation unit, **c** visual apparatus unit for phase separation observation, and **d** inside view of baffle mixture tank. (Color figure online)

routes. As most of the globally produced In being used in optical display devices, the industrial-scale In recovery from EOL e-wastes like LCD, and LED are a challenge for the circular economy notion of this metal. As reported in the literature, after leaching the In content in leach liquor is very small for industrial recovery, which is a genuine challenge and adversely affects the In recovery interests from these resources [12, 14, 15].

Indium and Sn recovery from ITO bearing EOL waste LCD and its optimization process has already been developed in our earlier research and reported elsewhere [12, 15]. Similarly, In recovery from LEDs industry waste as a part of the interest of different projects pursued previously has also been reported elsewhere [16]. As In and Sn concentration is low in the waste LCD glass, and similarly, In concentration

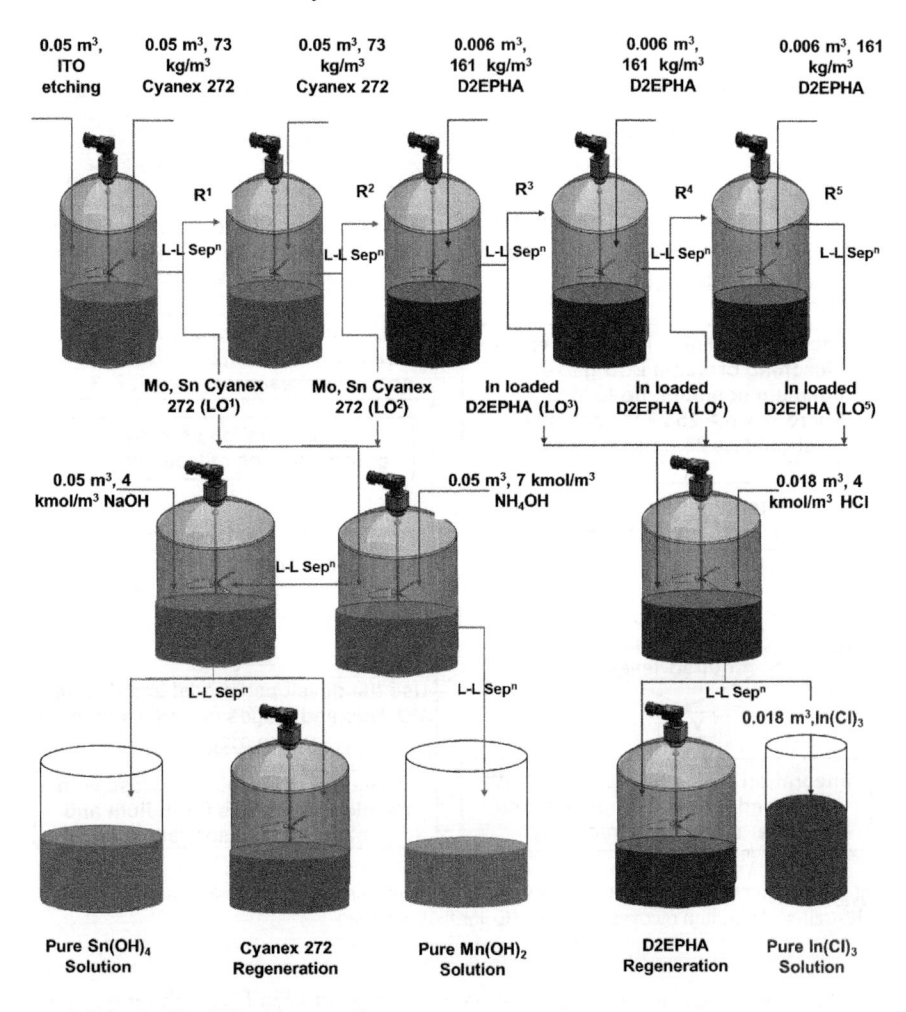

Fig. 4 Pilot-scale verification of developed bench-scale cross-current simulation for recovery of metals and treatment of ITO etching industry wastewater. (Color figure online)

is also significantly low in the LEDs, In valorization is a challenge even after efficient leaching that needs to be addressed during the process development through a cost-efficient approach [16]. All the challenges can be overcome by integrating the leaching optimization process [12, 15, 16] to the hydrometallurgical In recovery process developed for ITO etching wastewater as presented in Fig. 5. The waste LCD leaching optimization process can be integrated with the developed process through Route I (Fig. 5). Other In secondary resources can also be integrated through leaching optimization while applying Route II (Fig. 5). The prime challenges for integrating

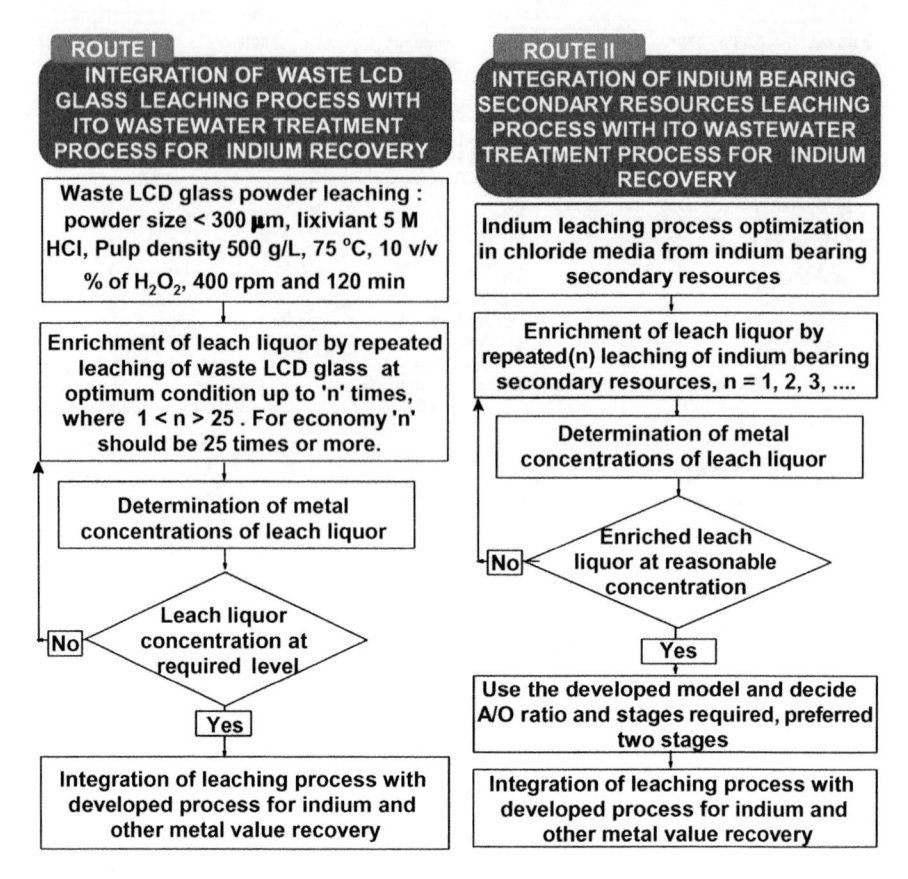

Fig. 5 Different routes for the integration of leaching optimization process to the developed industrial-scale indium recovery process. (Color figure online)

the above processes are (i) concentration difference between ITO etching wastewaters (ii) metal content, (iii) solvent concentration to be used, and (iv) stages required for the waste LCD leach liquor reported elsewhere, and metal content thereof [11–13, 15, 16].

Beneficiation and recovery of In from waste LCD glass has been optimized in our earlier investigation [12, 15]; the optimum condition for In leaching was powder size <300 mm, 5.0 M HCl, pulp density 500 g/L, 75 °C, H_2O_2 10%, reaction time 120 min, and stirring speed 400 rpm. The concentration of different species can be handled through the enrichment of In and Sn in leach liquor to the requisite concentration by Route I as reflected in Fig. 5. Clearly in Route I, repeated use of the same leach liquor against fresh LCD waste can enrich the In and Sn to the desired level so as to suitably integrate with the developed process presented in Fig. 4. Theoretically, 25 times of repeated leaching and recycling using the same leach liquor could attain the required level appropriate for downstream In metal recovery. Then, metal purification and recovery could be managed through the hydrometallurgical In recovery process from

ITO etching wastewater represented previously (Fig. 4). The proposed integration process through the route I is currently being further investigated.

Similarly, through Route II mentioned in Fig. 5, other secondary In bearing resources like waste LED can be handled. In Route II also, followed by leach-liquor enrichment, the impurity metal can be selectively scrubbed using Cyanex 272. For the scrubbing (extractive) Cyanex 272 concentration and volume can systematically be engineered using the developed model presented elsewhere [13]. Followed by such as a scrubbing sequence, the In can finally be purified using D2EPHA. Similarly, managing D2EPHA concentration and volume can systematically be engineered using the developed model presented elsewhere [13]. In either case once In is enriched and scrubbed, it can be recovered as In metal sponge through wet chemical reduction using Fe powder.

Conclusion

The developed value recovery process for ITO etching industry wastewaters can be used for the recovery of metal values from the EOL materials particularly LCDs and LEDs through the integration of the leaching optimization process. The integration of the processes can be engineered through two different routes, i.e., (i) multi-stage leaching of In bearing secondary wastes using the same leach liquor repeatedly followed by integrating the developed hydrometallurgical metal recovery process, and (ii) multi-stage leaching followed by manipulating the Caynex 272 and D2EPHA concentrations or process parameter presented in the developed process. Through the proposed approach, the semiconductor manufacturing industry and ITO recycling industry can address various pressing issues like (i) waste disposal, (ii) In recovery, (ii) circular economy, simultaneously. Integration of the process can also close the loop for the circular economy and can be part of cradle-to-cradle technology management which could be able to lower the futuristic carbon economy.

Acknowledgements The current research project is supported by the R&D Center for Valuable Recycling (Global-Top Environmental Technology Development Program) funded by the Ministry of Environment (Project No.: GT-11-C-01-020-0).

References

1. Kefeng Z (2018) Review and analysis of indium market detailed comparison of data between China, the United States, Japan and South Korea. https://news.metal.com/newscontent/100 896637/zhao-kefeng:-review-and-analysis-of-indium-market-detailed-comparison-of-data-between-china-the-united-states-japan-and-south-korea/. Accessed 07 Aug 2020
2. Grand View Research Inc. (2017) Indium market size worth $584.8 million by 202. Accessed 07 Aug 2020

3. U.S. Geological Survey (2016) Mineral commodity summaries 2016: U.S. Geological Survey. In. U.S. Department of the Interior, U.S. Geological Survey, Reston, Virginia
4. U.S. Geological Survey (2020) Mineral commodity summaries 2020: U.S. Geological Survey. In. U.S. Department of the Interior, U.S. Geological Survey, Reston, Virginia
5. U.S. Geological Survey (2018) Mineral commodity summaries 2018: U.S. Geological Survey. In: U.S. Department of the Interior, U.S. Geological Survey, Reston, Virginia
6. Communication from the commission to the european parliament, the council, the European economic and social committee and the committee of the regions on the 2017 list of critical raw materials for the EU. In: Internal Market, Industry, Entrepreneurship and SMEs. Brussels (2017)
7. U.S. Department of Energy (2011) U.S. Department of Energy Critical materials strategy
8. Bauer D, Diamond D, Li J, Sandalow D, Telleen P, Wanner B (2010) U.S. Department of Energy Critical Materials Strategy
9. Chung C-H, Ko Y-W, Kim Y-H, Sohn C-Y, Hye Yong C, Ko Park S-H, Jin Ho L (2005) Radio frequency magnetron sputter-deposited indium tin oxide for use as a cathode in transparent organic light-emitting diode. Thin Solid Films 491(1–2):294–297. https://doi.org/10.1016/j.tsf.2005.06.003
10. Kim H, Gilmore CM, Piqué A, Horwitz JS, Mattoussi H, Murata H, Kafafi ZH, Chrisey DB (1999) Electrical, optical, and structural properties of indium–tin–oxide thin films for organic light-emitting devices. J Appl Phys 86(11):6451–6461. https://doi.org/10.1063/1.371708
11. Swain B, Mishra C, Hong HS, Cho S-S (2015) Treatment of indium-tin-oxide etching wastewater and recovery of In, Mo, Sn and Cu by liquid–liquid extraction and wet chemical reduction: a laboratory scale sustainable commercial green process. Green Chem 17(8):4418–4431. https://doi.org/10.1039/c5gc01244a
12. Swain B, Mishra C, Hong HS, Cho SS (2016) Beneficiation and recovery of indium from liquid-crystal-display glass by hydrometallurgy. Waste Manag 57:207–214. https://doi.org/10.1016/j.wasman.2016.02.019
13. Swain B, Mishra C, Hong HS, Cho S-S, Lee SK (2015) Commercial process for the recovery of metals from ITO etching industry wastewater by liquid–liquid extraction: simulation, analysis of mechanism, and mathematical model to predict optimum operational conditions. Green Chem 17(7):3979–3991. https://doi.org/10.1039/c5gc00473j
14. Akcil A, Agcasulu I, Swain B (2019) Valorization of waste LCD and recovery of critical raw material for circular economy: a review. Resour Conserv Recycl 149:622–637. https://doi.org/10.1016/j.resconrec.2019.06.031
15. Swain B, Lee C, Hong H (2018) Value recovery from waste liquid crystal display glass cullet through leaching: understanding the correlation between indium leaching behavior and cullet piece size. Metals 8(4). https://doi.org/10.3390/met8040235
16. Swain B, Mishra C, Kang L, Park KS, Lee CG, Hong HS (2015) Recycling process for recovery of gallium from GaN an e-waste of LED industry through ball milling, annealing and leaching. Environ Res 138:401–408. https://doi.org/10.1016/j.envres.2015.02.027

Recovery of Lithium (Li) Salts from Industrial Effluent of Recycling Plant

Archana Kumari, Pankaj Kumar Choubey, Rajesh Gupta, and Manis Kumar Jha

Abstract To cope up with the supply–demand gap of lithium (Li) an essential energy element, the recycling of waste industrial effluent (generated after cobalt recycling from waste Li-ion batteries) is targeted. In industry, after the recovery of Co, Cu, Ni, and graphite from one ton of black cathodic material of Li-ion batteries about 8 m^3 of waste effluent containing 5–10 g/L Mn and 1–3 g/L Li is generated. Systematic precipitation studies were carried out using saturated alkaline solution varying Eh/pH of the effluent. Settling time 30 min and pH ~12 were found to be optimum conditions for maximum precipitation of Li (~90%) as salt. Precipitation studies for Mn/ Li with scientific validation were also carried out and discussed. The process developed has tremendous potential to be commercialized in industry after scale-up studies.

Keywords Lithium-ion batteries (LIBs) · Lithium (Li) · Precipitation · Hydrometallurgy

Introduction

Li, an essential energy element, has been extensively used in lithium-ion batteries, computer products, communication devices and other fields. Li and its compounds are usually produced from their primary resources, which is not sufficient to cope up with the amount of Li required. The limited supply of Li is not forbidding its utilization in battery production and is constantly escalating [1–4]. The requirement of Li by the battery industries is estimated to increase from 43% in 2017 to 65% in 2025 [5]. Almost one-third of Li present in the earth crust will be consumed in making electric vehicle batteries by 2050 and ultimately the Li resources are predicted

---

A. Kumari · P. K. Choubey · M. K. Jha (✉)
Metal Extraction and Recycling Division, National Metallurgical Laboratory (CSIR), Jamshedpur, India
e-mail: mkjha@nmlindia.org

R. Gupta
M/s Evergreen Recyclekaro India Pvt. Ltd., Mumbai, India

to be entirely exhausted by 2080 [6]. Thus, to cope up with the supply–demand gap of Li, it is essential to ensure a reliable alternative source of Li. Huge amount of effluent containing substantial amount of Li is generated by battery recycling industries, where after the recovery of Co, Cu, Ni, and graphite from the black cathodic material of LIBs, the effluent is discarded. It is expected to have loss of Li in drained/ treated effluent during the final disposal. Moreover, variation in the concentration of Li is often noticed due to complex and heterogeneous nature of scrap LIBs which has limited the research work on laboratory scale rather than scale-up or commercial scale. The effluent generated during processing of LIBs showed the presence of other metals (Mn, Co, Cu, Ni) as impurities, which when disposed to the environment can cause severe damage to the aquatic life. However, recuperation of Li from industrial effluent will not only assuage the load on natural resources but also solve the problems caused to the ecosystem. Hydrometallurgical processes consisting of solvent extraction, precipitation, or adsorption have been globally applied for the recovery of Li from effluents. Researchers worked for the recovery of Li from various effluent. Table 1 summarizes different routes adopted to recover Li from wastewater [7–11]. Although studies have been carried out to recover Li from various effluent

Table 1 Different routes utilized to recover Li from wastewater

Process used	Remarks	References
Electrometallurgy	Li recovery and organic pollutant removal from industrial wastewater was proposed where an electrochemical system containing a Li-recovering electrode and an oxidant-generating electrode was used to simultaneously recover Li and decompose organic pollutants. This recovers 98.6 mol% Li and reduces organic pollutants by 65%	Kim et al. [7]
Adsorption	A magnetic Li-ion imprinted polymer, Fe_3O_4@SiO_2@IIP was synthesized using novel crown ether to recover Li from wastewater. At optimum pH = 6 was for adsorption and polymer takes 10 min to reach complete equilibrium. About 89.8% of the Li was recovered	Luo et al. [8]
Leaching	Selective recovery of Li from spent LIBs by coupling advanced oxidation and chemical leaching processes was studied. Loss of Li was only 2.06%. Li_2CO_3 of purity 99.0% was finally obtained	Lv et al. [9]
Precipitation	Two-stage precipitation process using Na_2CO_3 and Na_3PO_4 as precipitants was developed to recover Li as 74.72% raw Li_2CO_3 and 92.21% pure Li_3PO_4, respectively from effluent	Guo et al. [10]
Precipitation	A two-step precipitation of Li_2CO_3 using CO_2 as a precipitation agent from Li-containing alkaline wastewaters was studied. At 95 °C, more than 99.5% sparingly soluble Li_2CO_3 was produced in the second step, whilst most impurities remain in the solution	Jandová et al. [11]

of recycling plant, much consideration has been made on the various methods used for recycling Li rather than its effective recovery from low concentration solution. Moreover, lack of selectivity still persists. Keeping in view of the above, CSIR-NML, Jamshedpur has developed feasible technology for battery recycling and transferred the same to different Indian industries. Based on this, the present paper reports the development of a complete process flowsheet to recover Li and Mn as a valuable product from effluent generated after Co, Cu, and Ni recycling from waste LIBs using precipitation process.

Systematic and scientific precipitation studies were carried out and after studying various process parameters, optimal condition for the maximum extraction of Li and Mn from the effluent was obtained. The metal-free effluent generated containing carbonate ions will be re-utilized. Developed process is novel and has potential approach for Li recovery from effluent generated by the recycling industries.

Materials and Method

Materials

Present work is carried out using the effluent generated after the recovery of Co, Cu, Ni, Fe, and graphite during LIBs recycling to recover Li and Mn. The effluent was chemically analyzed and found to contain about 2.31 g/L Li and 5.52 g/L Mn. The pH of the effluent generated was recorded to be ~3. Sodium hydroxide (NaOH) and sodium carbonate (Na_2CO_3), both supplied by Merck, India was used as precipitant to precipitate Mn and Li, respectively, from the generated effluent. Both the precipitating reagents were diluted to the required concentration using de-ionized water. Apart from these, H_2SO_4 used during this study (for pH adjustment) was of analytical grade (AR) supplied by Rankem, India. All chemicals used for the experimental studies were used without further purification. Samples to be analyzed were diluted using de-ionized water.

Procedure

Based on our group (CSIR-NML, Jamshedpur) experience in the area of LIBs recycling, a complete process flowsheet including pre-treatment—leaching—solvent extraction—precipitation is presented in this paper with detailed discussion regarding Li and Mn recovery. Waste LIBs were received from a recycling industry, M/s Evergreen Recyclekaro Pvt. (India) Ltd. situated at Mumbai, India. Spent LIBs were initially discharged, crushed (using scutter crusher) and further beneficiated to separate the black cathodic material. The typical analysis of metals present in this black powder is presented in Table 2.

Table 2 The composition of metals present in black powder of LIBs

Elements	Co	Mn	Li	Cu	Ni	Fe
Wt.%	15.16	6.30	2.36	1.29	1.95	0.35

Variation in the composition of metals in the black powder is generally observed due to quality and type of LIBs collected and it may range between 10–20% Co; 0.5–1.5% Fe; 1.5–3% Ni; 2.1–3.5% Li; 7.5–9.5% Mn, and 4.1–6% Cu. The leaching of metals from the black cathodic material of scrap LIBs were carried out in a three-necked closed Pyrex glass reactor fitted with a condenser using a hot plate with a temperature controlling sensor. Magnetic stirrer was used for mechanical agitation. The obtained leach liquor was subjected to solvent extraction resulting in maximum extraction of Cu and Ni using suitable extractant. Further, selective separation of Co takes place via precipitation technique. Solvent extraction and precipitation process result in complete extraction of Co, Ni, and Cu leaving Mn and Li in the raffinate. This raffinate when generated in huge quantity in recycling industries mostly discarded due to lack of viable technologies for their processing. In order to prevent the wastage of Li and Mn left in the effluent generated were collected, mixed properly and experiments for their precipitation were carried out. After selecting the precipitating reagent, optimization of suitable conditions for batch scale studies as well as validation of the same was also made so as to collect salts of Li and Mn. Satisfactory mass balance was obtained for representative test samples. The content of metals in the sample as well as in the solution generated during the experiments was analyzed using atomic absorption spectrometer (AAS) (Perkin Elmer model, Analyst 200; USA). Metrohm Basic Titrino 797 with glass pH combination electrode and automatic temperature correction was used for pH measurement.

Results and Discussion

Generation of Effluent Containing Li and Mn

Scrap LIBs were initially ensured to be completely discharged and beneficiated to separate the plastic part that floats on top, black cathodic powder which remains suspended and metallic content settles at the bottom. Among these materials, the black cathodic material mainly containing Co, Li, Mn, Cu, Ni, and graphite was used for the experimental purpose. The beneficiated black cathodic material obtained was leached using suitable concentration of sulfuric acid (H_2SO_4) in presence of an oxidizing agent, hydrogen peroxide (H_2O_2) at an elevated temperature for some time to get maximum amount of metals in the solution. It was observed that H_2SO_4 with H_2O_2 proved to be the positive lixiviant for leaching of metals. The whole slurry was filtered and the residue was washed for further analysis. More than 98% leaching of metals was achieved and the obtained leach liquor majorly contained about 15.03 g/L

Recovery of Lithium (Li) Salts … 95

Lithium ion batteries (LIBs)
↓
Physical pre-treatment
↓
Hydrometallurgical process
↓

Recovery of metals (Cu, Ni, Co) & Graphite   Effluent generated (Li & Mn)

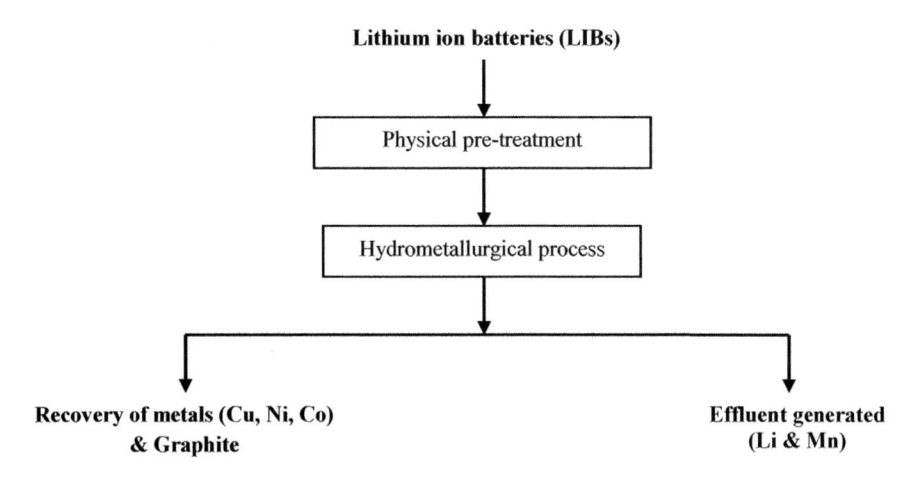

Fig. 1 Process flow for the generation of effluent containing Li and Mn. (Color figure online)

Co; 2.2 g/L Li; 1.3 g/L Cu; 1.87 g/L Ni; and 5.5 g/L Mn. Further, Cu, Ni, and Co were selectively separated using solvent extraction technique. Major amount of Co, Ni, and Cu, as well as their metallurgical advantages, attracted industrialists towards their extraction while Li along with Mn gets disposed off as effluent (Fig. 1). Lack of cost-effective process to recover Li and Mn results in their loss. Thus, huge amount of effluent was generated which require proper treatment.

Selective Precipitation of Mn from the Effluent

The effluent obtained after extraction of Co was found to contain 2.31 g/L Li and 5.52 g/L Mn. The pH of the effluent was ~3.5. Due to low concentration of Li, the effluent was concentrated 10 times and analyzed. It was found that the content of Li and Mn increased up to 20.9 g/L and 53.1 g/L, respectively, and pH was ~4.5. For selective precipitation of Mn from this concentrated solution, optimization of the precipitation parameters was carried out including effect of pH, settling time, and temperature. Using 50% NaOH solution at room temperature, the pH of the effluent was increased up to 11, provided with constant stirring and maintaining settling time of 30 min. It was observed that ~pH 10.5; ~99.99% of Mn present in the effluent get precipitated in two stages (Fig. 2). The reaction taking place is represented by the equation below:

$$MnSO_4 + 2NaOH = Mn(OH)_2 \downarrow + Na_2SO_4 \qquad (1)$$

The amount of NaOH required for increasing the pH was also calculated and found that for 200 mL of effluent (containing 53.1 g/L Mn), ~2.5 g of NaOH is required to reach pH 10.5 and completely precipitate Mn from the effluent. It was noticed that

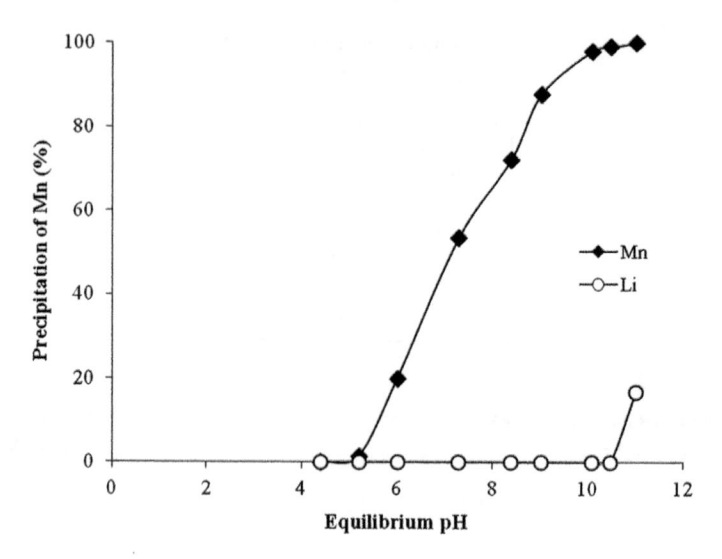

Fig. 2 Effect of pH on precipitation of Mn (Aqueous: 53.1 g/L Mn and 20.9 g/L Li; Reagent: 50% NaOH; Temp.: 30 °C; Time: 30 min; Stirring speed: 300 rpm)

beyond pH 11, minor amount of Li present in the raffinate get co-precipitated. Thus, at pH 10.5, Mn was selectively separated. The effluent was further filtered and a black precipitate of Mn(OH)$_2$ was collected. This precipitate of Mn(OH)$_2$ was dried to get MnO$_2$ of purity 99%. After Mn removal, the raffinate (analysed to be free from Mn) was ready for Li recovery.

Chemical Precipitation for Li Recovery

After successful recovery of Mn from the effluent, focus was made towards the Li recovery. Analysis of the raffinate (free from Mn) showed the presence of 20.88 g/L Li. Li was present as liquid phase, which was selectively precipitated as solid phase by the chemical reaction during this precipitation process. In this case, Na$_2$CO$_3$ was used to precipitate Li according to the equation:

$$Li_2SO_4 + Na_2CO_3 = Li_2CO_3 + Na_2SO_4 \tag{2}$$

After carrying out the precipitation experiments at different conditions, it was found that temperature plays an imperative role during Li precipitation. The fact is that the solubility of Li$_2$CO$_3$ decreases with increase in temperature and thus, to reduce the loss of Li due to solubility, it is beneficial to precipitate Li at elevated temperature. Consequently, after optimizing other parameters, the effect of temperature was also studied at room temperature (30 °C), 50 and 90 °C as presented in Fig. 3. It was observed that elevated temperatures favored the rate of Li precipita-

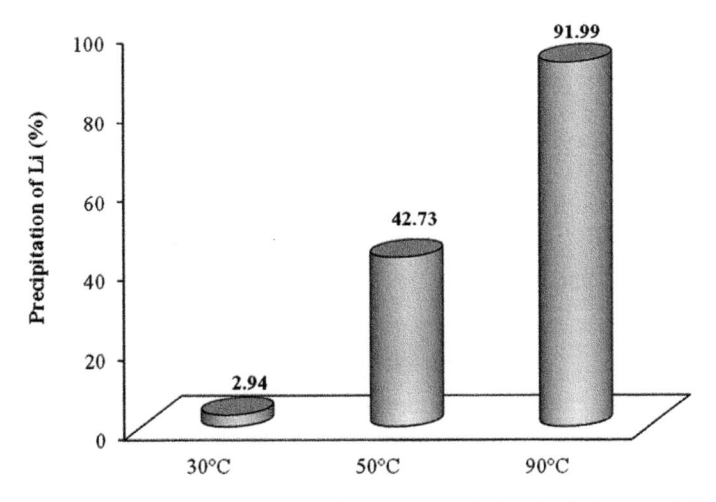

Fig. 3 Effect of temperature on precipitation of Li. (Aqueous: 20.88 g/L Li; Reagent: 50% Na_2CO_3; Time: 15 min; Stirring speed: 300 rpm). (Color figure online)

tion. In 15 min time, ~90% precipitation of Li as carbonate takes place maintaining the solution temperature 90 °C. As far as the requirement of Na_2CO_3 is concerned, it was noticed that 200 mL of 20.88 g/L Li-containing raffinate needs about 50 g of Na_2CO_3 for complete Li precipitation. The salt of Li_2CO_3 get precipitated. The obtained Li_2CO_3 was washed with hot water as water-washing decreases the possibility for the presence of impurities (containing Na^+) in Li_2CO_3 produced. The purity of obtained Li_2CO_3 salt was 99%.

In order to validate the result of Li precipitation, standard Eh–pH diagram was drawn using HSC software and presented in Fig. 4. The experimental data were validated with the theoretical Eh–pH diagrams showing the recovery of Li as carbonate usually starts at pH 9 but as optimized, carbonate of Li occurs in solution having pH range above 11.

Processing of Wastewater

It is essential to process the wastewater left after Mn and Li recovery for re-utilization or water treatment to check its suitability for disposal in the environment. If allowed to be discharged in the environment, it is essential to maintain pH of the wastewater between pH 6 and 7 for disposal. In this condition, the effluent left after Li recovery contains Na_2CO_3 and thus, the temperature was decreased to 4 °C for more than 24 h and crystals of Na_2CO_3 were collected and water was used further for the hydrometallurgical process. The content of metal ions in ppb level is required to be checked and TDS to be maintained in the range 150–200. The treated water can be utilized/recycled in the industry.

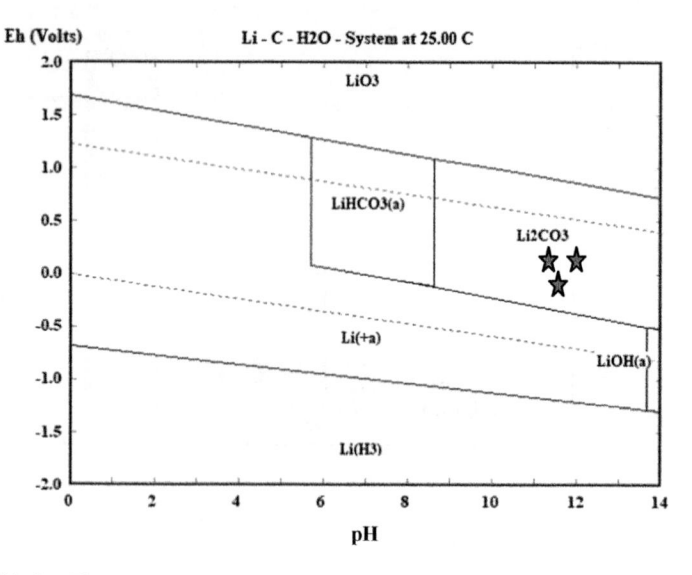

Fig. 4 Validating Eh–pH diagrams with experimental data for Li precipitation (=Li$_2$CO$_3$). (Color figure online)

Conclusions

Based on the obtained experimental results for separation and recovery of Li and Mn from the effluent generated after Co, Cu, Fe, and Ni extraction, the following conclusions can be drawn:

- Effluent containing Mn and Li were precipitated using NaOH at room temperature and Na$_2$CO$_3$ at 90 °C, respectively without using any additive.
- About 99.99% of Mn and Li were found to be selectively precipitated with purity ~99% at pH 10.5 in two stages and 12.34 in single stage, thus chances of co-precipitation was reduced.
- Temperature plays a very essential role where the solubility of Li$_2$CO$_3$ decreases with increase in temperature and thus, it is beneficial to precipitate Li at elevated temperature.
- Filtration separates the hydroxides of Mn and carbonate of Li leaving metal depleted wastewater.
- The developed process will be economical as it consumes comparatively less amount of energy and time than the processes reported previously. The complete process flowsheet is shown in Fig. 5. Lab-scale data shows feasibility of the process; however, some scale-up studies/ pilot scale studies are required before commercialization.
- After Li and Mn extraction, the effluent generated can be re-utilized in industry after the treatment and maintaining TDS and pH.

Effluent (containing Mn & Li)

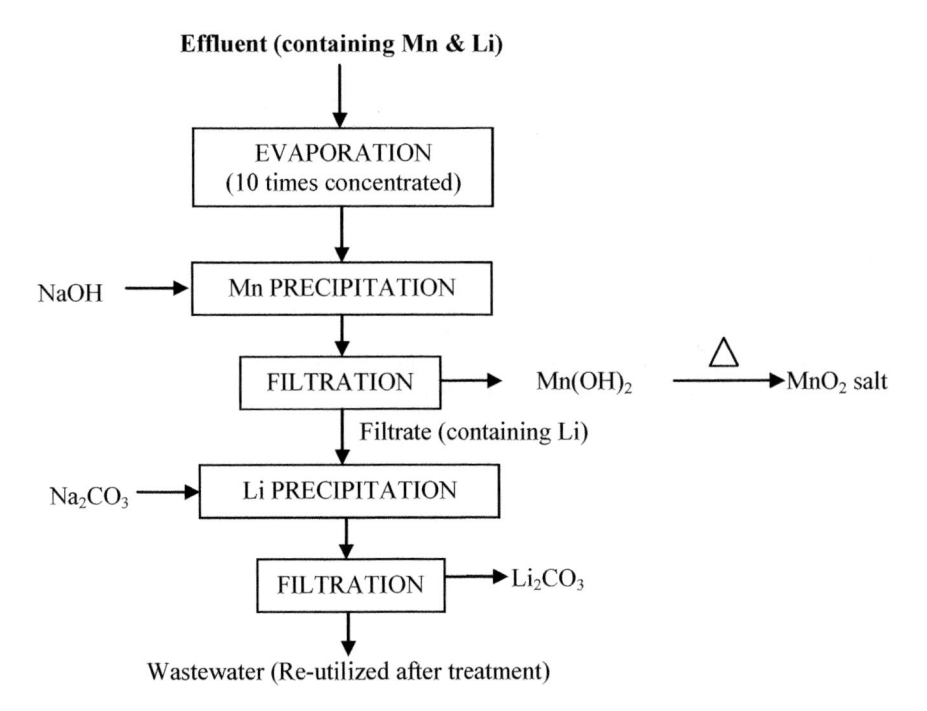

```
         ↓
   EVAPORATION
(10 times concentrated)
         ↓
NaOH → Mn PRECIPITATION
         ↓                        Δ
     FILTRATION → Mn(OH)₂ ────→ MnO₂ salt
         ↓
  Filtrate (containing Li)
         ↓
Na₂CO₃ → Li PRECIPITATION
         ↓
     FILTRATION → Li₂CO₃
         ↓
Wastewater (Re-utilized after treatment)
```

Fig. 5 A complete process flowsheet for the recovery of Li and Mn from the effluent generated after LIBs processing

Acknowledgements This paper is based on the joint collaborative research work carried out between CSIR-National Metallurgical Laboratory (CSIR-NML), Jamshedpur, India and M/s Evergreen Recylekaro India Pvt. Ltd., Mumbai, India. Authors are thankful to the Director, National Metallurgical Laboratory for giving permission to publish the paper. M/s Evergreen Recylekaro India Pvt. Ltd., Mumbai, India has also been acknowledged for the financial support provided by them.

References

1. Lv WG, Wang ZH, Cao HB, Sun Y, Zhang Y, Sun Z (2018) A critical review and analysis on the recycling of spent lithium-ion batteries. ACS Sustain Chem Eng 6(2):1504–1521
2. Zheng X, Zhu Z, Lin X, Zhang Y, He Y, Cao H, Sun Z (2018) A mini-review on metal recycling from spent lithium ion batteries. Engineering 4:361–370
3. Zheng XH, Gao WF, Zhang XH, He MM, Lin X, Cao HB, Zhang Y, Sun Z (2017) Spent lithium-ion battery recycling-Reductive ammonia leaching of metals from cathode scrap by sodium sulphite. Waste Manage 60:680–688
4. Swain B (2017) Recovery and recycling of lithium: a review. Sep Purif Technol 172:388–403
5. Zhao C, He M, Cao H, Zheng X, Gao W, Sun Y, Zhao H, Liu D, Zhang Y, Sun Z (2020) Investigation of solution chemistry to enable efficient lithium recovery from low-concentration lithium-containing wastewater. Front Chem Sci Eng 14(4):639–650

6. Yang SX, Zhang F, Ding HP, He P, Zhou HS (2018) Lithium Metal Extraction from Seawater. Joule 2(9):1648–1651
7. Kim S, Kim J, Kim S, Lee J, Yoon J (2018) Electrochemical lithium recovery and organic pollutant removal from industrial wastewater of a battery recycling plant. Environ Sci: Water Res Technol 4:175–182
8. Luo X, Guo B, Luo J, Deng F, Zhang S, Luo S, Crittenden JC (2015) Recovery of lithium from wastewater using development of Li ion-imprinted polymers. ACS Sustain Chem Eng 3(3):150209143423000
9. Lv W, Wang Z, Zheng X, Cao H, He M, Zhang Y, Yu H, Sun Z (2020) Selective recovery of lithium from spent lithium-ion batteries by coupling advanced oxidation processes and chemical leaching processes. ACS Sustain Chem Eng 8(13):5165–5174
10. Guo X, Cao X, Huang G, Tian Q, Sun H (2017) Recovery of lithium from the effluent obtained in the process of spent lithium-ion batteries recycling. J Environ Manage 198(1):84–89
11. Jandova J, Dvorak P, Kondas J, Havlak L (2012) Recovery of lithium from waste materials. Ceramics—silikaty 56(1):50–54

Extraction of Platinum Group Metals from Spent Catalyst Material by a Novel Pyro-Metallurgical Process

Ana Maria Martinez, Kai Tang, Camilla Sommerseth, and Karen Sende Osen

Abstract The extraction of platinum group metals (PGM) contained in waste automobile catalyst monolithic honeycomb was investigated by a novel approach that combines a pyro-metallurgical and electrolysis step. The first step aims to both up-concentrate the amount of PGMs by using a metal collector, as well as to prepare the conductive material to be used as anode in the electrolysis step. The electrolysis step is carried out in a molten chloride electrolyte, where the PGMs remain as metallic residue, and the refined metal is further reused in the pyro-metallurgical step. Optimization of the pyro-metallurgical process led to 82–100% metal recovery rates, while the PGM recovery rates were close to 100%. Furthermore, the electrolyte composition and working temperature, as well as cell design of the subsequent electrolytic method, were adjusted. The process was assessed in a lab-scale electrolysis reactor, where PGMs could be extracted selectively at a current efficiency of around 70%.

Keywords Platinum group metals · PGM · Secondary resources · Spent catalyst · Recycling · Molten salts · Pyro-metallurgy

Introduction

Platinum group metals (PGMs) are six chemically very similar elements: i.e., platinum (Pt), palladium (Pd), rhodium (Rh), iridium (Ir), ruthenium (Ru), and osmium (Os). Mineable deposits are very rare and found in relatively few areas of the world. South Africa dominates the PGM world production with 58%, Russia accounts for a further 26%, most of it as a co-product of nickel mining. Moreover, the world demand is steadily increasing mostly due to the high loads of PGM currently used in automobile catalysts, thus complying with the increasingly restrictive emissions legislation.

A. M. Martinez (✉) · K. Tang · C. Sommerseth · K. S. Osen
SINTEF, 7034 Trondheim, Norway
e-mail: anamaria.martinez@sintef.no

To decouple from unstable suppliers, secondary material streams must be fostered. Moreover, the recovery of these valuable elements from by-products and waste streams is in good harmony with the goals of the Circular Economy towards zero-waste societies. Efficient and selective extraction of PGM from different matrices (ores or wastes) is challenging, mainly due to their chemical properties (inertness). Ways of processing and extracting them have been, and still are, the subject of many investigations through decades [1].

Within the frame of the EU-financed project PLATIRUS (GA 730224), SINTEF has investigated the possibility of recovering PGM from spent automobile catalysts by different innovative methods involving molten salts [2]. In this work, the approach consists of a pyro-metallurgical treatment of the waste catalyst material (ceramic monolithic honeycomb) followed by an electrorefining process from a molten salt electrolyte.

Pyro-metallurgical processes provide the best conditions of pre-concentration of PGMs from the very diluted waste flows [3], thus being the most frequently used methods for extracting PGM from spent automotive, petrochemical catalysts and PGM sweeps. The benefits of the pyro-metallurgical processing methods include large PGM recovery rates and high throughput. However, challenges are still encountered, and investigations are still focused on further enhancing the affinity of PGM particles for the melted collector metal (usually Cu, Fe, or Ag), and the viscosity of the slag phase.

Moreover, the subsequent electrorefining process from a molten salt media allows the extraction of the collector base-metal from the PGM-containing anode with better selectivity and lower energy consumption than state-of-art refining process in aqueous solutions. This is mainly owed to the fact that, commonly, the kinetics of the electrode charge-transfer reactions in molten salts are fast due to the high operational temperatures [4]. Moreover, the stability of lower valences of metallic species in molten chloride media implies a lower voltage needed (lower energy) in the analogous electrorefining process in aqueous solutions where higher valences are the solely stable species.

In this work, SINTEF has investigated the possibility of extracting the PGM elements from the spent catalyst material by first converting the waste in a suitable and conductive anode with up-concentrated PGM content by means of a pyro-metallurgical process. Subsequently, the PGMs were selectively extracted by means of an electrolysis process from a molten salt media. Optimization of the pyro-metallurgical process led to PGM recovery rates close to 100%. Furthermore, and after adjusting electrolyte composition, working temperature, and cell design, the subsequent electrolytic method was assessed in a lab-scale electrolysis reactor, where the metal collector was refined and recovered for further use. In this way, the PGMs could be selectively separated at a current efficiency of ca. 70%.

Experimental

Experimental Setup

The end-of-life automobile catalyst monolithic honeycomb material was provided in the form of powder by Monolithos Catalysts Ltd [5]. Characterization and analysis of the indialite/cordierite ceramic monolith material is reported elsewhere [2]. Typical PGM (Pt, Pd, Rh) content is of ca. 2500 ppm.

The pyro-metallurgical pre-treatment was carried out in a large induction furnace capable of accommodating 4 kg of waste catalyst material per trial. Due to the nature of very dilute PGMs content in the monolithic honeycomb waste, Cu metal was used as collector of the metal phase during the thermal treatment. The slag composition was modified to get the best conditions (viscosity) for metal/slag separation by adding CaF_2 and/or CaO fluxes. Graphite crucible and furnace lids allowed a CO/CO_2 atmosphere in the furnace during the slag pre-treatment. After the trials, the crucible was cut, and the metal phase separated from the slag. The metal fraction obtained was weighted, and further characterized by SEM-EPMA and analyzed by ICP-MS.

The experimental cell in the electrolysis step consisted of a programmable vertical tight furnace with a mullite liner that housed a glassy carbon crucible ($\emptyset_{int}$ = 70 mm) used as electrolyte container. A water-cooled lid supported a metal top plate that allowed the insertion of the anode and cathode, as well as thermocouple for temperature control. The system was kept under inert atmosphere using Ar (purity of 99.999%).

The electrolyte was the LiCl–KCl mixture at the eutectic composition, i.e., 58.2 mol% LiCl and 41.8 mol% KCl. All salts and the catalyst waste material were dried at 200 and 100 °C, respectively, for at least 48 h prior to use. The working temperature was kept to 470 °C monitored continuously using a thermocouple type S (Pt–Pt 10% Rh) shielded by a closed-end alumina tube and connected to a multichannel Keithley 2000 Multimeter.

The metallic phase from the pyro-metallurgical up-concentration step was introduced inside a small alumina crucible and on top of a graphite disk, which was used to establish the electric contact using a W wire protected from the electrolyte by an alumina tube. This anode arrangement was polarized against a 6 mm diameter Cu cathode rod connected to a steel current collector. A reference electrode was used to control the anode and cathode voltages. This was based on the AgCl/Ag system and consisted of a silver wire (1 mm diameter) dipped into a silver chloride solution (0.75 mol kg^{-1}) in the LiCl–KCl electrolyte, all housed in a closed-end mullite tube.

Methodology

Firstly, the waste catalyst powder was mixed with the proper amounts of fluxes (CaF_2 or CaO), and with or without different amounts of metal collector (Cu). Then, the mixture was placed in a graphite container, which was introduced in the high-temperature induction furnace. The effect of temperature and holding time was investigated. After the heat treatment, the furnace was cooled down to room temperature, the crucible extracted and subsequently cut for further SEM-EPMA analysis. Samples of the metallic phase were analyzed by ICP-MS/OES (Inductively Coupled Plasma Mass Spectrometry/Optical Emission Spectrometry).

The recovered PGM-containing Cu metallic phase was further transferred to the electrorefining cell where it was used as anode.

The electrorefining experiments were performed by applying a constant anode voltage (potentiostatic mode versus AgCl/Ag reference electrode) using an Autolab potentiostat. All voltages, as well as the resulting currents, were logged by using a Keithley 2000 Multimeter.

Samples of the molten bath were taken prior to and after the electrorefining trials, using a quartz tube with a quartz frit to avoid solid particles. The samples were cooled down in a desiccator and stored until analyzed by ICP-SFMS (Inductively Coupled Plasma Sector Field Mass Spectrometry).

Once the experiment was finished, the furnace was cooled down to room temperature under argon atmosphere, and the next day the cell was opened and investigated. The Cu cathode and anode lump were recovered and analyzed by ICP-MS/OES after washing the entrapped salt with water.

Results and Discussion

Optimization of the Pyro-Metallurgical Step

Metal-rich droplet losses in slags due to insufficient phase separation remain an important issue in pyro-metallurgical processes [6]. Therefore, the design of an appropriate slag system is the most important task when developing an efficient process for the recovery of PGM from waste catalyst material. In this respect, viscosity seems to be the most important physical property of the slag, which depends primarily upon the working temperature and composition of the slag. Moreover, interfacial energy/tension between the PGM micro-particles and the molten slag may also be of great importance to promote separation of the metal phase, and so achieving best recovery efficiencies of PGM from the waste material.

The catalyst waste material consists mainly of cordierite ($Mg_2Al_4Si_5O_{18}$), zeolite (typically, $Na_2Al_2Si_3O_{10} \cdot 2H_2O$) and corundum ($Al_2O_3$) or other oxides like ceria (CeO_2) and zirconia (ZrO_2) acting as catalyst layers that support fine PGM particles.

Then, most part of the spent catalyst material will act as slag in the pyro-metallurgical process.

It is well known that the viscosity of the slag increases with increasing concentrations of network formers (e.g., SiO_2 and Al_2O_3), whereas both network breakers (e.g., CaO, MgO, MnO, and FeO) and fluxes (e.g., Na_2O, K_2O, Li_2O, CaF_2, and B_2O_3) decrease viscosity [7]. In addition, increasing working temperatures reduces the viscosity of the slag.

On the other hand, the affinity of PGM particles for the melted collector metal is a key factor in the pyro-metallurgical process. Cu was chosen in this study due to the well-known mutual solubility between PGM and Cu. However, the largest disadvantage of using Cu as metal collector is its price. Therefore, it is important to recover the Cu for re-use.

Furthermore, large bubbles are produced during the gas-blowing stages during the pyro-metallurgical treatment. Therefore, a sedimentation step is necessary before phases can be tapped. The settling time of this step needs to be long enough to allow the different phases to separate, but to avoid large energy consumption and production times it cannot be too long.

Taking the above into consideration, the following parameters and values were tested:

(i)   Slag composition, i.e., CaO and CaF_2 additions, 2.5, 5, 10, and 20 wt%.
(ii) Atmosphere conditions, i.e., inert (Ar), air, and CO/CO_2.
(iii) Temperature, i.e., 1500–1650 °C.
(iv) Holding time, i.e., 1–3 h.
(v)   Metal additive and amount, i.e., 0, 2.5, 10, and 15 wt% Cu with respect to catalyst waste.

Table 1 summarizes the parameters used, and the results obtained in terms of metal recovery rates and PGM extraction efficiencies in different pyro-metallurgical tests. Figure 1 shows some examples of micrographs from the crucible after the pyro-metallurgical treatment.

The results showed that the addition of Cu metal, as PGMs collector, increased the recovered metal droplet size, then accelerating the settling velocity of metal droplets, thus increasing the recovery rate. Moreover, metal recovery is enhanced when increasing density difference between metal and slag, as well as when decreasing the viscosity of the molten slag by adding CaO. In addition, the effect of reactor wall on the PGM recovery was also studied experimentally. It was found that a 250% increase in diameter of the crucible leads to an improvement of ca. 5% in the overall PGMs recovery.

When investigating the results obtained, the optimal parameters for the pyro-metallurgical pre-treatment of the ceramic catalyst material can be summarized as follows:

(i)   10–15 wt% CaO addition. The slag viscosity obtained is optimal, so there is no need of adding CaF_2, which will increase the hazardous nature of the slag by-product.

Table 1 Pyro-metallurgical tests, parameters, and results obtained in terms of metal phase recovered and PGM amount extracted

Exp#	CaO (wt%)	Cu (wt)	T (°C)	Holding time (min)	Metal recovery (%)	Pt recovery (%)	Pd recovery (%)	Rh recovery (%)	Overall PGM recovery (%)
1	10	11.2	1600	90	84.2	88.54	26.09	75.91	66
2	20	10.5	1588	60	88.5	78.92	78.21	59.44	72
3	20	15	1588	60	88.2	71.97	75.79	57.70	66
4	20	10	1651	60	100.0	100	100	100	100
5	20	15	1650	60	86.0	66.23	109.57	76.53	68

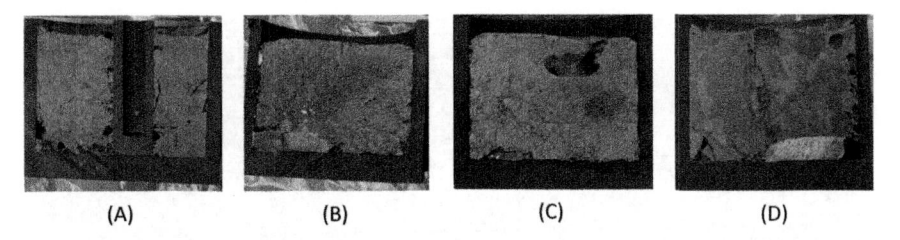

Fig. 1 Examples of crucibles after the pyrometallurgical treatment according to Table 1. **a** Exp #2; **b** Exp #3; **c** Exp #4; **d** Exp #5. (Color figure online)

(ii) CO reducing atmosphere, obtained by using graphite linings/containers.
(iii) Working temperature in the range of 1600–1650 °C.
(iv) Holding time, 1–1.5 h.
(v) Cu additive, 10 wt%, with respect to the catalyst sample and PGM contents.

The choice of the concluding parameters is based upon the recovery of PMGs, energy efficiency as well as cost-effectiveness of the process.

SEM micrographs of the metal and slag phase using the optimal conditions as shown in Fig. 2. EPMA analysis showed that the metal phase, mainly consisting of Cu, contained also significant amounts of Fe, as well as some Si (cf. Fig. 3). Moreover, the amount of metal in the slag phase was negligible, thus confirming the good metal separation and recovery.

The pyro-metallurgical step was able to pre-concentrated the PGMs by ca. 10 times from the waste feedstock to the generated metallic phase, with an energy consumption of ca. 5.5 kWh kg^{-1} Cu recovered.

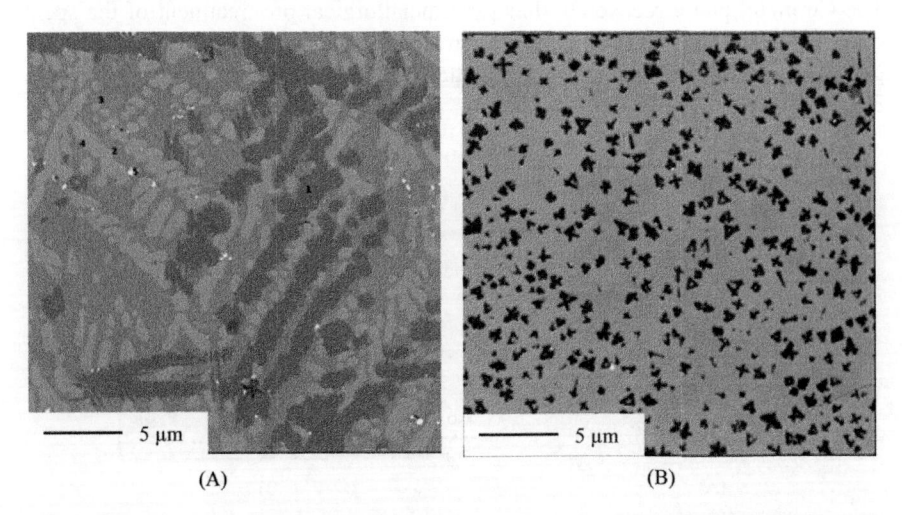

Fig. 2 SEM micrographs obtained using the optimal conditions for the pyro-metallurgical process. **a** Metal phase; **b** Slag phase. (Color figure online)

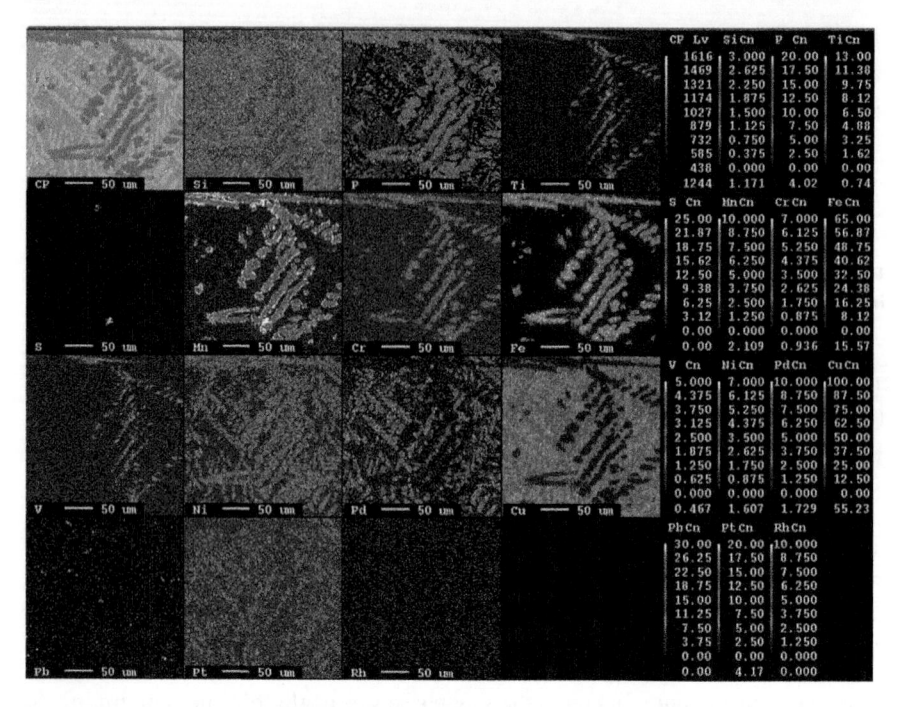

Fig. 3 EPMA analysis of the metal phase from Fig. 2a. (Color figure online)

Optimization of the Electrorefining Step

The Cu metal phase recovered after pyro-metallurgical pre-treatment of the spent catalyst (cf. Fig. 4a) was transferred to an electrolysis cell where it was anodically polarized against a Cu cathode. Schematics of the electrorefining cell is shown in Fig. 4b.

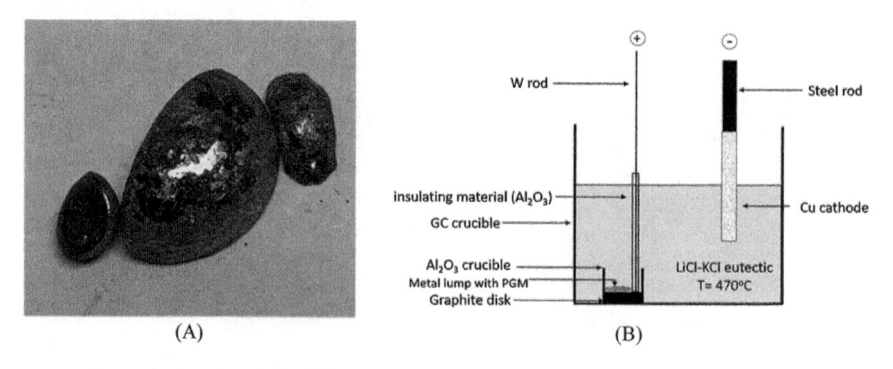

Fig. 4 a Examples of metal phase (Cu + PGM) recovered after pyro-metallurgical step. **b** Schematics of the experimental electrorefining setup. (Color figure online)

ICP-MS analysis of the metal lump showed that the most relevant impurity that could affect the electrorefining process is Fe (cf. Table 2). It is expected that in the cases where the content of Si is quite significant (3 wt%), it will leave the electrolyte to the gas phase as $SiCl_4$ gas. P is another significant impurity in the metal lump, as the P content in the catalyst waste is up-concentrated in the metal phase. If P is anodically oxidized, it will be collected in the gas phase in the form of PCl_3 gas.

The standard potentials of the most relevant electrochemical systems are gathered in Table 3. Despite the values do not take into account the solvation effects of the dissolved metal ions by the Cl^- ions from the chloride electrolyte, it is possible to predict that Cu, Fe, Si, and P could be anodically extracted, as Cu(I) and Fe(II) ions and PCl_3 and $SiCl_4$ gas, while Pt being still in metallic form in the anode. In the LiCl–KCl electrolyte, the formation of potassium hexachloro-platinate complex is very stable, so the anodic dissolution giving K_2PtCl_6 is more favoured than that of the Pt-chlorocomplexes (200 mV, cf. Table 3). Despite this, the standard potentials of Pt, Pd, or Rh dissolution are far more anodic from that of the Cu system.

The electrochemical behavior of PGM-containing Cu metal lump was studied by linear sweep voltammetry (cf. Fig. 5a). As expected, its anodic dissolution corresponded to that of metallic copper [9], though some extra ohmic resistance could be observed due to the presence of an oxide layer in the Cu-PGM metal lump. The results showed the optimal anodic potentials to be applied in the electrorefining process.

Electrorefining tests were carried out in the eutectic LiCl–KCl mixture at ca. 450 °C containing ca. 1.5 wt% CuCl, using the experimental setup shown in Fig. 4b. A series of potentiostatic electrolysis trials were systematically run, by anodic potentials applied ranging from −0.2 to +0.3 V versus AgCl/Ag reference system. The current values registered during the trials are shown in Fig. 5b.

Electrolyte samples, as well as the metal lumps before and after the electrolysis trials, were analyzed by ICP-MS. The results indicated that a significant weight loss of the Cu-PGM lump was detected when the electrolysis process was carried out at an applied anodic potential of at least +0.2/+0.3 V versus the AgCl/Ag reference system, where significant current was obtained (cf. Fig. 5b).

BSE micrographs of the metal lumps after the electrorefining clearly showed an up-concentration of the PGM-rich areas (bright areas in Fig. 6b), compared to the micrographs of the Cu-PGM sample before the direct electrolysis process (cf. Fig. 6a). These results confirm the PGM up-concentration in the anode after the direct electrolysis process.

Analysis of the Cu deposit at the cathode and the amount of Cu in the electrolyte before and after the electrorefining process showed that Cu was selectively extracted from the Cu-PGM anode, the current efficiency of the process being ca. 70%, which is not so bad considering the small laboratory cell. Considering that the optimal cell voltage applied is in the order of 0.4–0.5 V (cf. Fig. 5b insert), the energy consumption is estimated to be in the order of 0.2–0.3 kWh/kg^{-1} Cu refined. Considering an average PGM content of 2500 ppm in the waste catalyst, the energy consumption related to the PGM extraction is of 7 kWh/kg^{-1} PGM.

Table 2 ICP-MS of the metal lump obtained in the different experiments, as shown in Table 1

Exp #	Cu (wt%)	Pt (ppm)	Pd (ppm)	Rh (ppm)	Fe (wt%)	Si (ppm)	Cr (ppm)	Ni (ppm)	P (wt%)	Ti (ppm)
1	91.1	2273	13686	2253	2.78	100	n.a	700	2.23	40
2	95.9	2509	17100	1958	1.93	5000	107	1033	1.58	6
3	102.5	1627	10436	1272	1.48	3000	85	730	1.29	3
4	81.4	3722	19855	3693	3.56	30300	399	1043	2.68	2
5	95.8	2416	9864	1733	3.32	29300	482	761	1.55	20

Table 3 Standard potentials of most relevant electrochemical systems determined from theoretical thermodynamic values of the pure chloride substances using HSC software [8]

System	E^0 /V versus Cl_2 /Cl^- @ 450 °C
$Fe(III) + e^- = Fe(II)$	0.099
$Cl_2 (g) + 2e^- = 2Cl^-$	0
$Cu(II) + e^- = Cu(I)$	−0.181
$Pt(IV) + 5e^- = Pt (s)$	−0.212
$PdCl_6^{2-} + 4e^- = Pd(s) + 6Cl^-$	−0.269 (in the presence of KCl)
$Pd(II) + 2e^- = Pd(s)$	−0.328
$Rh(III) + 3e^- = Rh(s)$	−0.357
$P(V) + 5e^- = P$	−0.396
$PtCl_6^{2-} + 4e^- = Pt(s) + 6Cl^-$	−0.403 (in the presence of KCl)
$P(III) + 3e^- = P$	−0.826
$Cu(I) + e^- = Cu$	−1.017
$Fe(II) + 2e^- = Fe$	−1.241
$Si(IV) + 4e^- = Si$	−1.469

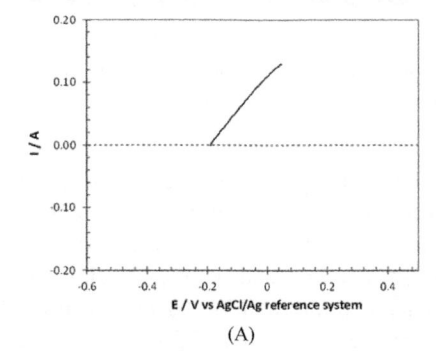

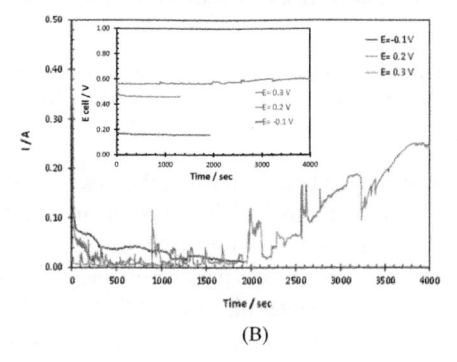

Fig. 5 a Example of linear sweep voltammogram obtained using the Cu-PGM metal lump as working electrode in the eutectic LiCl–KCl mixture at 450 °C. Sweep rate 100 mV s^{-1}. **b** Current reads during direct (potentiostatic) electrolysis trials at different anode voltages applied (given vs. AgCl/Ag reference system). The cell voltages obtained in each case are showed in the insert. (Color figure online)

Summary and Concluding Remarks

The extraction of platinum group metals (PGM) contained in waste automobile catalyst monolithic honeycomb was investigated by a novel approach that combines a pyro-metallurgical and electrolysis step.

Optimization of the pyro-metallurgical step gave the best conditions to achieve the best efficiencies in PGM recovery. It was found that the recovery of the metal fraction is enhanced when increasing the density difference between metal and slag,

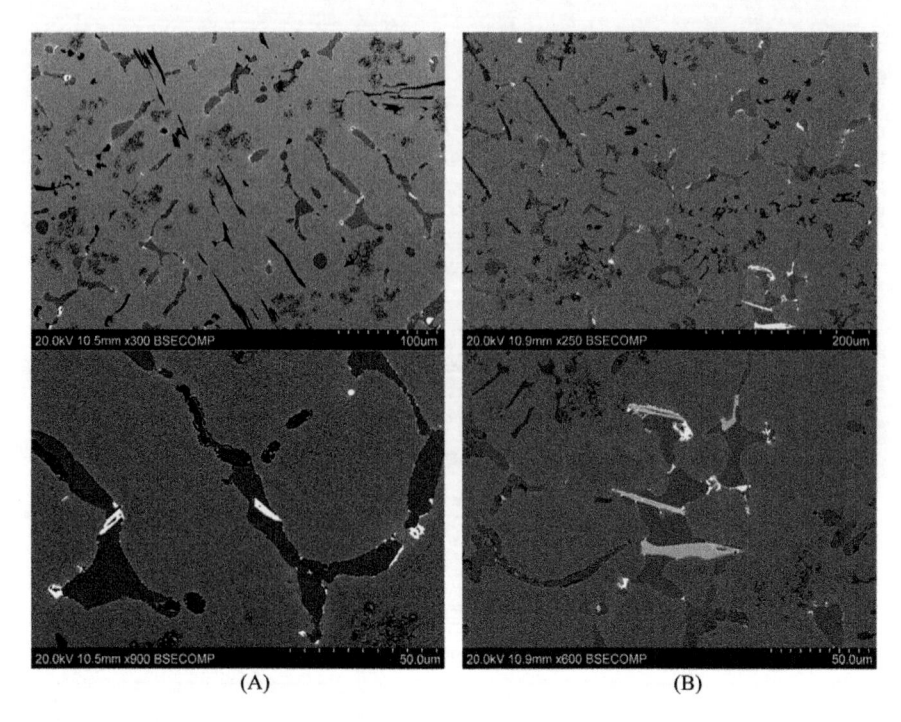

Fig. 6 BSE images of the Cu-PGM metal lumps at different magnifications, **a** before, and **b** after the electrolysis process at applied anode potential of +0.3 V versus AgCl/Ag reference system

as well as when decreasing the viscosity of the molten slag. Moreover, the addition of Cu metal, as PGMs collector, increased the recovered metal droplet size, then accelerating the settling velocity of metal droplet, and increasing the recovery rate. The optimal process parameters included 10–15 wt% CaO and 10 wt% Cu additions to the waste material, 1600–1650 °C working temperature, and 1–1.5 h holding time. The energy consumption of the process is estimated to be of ca. 5.5 kWh kg^{-1} Cu recovered.

The subsequent electrolysis process is carried out from a molten salt electrolyte, thus allowing the electrorefining of Cu from the PGM-containing metal with better selectivity and kinetics as well as lower energy consumption than in state-of-art processes using aqueous solutions. This is due to the fact that Cu(I) species are stable in the molten salt electrolyte, thus the voltage (and so energy) needed in the electrorefining process is lower than that in an analogous aqueous solution process where Cu(II) are the solely stable species. Experimental trials in a small laboratory-scale cell demonstrated that Cu could be extracted selectively at a current efficiency of ca. 70%. The energy consumption of the process was estimated to be in the range of 0.2–0.3 kWh/kg^{-1} Cu refined which amounts to ca. 7 kWh/kg^{-1} PGM, assuming an overall PGM content (Pt, Pd, Rh) in the waste automobile catalyst monolithic honeycomb of 2500 ppm.

Acknowledgements This work has received funding from the European Union's H2020 Programme under GA No. 730224 (www.platirus.eu).

References

1. Sinisalo P, Lundström M (2018) Refining approaches in the PGM processing value chain. a review. Metals 8:203–215
2. Martinez AM, Osen KS, Støre A (2020) Recovery of PGM from secondary sources by selective chlorination from molten salt media. Rare Metal Technol 221–233
3. Peng Z et al (2017) Pyro-metallurgical recovery of platinum group metals from spent catalysts. JOM 69:1553–1562
4. Kisza A, Kazmierczak J (1991) Kinetics of electrode reactions on metallic electrodes in pure molten chlorides. Chem Papers 45(2):187–194
5. https://monolithos-catalysts.gr/en/
6. Bellemans I, De Wilde E, Moelans N, Verbeken K (2018) Metal losses in pyro-metallurgical operations—a review. Adv Colloid Interface Sci 47–63
7. Mills K (2014) How mold fluxes work. In:Seethraman S (ed) Treatise on process metallurgy, vol 3. Industrial processes. Elsevier 435–475
8. HSC Chemistry v7.11, © Outotec
9. Castrillejo Y, Bermejo MR, Martínez AM, Pardo R, Sánchez S, Picard G (1999) Electrochemical behaviour of copper ions in molten equimolar $CaCl_2$-NaCl mixture at 550 °C. In: Gaune-Escard M (ed) Advances in molten salts, 32–49. ISBN 1–56700–142–4

Developed Commercial Processes to Recover Au, Ag, Pt, and Pd from E-waste

Rekha Panda, Manis Kumar Jha, Jae-chun Lee, and Devendra Deo Pathak

Abstract Due to the supply gap towards increasing demand as well as loss of precious metals by illegal recycling, present research reports application-oriented processes developed at CSIR-NML, India to recover precious metals from small components of e-waste containing ~0.1–0.8% Ag, ~0.03–0.9% Au, ~0.01–0.02% Pd, ~0.0003–0.0005% Pt, and related effluent. Firstly, ~99.99% Au was recovered from plated e-waste using the process of selective leaching followed by charcoal adsorption and heat treatment, whereas the second process consists of dismantling, physical/ chemical pre-treatment of e-waste followed by hydrometallurgical processing to recover 99% Ag, 99.9% Au, 95% Pd, and 90% Pt. Apart from the above, leaching and selective precipitation were used to recover ~95% Ag from waste computer keyboards. The effluent generated during the e-waste processing was found to contain ~8–10 mg/L Au, which was also recovered using ion-exchange technique. All processes presented are scientifically validated and commercially viable after scale-up studies.

Keywords Precious metals · Recycling · E-waste · Hydrometallurgy

Introduction

Precious metals (PMs) naturally occurring elements with high monetary value are usually present in uncombined form in rocks or placer deposits of water bodies as well as in alluvial sands. They are regarded as noble metals owing to their remarkable

---

R. Panda · M. K. Jha (✉)
Metal Extraction and Recycling Division, CSIR-National Metallurgical Laboratory, 831007, Jamshedpur, India
e-mail: maniskrjha@gmail.com; mkjha@nmlindia.org

J. Lee
Mineral Resources Research Division, Korea Institute of Geosciences and Mineral Resources (KIGAM), Daejeon, South Korea

R. Panda · D. D. Pathak
Department of Applied Chemistry, Indian Institute of Technology (ISM), Dhanbad 826004, India

resistance towards the external environment. Gold (Au), silver (Ag), platinum (Pt), and palladium (Pd) belong to the group of precious metals. Particularly, Au and Ag have been highly prized by mankind since ancient times, whereas Pt and Pd have been recently discovered [1]. In ancient times, precious metals were mostly used for fabrication of religious decorative articles, jewellery, and currency due to their beauty, stability, and scarcity but in the present scenario they play a key role in the industrialized world resulting in manifold applications from industrial to advanced material science owing to their incomparable chemical and physical properties. PMs are widely being used in electrical and electronic manufacturing industries to produce electronic components such as printed circuit boards (PCBs) of computers, mobile phones, parts of medical and telecom equipment, smart-cards, sensors. Specially, Au, Ag, Pt, and Pd play important role in the modern electronic devices. On global aspect, the net consumption Au, Ag, and Pd by the electronics manufacturing sector is about 20%, 35%, and 16%, respectively, out of the total quantity mined annually [2]. The worldwide demand of Au in the field of electronics was around 254 tons in the year 2015 [3]. About 130 kg of Cu, 3.5 kg of Ag, 0.34 kg of Au, and 0.14 kg of Pd could be produced by recycling one ton of mobile phones which is almost 10–100 times greater than their respective ores [4, 5].

In the present scenario, the financial and technological advancements along with modern lifestyle and high living standard of people have initiated availability of cheaper and improved electronic products in the market, which has led to incredible increase in the volume of electronic goods being procured by the consumers. It is projected that annual revenue of $ 62.5 billion is to be generated by the end of 2020, according to the report of global e-waste management market [4]. Replacement of older models by new ones has resulted in the generation of huge quantities of electronic waste (e-waste) at their end of life which needs to be properly managed [6]. E-waste is also termed as urban ore as they are generated in urban area, whereas their collection is called urban mining. The e-waste containing significant amount of PMs proposes an important recycling potential for the secondary supply of precious metals. The price of precious metal and cost of exploration is so high that urban mining could be a key solution to meet their present demand as this can save the exploration time and cost as well as high grade of metals compared to conventional mining could be produced [7]. The depleting conventional reserves of precious metals coupled with increase in their demand have imposed a challenging task ahead for sustainable development. In the current scenario, e-waste recycling industries are in need of processes to recycle precious metals from obsolete electronic devices.

India has been progressively emerging as an industrialized and economically advancing country in the world with major producer of many important metals and industrial minerals but the status of precious metals production is rather dismal. Consequently, importance has been felt to research on precious metals recovery from secondary resources, especially e-waste, using economically feasible technologies. Therefore, it is necessary to exploit alternative resources and develop technologies to meet the future demands of precious metals for their various applications.

Recycling of e-waste to recover precious metals can be achieved using pyro-, hydro, electro-metallurgical, or hybrid processes. Pyrometallurgy is a conventional method which needs more investment but recovery is difficult in view of high purity along with the loss of precious metals. Moreover, it also leads to emission of toxic gases into the environment, whereas hydrometallurgical processes are gaining much attention to recover precious metals from e-waste due to its high selectivity, low toxic emission, and high purity product formation. Various pre-treatment methods (roasting, pyrolysis, pre-leaching, etc.) have been reported to remove organic substances, which hinders the leaching of precious metals from source material [8]. After pre-treatment, leaching is carried out for maximum dissolution of metals. Several authors have reported the recovery of precious metals using cyanide and non-cyanide leachants. Cyanide is the most common leachant and has been employed for more than a century for dissolution of gold owing to its high selectivity and stability of dicyanoaurate complex [9, 10]. Other non-cyanide lixiviants such as thiosulphate [11, 12] and aqua regia [13] have also been reported for effective leaching of precious metals. Recovery of gold from waste solution using precipitation [14], solvent extraction [15, 16], adsorption, and ion exchange [3, 17, 18], etc., have also been reported.

From the extensive literature review made, it was concluded that there is still a lack of feasible and indigenous technology to recover precious metals from e-waste in environmental friendly manner. Keeping in view, the significant conflict between availability of natural resources of precious metals, lack of viable technology, as well as its loss by illegal recycling practices, CSIR-NML in India has been continuously engaged in making sincere R & D efforts to develop application-oriented e-waste recycling processes using pyro-/hydrometallurgical/combination of both. The work developed and reported by our group is clearly different than the other researchers which mostly reported basic scientific studies whereas in present work emphasis has been made on industrial viability. The flow-sheets developed at CSIR-NML to recover precious metals (Au, Ag, Pd, and Pt) as value-added marketable products from e-waste are scientifically validated, economical, and eco-friendly that will not only lead to proper e-waste management but also produce high-value products which will add to the economy of the nation.

Developed Application-Oriented Processes at CSIR-NML

Materials

The outer surface of PCBs and its populated small components containing significant amount of precious metals (Au, Ag, Pt, and Pd) from different types of e-waste comprising of personal computers, laptops, mobile phones, printers, television, etc., were used as raw material for process development. PCBs with gold coating on surface, ICs, MLCCs, Mylar sheet of keyboards and dilute effluent left after processing of e-waste were used for experimental purpose.

Minerals acids such as hydrochloric acid (HCl), sulfuric acid (H$_2$SO$_4$), and nitric acid (HNO$_3$) were procured from Merck, India, whereas salts viz. sodium metabisulphite (SMB), potassium chloride (KCl), sodium chloride (NaCl), and sodium hydroxide (NaOH) used for precipitation studies were supplied by Rankem, India. Organic extractants 2-hydroxy-5-nonylacetophenone oxime (Trade name LIX 84/ LIX 84IC) and mixture of 5-nonylsalicylaldoxime and 2-hydroxy-5-nonylacetophenone oxime in ratio 1:1 (Trade name LIX 984 N) used for solvent extraction were procured from M/s Cognis Corporation, Ireland. Apart from this, different dilute solutions were prepared using de-ionized water.

Methodology

Hydrometallurgical route was followed for the development of feasible flow-sheet to recover Ag, Au, Pt, and Pd from e-waste. Initially, all scraps were dismantled and parts such as PCBs, hard disks, monitors, plastics, batteries were separated. The PCBs were further depopulated using thermal treatment to de-solder and liberate the small components such as integrated circuits (ICs), multilayered ceramic capacitors (MLCCs), processors, diodes, transistors, etc., mounted on its surface. The selected samples were pulverised in a high-impact pulveriser to get mixture of metals, plastics, ceramics, and epoxy, which was further processed for physical beneficiation to separate the metallic and non-metallic parts using wet gravity separation technique. The metallic concentrate containing precious metals were leached in a three-necked pyrex reactor well fitted with a condenser in order to prevent the evaporation of gases evolved at high temperatures. Hot plate was used for providing heat to the leaching system, whereas magnetic stirrer was used for proper agitation of the samples. The leach liquor obtained after filtration was further processed using advanced separation techniques (solvent extraction/ precipitation/ ion exchange) to collect pure metal solution from which value-added marketable products of precious metals could be produced. The leached residue was washed, dried in a vacuum oven, and analyzed to obtain the mass balance for each set of experiment. The systematic approach used for recovery of precious metals from e-waste is presented in Fig. 1. The chemical composition of all raw materials used as well as solution prepared were analyzed using an atomic absorption spectrometer (AAS) (Perkin Elmer Model, Analyst 200; Make: USA) and inductively coupled plasma-optical emission spectroscopy (ICP-OES) (VISTA-PMX, CCD Simultaneous; Make-Australia).

E-waste / urban ore

↓

STEP 1: DISMANTLING, DEPOPULATION & CLASSIFICATION

↓

Components containing precious metals (PMs)

↓

STEP 2: PRE-TREATMENT/ PHYSICAL BENEFICIATION

↓

Metallic concentrate (PMs and impurities)

↓

STEP 3: LEACHING OF METALLIC CONCENTRATE

↓

Leach liquor containing PMs

↓

STEP 4: PURIFICATION OF LEACH LIQUOR

↓

Purified solution of PMs

↓

STEP 5: PRECIPITATION/ CEMENTATION/ ELECTROWINNING

↓

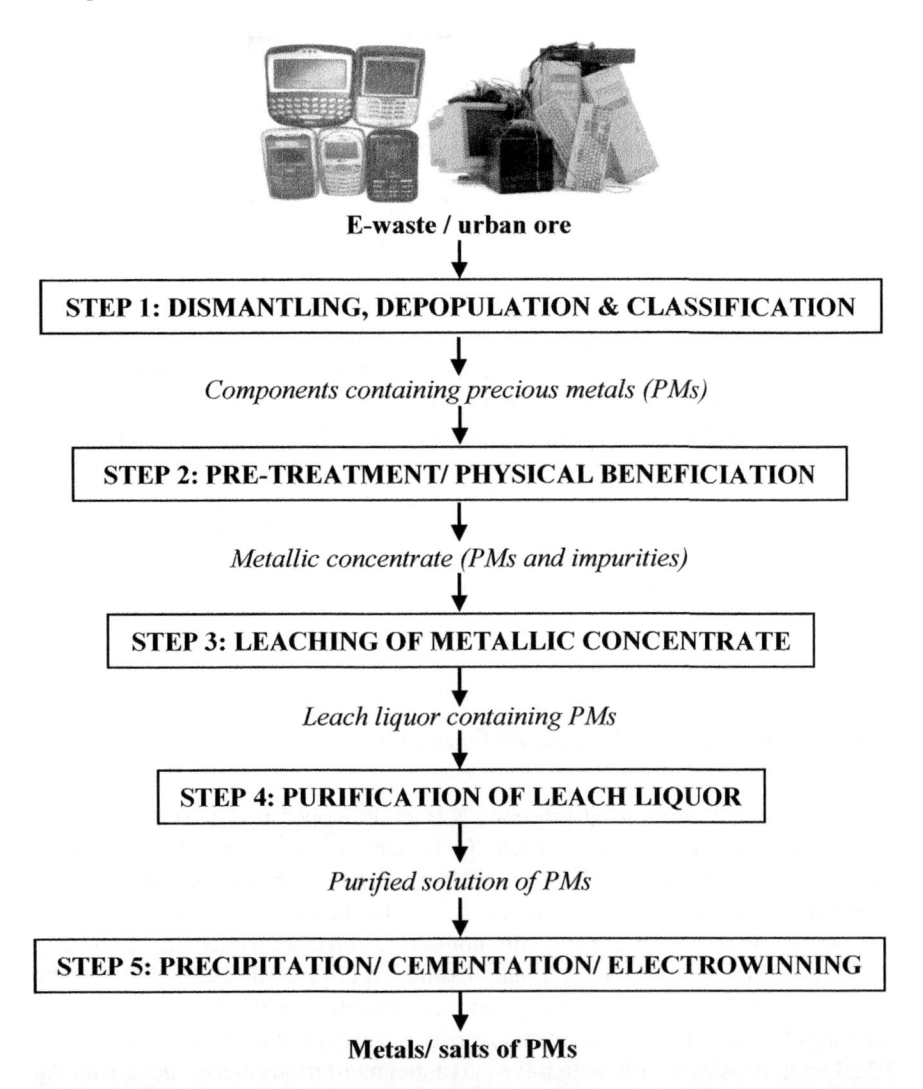

Metals/ salts of PMs

Fig. 1 Systematic approach for the recovery of precious metals from e-waste. (Color figure online)

Hydrometallurgical Flow-Sheets Developed for the Recovery of Precious Metals

Recovery of Au from Outer Surface of PCBs of Various Electronic Equipments

Panda et al. [3] developed a suitable hydrometallurgical process for the separation/ recovery of precious metals particularly Au from the waste PCBs of used and discarded electronic equipments has been developed and transferred to Indian recycling companies. Depending upon the type and composition of the source materials, two processes have been developed. First one cyanidation of gold followed by charcoal adsorption and second one being the selective dissolution of non-ferrous metals to get pure gold as residue. The outer surface of PCBs plated with gold was analysed and found to contain ~0.01–0.03% Au. The gold-plated PCBs were leached in cyanide medium for maximum dissolution of gold followed by the process of charcoal adsorption where all gold gets adsorbed onto the activated charcoal [3]. Further, the gold adsorbed charcoal was burnt at elevated temperature (~1350 °C) to obtain gold metal. In the second process, selective dissolution of other metals (impurities) present in the PCBs was carried out leaving gold in the residue. By this method all non-ferrous metals got dissolved except gold. The residue containing gold obtained was melted at ~1350 °C to obtain pure gold metal. Both the processes were carried out in closed system following proper safety measures. The effluent generated during the experiments was properly treated by electrolytic/chemical oxidation to completely decompose the cyanide content, whereas the gases NOx evolved during nitric dissolution was passed through scrubber. Both the processes are feasible depending upon the composition of the raw material. Gold of high purity was obtained using further purification techniques.

Recovery of Ag, Au, Pd, and Pt from ICs

A feasible hydrometallurgical process has been developed by Panda et al. [19] to recover precious metals from the waste ICs present in e-waste (Fig. 2). The process has also been transferred to Indian recycling company. The beneficiated metallic concentrate obtained from ICs was analysed and found to contain ~0.6% Ag, ~0.1% Au, ~0.01% Pd, ~0.0005% along with non-ferrous metals (mainly Cu). Two stage leaching was carried out for maximum dissolution of precious metals followed by their purification and recovery using advance separation techniques. In first stage, leaching of the metallic concentrate was carried out using 4 M HNO_3 at 90 °C keeping 50 g/L pulp density for 1 h to dissolve maximum non-ferrous metals along with Ag leaving other precious metals (Au, Pd, and Pt) in the leached residue [19].

HNO_3 was found to be suitable leachant for non-ferrous metals (Cu, Ni, etc.) and Ag. About 98% leaching of metals was achieved under the above experimental condition. Whereas in second stage, the leached residue containing Au, Pd, and Pt was dissolved in mixture of hydrochloric acid and sodium hypochlorite to get more than 95% precious metals in solution. Further, advance separation techniques were used to obtain purified solution of precious metals from which metals/ salts were recovered as marketable products. The leach liquor obtained in the first stage was found to contain Ag, Cu, and Ni. Initially, it was processed for the precipitation of Ag followed by the selective separation of Cu and Ni by solvent extraction technique using organic extractant LIX 984 N. Almost all Ag present in the solution was recovered as silver chloride (AgCl) using NaCl and 10 min mixing time. Solvent extraction was carried

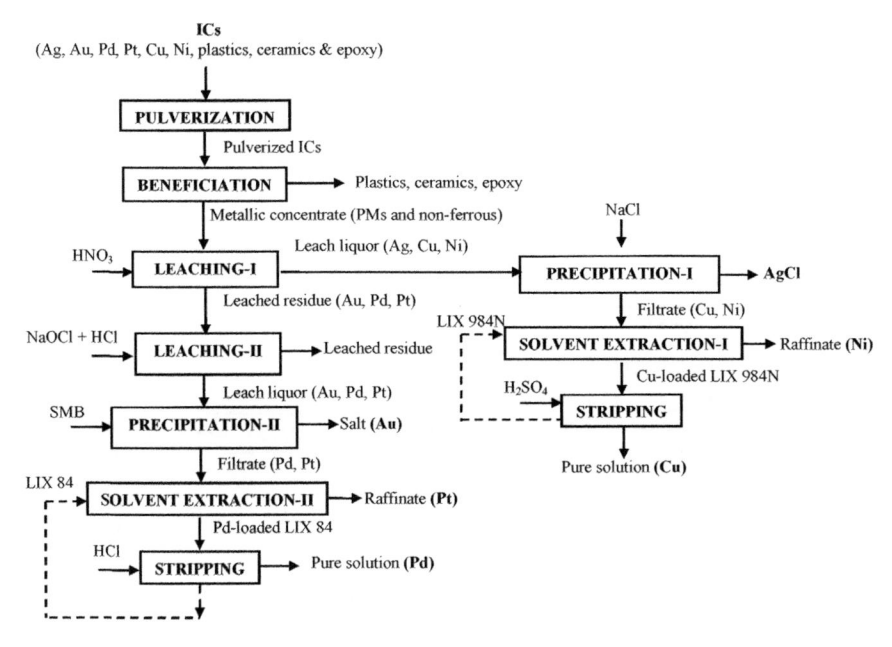

Fig. 2 Applied hydrometallurgical process to recover precious metals from the waste ICs [6, 19]

out using organic extractant LIX 984 N at pH of 2.97 maintaining phase ratio (O/A) of 1:1 at 10 min residence time. About 99.99% of Cu was extracted from the filtrate leaving Ni in the raffinate [6]. The leach liquor obtained in the second stage containing Au, Pt, and Pd was precipitated using sodium metabisulphite (SMB) to get salt of Au. The precipitate of Au obtained was filtered and separated, whereas Pt and Pd remained in the solution. Further, solvent extraction method was used to selectively extract Pd using extractant LIX 84 leaving Pt in the raffinate. The Pd-loaded LIX 84 was stripped using HCl to get pure solution of Pd. Salts of Pd and Pt were obtained from the stripped solution and raffinate, respectively, using evaporation technique.

Recovery of Ag and Pd from MLCCs

An application oriented process flow-sheet to recover Ag and Pd from MLCCs present in PCBs of e-waste has also been developed by Panda et al. [20] (Fig. 3). The depopulated MLCCs from PCBs were pulverized, chemically digested, and analyzed. About 1.08% Ag, 0.14% Pd along with 1.76% Cu and 11.1% Ni were found to be present in the pulverized MLCCs. As Ni was the major impurity present in the pulverized sample, therefore, it was removed by selective leaching using 2 M HCl at 75 °C and pulp density 100 g/L in two stages. The obtained leached residue was washed, dried, and further leached using 4 M HNO_3, 80 °C, pulp density 100 g/L,

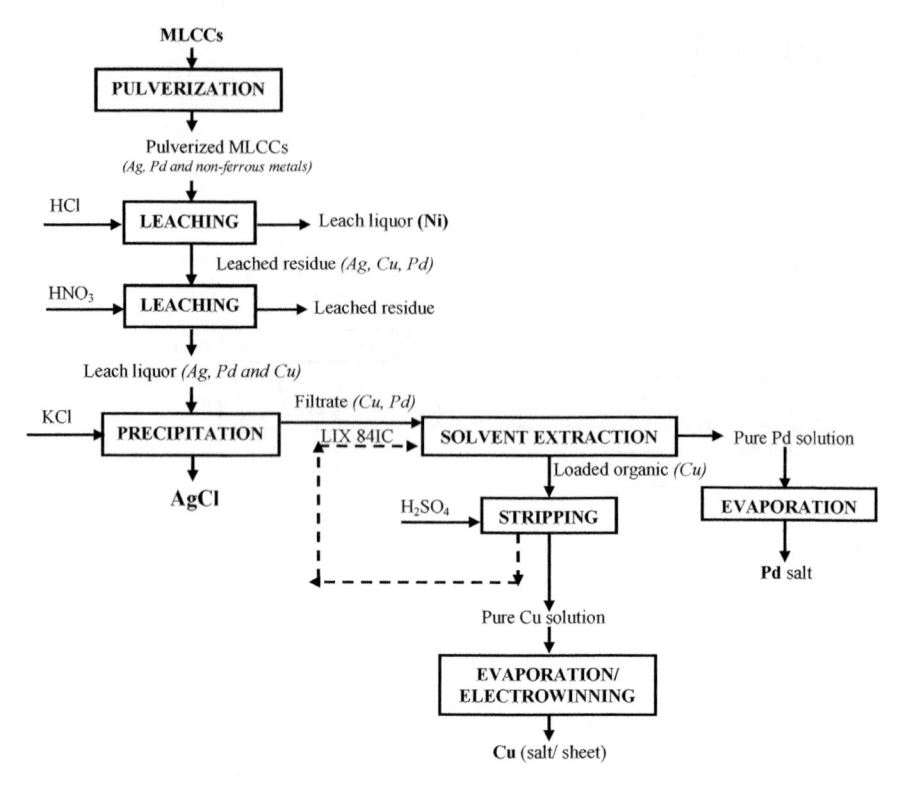

Fig. 3 Developed hydrometallurgical process to recover precious metals from the waste MLCCs [20]

and mixing time 1 h to get 99.99% of both Ag and Pd in solution. The leach liquor containing Ag, Pd, and Cu was put to precipitation studies using KCl to selectively precipitate Ag as AgCl. The Ag depleted solution was put to solvent extraction study using LIX 84IC to selectively extract Cu from the solution leaving Pd in the filtrate. The Cu-loaded organic was stripped using diluted H_2SO_4 to get pure solution of Cu. Both the pure solution of Cu and Pd obtained were evaporated separately to get their pure salts [20].

Recovery of Ag from Mylar Sheet of Keyboards

Keyboards of personal computers are one of such electronic good that contains significant amount of Ag (~2%). These keyboards after their end-of-life are treated as waste and become the part of waste stream. Keeping in view the increasing demand, stringent environmental regulations and limited natural resources of Ag, a hydrometallurgical process flow-sheet has been developed at CSIR-NML to recover Ag from scrap

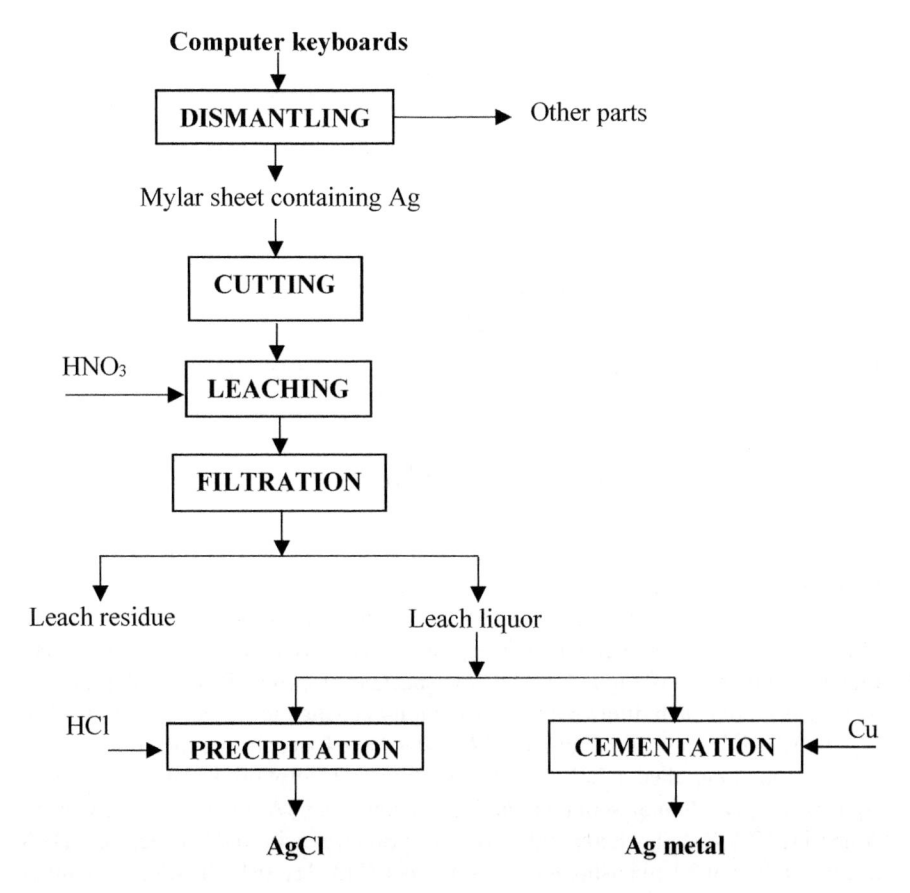

```
Computer keyboards
        │
        ▼
   DISMANTLING ──────▶ Other parts
        │
        ▼
Mylar sheet containing Ag
        │
        ▼
     CUTTING
        │
HNO₃    ▼
  ────▶ LEACHING
        │
        ▼
    FILTRATION
    │        │
    ▼        ▼
Leach     Leach liquor
residue       │
         ┌────┴────┐
HCl      ▼         ▼        Cu
 ──▶ PRECIPITATION  CEMENTATION ◀──
         │         │
         ▼         ▼
        AgCl    Ag metal
```

Fig. 4 Novel process flow-sheet for the recovery of Ag from Mylar sheet

computer keyboards (Fig. 4). Initially, the scrap keyboards of personal computers were dismantled and the transparent plastic thin film (Mylar sheet) containing Ag was separated. The Mylar sheet was cut into small pieces and further leached using HNO_3 at elevated temperature (90 °C) in 60 min contact time for maximum dissolution of Ag. More than 95% Ag leaching was achieved under above mentioned experimental condition. From the obtained leach liquor, Ag was recovered as high purity marketable product (salt/ metal) using precipitation/ cementation technique. The developed process flow-sheet is economical, environment friendly and has potential to be translated in industry.

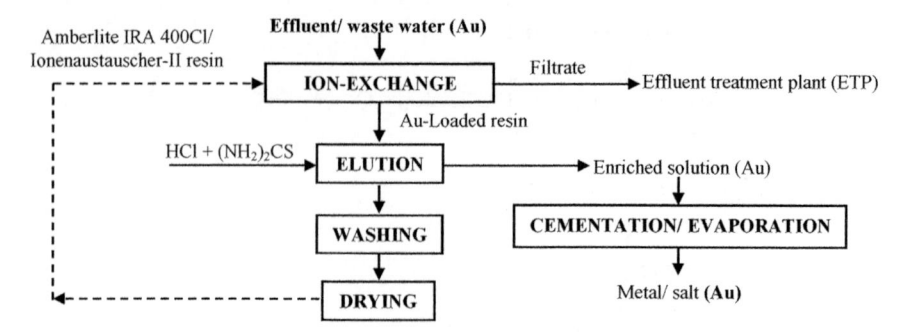

Fig. 5 Developed process flow-sheet to recover Au form dilute solution using adsorption technique [3, 21]

Recovery of Au from Wastewater/Effluent

Apart from the above-described flow-sheets related to recovery of precious metals from e-waste, R & D efforts have also been made to recover precious metal specially Au from wastewater/ effluent generated during the processing as well as recycling of e-waste. The effluent generated during the e-waste processing was analyzed and found to contain ~8–10 mg/L Au. The flow-sheet developed by Panda et al. [3, 21] to recover Au from such dilute solution using ion-exchange technique is presented in Fig. 5. Two different resins Amberlite IRA 400Cl and Ionenaustauscher-II have been used to obtain enriched solution of Au. Maximum adsorption of Au took place at equilibrium pH 9.6 in 30 min maintaining aqueous/resin (A/R) ratio 25 mL/g using Amberlite IRA 400 Cl, whereas 99% Au adsorption was achieved between eq. pH ~8 in contact time of 30 min using Ionenaustauscher-II [3, 21]. In both the cases, elution of the loaded resin was achieved using mixture of hydrochloric acid and thiourea in contact time of 30 min. About 88 times enriched solution of Au was obtained, from which value-added marketable products (salt or metal) could be produced.

Conclusion

Based on the various processes developed at CSIR-NML for the recovery of precious metals from different segments of e-waste, the following conclusion has been drawn:

- E-waste is the potential source for the recovery of precious metals (Au, Ag, Pt, and Pd).
- Generation of ample amount of e-waste annually, as well as lack of efficient recycling technology, lead to loss of valuable metals.
- Recycling of PCBs, ICs, MLCCs, and their related effluent to recuperate of Au, Ag, Pt, and Pd should be the prime concern towards zero waste.

- The processes developed to recover precious metals will minimize the gap between its demand and supply as well as curtail the load on import.

Acknowledgements The authors are thankful to the Director, CSIR-National Metallurgical Laboratory, Jamshedpur, for giving permission to publish this paper. One of the authors Ms. Rekha Panda would like to extend her sincere gratitude to CSIR, New Delhi (Grant: 31/10 (64)/ 2017-EMR-I), for providing Senior Research Fellowship to carry out this research work. Authors are also thankful to Indian Industries and Indo-Korean Collaboration.

References

1. Grimwade M (2009) Introduction to precious metals. Brynmorgen Press, Brunswick, USA, Metallurgy for Jewelers and Silversmiths
2. Khaliq A, Rhamdhani MA, Brooks G, Masood S (2014) Metal extraction processes for electronic waste and existing industrial routes: a review and Australian perspective. Resources 3:152–179
3. Panda R, Dinkar OS, Jha MK, Pathak DD (2020) Recycling of gold from waste electronic components of devices. Korean J Chem Eng 37(1):111–119
4. Rao MD, Singh KK, Morrison CA, Love JB (2020) Challenges and opportunities in the recovery of gold from electronic waste. RSC Adv 10:4300–4309
5. Jeon S, Ito M, Tabelin CB, Pongsumrankul R, Tanaka S, Kitajima N, Saito A, Park I, Hiroyoshi N (2019) A physical separation scheme to improve ammonium thiosulfate leaching of gold by separation of base metals in crushed mobile phones. Miner Eng 138:168–177
6. Panda R, Jha MK, Pathak DD, Gupta R (2020) Recovery of Ag, Cu, Ni and Fe from the nitrate leach liquor of waste ICs. Minerals Engineering 158:106584. https://doi.org/10.1016/j.mineng.2020.106584
7. Baba H (1987) An efficient recovery of gold and other noble metals from electronic and other scraps. Conserv Recycl 10(4):247–252
8. Panda R, Jha MK, Pathak DD (2018) Commercial processes for the extraction of platinum group metals (PGMs). In: Rare metal technology 2018, the minerals, metals & materials society, pp 119–130
9. Quinet P, Proost J, Van AL (2005) Recovery of precious metals from electronic scrap by hydrometallurgical processing routes. Miner Metall Process 22(1):17–22
10. Kulandaismy S, Rethinaraj JP, Adaikkalam P, Srinvasan GN, Raghavan M (2003) The aqueous recovery of gold from electronic scrap. J Metals 8:35–41
11. Hung VH, Lee JC, Jeong J, Hai HT, Jha MK (2010) The aqueous recovery of gold from electronic scrap. J Hazard Mater 17:1115–1119
12. Abbruzzese C, Fornari P, Massidda R, Veglio F, Ubaldini S (1995) Thiosulphate leaching for gold. Hydrometallurgy 39:265–276
13. Sheng PP, Etsell TH (2007) Recovery of gold from computer circuit board scrap using aqua regia. Waste Manage Res 25(4):380–383
14. Chmielewski AG, Urbanski TS, Migdal W (1997) Separation technologies for metals recovery from industrial wastes. Hydrometallurgy 45(3):333–344
15. Byoung HJ, Yi YP, Jeon WA, Seong JK, Tam T, Myong JK (2009) Processing of high purity gold from scraps using diethylene glycol di-N-butyl ether (dibutyl carbitol). Hydrometallurgy 95:262–266
16. Marzenko Z, Kowalski T (1984) The extraction of gold(III) from nitric acid medium. Anal Chim Acta 156:193–199

17. Tasdelen C, Aktas S, Acma E, Guvenilir Y (2009) Gold recovery from dilute gold solutions using DEAE-cellulose. Hydrometallurgy 96:253–257
18. Nguyen NV, Lee JC, Jha MK, Yoo KK, Jeong JK (2009) Copper recovery from low concentrate waste solution using DowexG- 26 resin. Hydrometallurgy 97:237–242
19. Panda R, Jha MK, Dinkar OS, Pathak DD (2020) Reclamation of precious metals from small electronic components of computer hard disks. In: Rare metal technology 2020, the minerals, metals & materials society, pp 243–250
20. Panda R, Dinkar OS, Jha MK, Pathak DD (2020) Hydrometallurgical processing of waste multilayer ceramic capacitors (MLCCs) to recover silver and palladium. Hydrometallurgy 197:105476
21. Panda R, Dinkar OS, Jha MK, Pathak DD (2020) Novel approach for selective recovery of gold, copper, and iron as marketable product from industrial effluent. Gold Bull 53:11–18

Part III
REEs

Innovative Reactors for Recovery of Rare Earth Elements (REEs)

Alison Lewis, Jemitias Chivavava, Jacolien du Plessis, Dane Smith, and Jody-Lee Smith

Abstract Interest in the recovery of Rare Earth Elements (REEs) has increased significantly in the last few years. There has been a concomitant increase in research and in process development for REE recovery [1]. Antisolvent crystallization has the potential to recover REE from solution at high yields and with minimal waste. However, antisolvent addition generally results in uncontrolled primary nucleation and very small product crystals. A better approach could be to carry out the crystallization in fluidized bed reactors. Therefore, our approach in this work was to focus on the development of a novel process for the recovery of REE by combining antisolvent crystallization and a fluidised bed process. Thermodynamic modelling showed that, when ethanol is added to a $Nd_2(SO_4)_3$ or $Dy_2(SO_4)_3$ solution as an antisolvent, the only solid products formed were the REE sulphate salts. Since the solubilities of the REE sulphate salts at any of the Organic/Aqueous (O/A) ratios are of similar orders of magnitude to those of salts that have been successfully recovered in a fluidised reactor process, an antisolvent, fluidised reactor process is potentially suitable for REE sulphate salts. Batch experiments showed that the yields are sufficiently high for a viable process. At the same time, the micrographs show that the nature of the formed crystals are such that they are likely to form uniform and robust coatings on seed particles in and fluidised bed reactor process. Therefore, our preliminary conclusion is that this REE system is well suited for further investigation in a combined antisolvent crystallization and fluidised bed process.

Keywords Crystallization · Fluidised bed reactor · Rare earth elements · Sustainability

A. Lewis (✉) · J. Chivavava · J. du Plessis · D. Smith · J.-L. Smith
Crystallization and Precipitation Research Unit, Chemical Engineering Department, University of Cape Town, Cape Town, South Africa
e-mail: Alison.Lewis@uct.ac.za

Introduction and Background

The Rare Earth Elements (REEs), which include the lanthanides, yttrium, and scandium, play an important role in many fields of advanced materials science [2]. In addition, due to their vital role in renewable energy technologies such as wind turbines, batteries, and electric cars [3], as well as in permanent magnets, lamp phosphors, catalysts, and rechargeable batteries, they are becoming progressively more important in the transition to a green economy [1].

Sources of REEs include raw deposits, waste materials as well as, potentially, seawater [4]. Despite the fact that the REEs have been identified as critical raw materials [5], typically only around 1% are recycled, with the rest being sent to waste and thus being removed from the materials cycle [6]. Because of limited primary resources on their territories, many countries will be forced to rely on recycling of REEs from pre-consumer scrap, industrial residues, and REE-containing End-of-Life products [7] as well as imports.

However, there is a significant but currently unrealised potential for recycling REE from end-uses such as permanent magnets, fluorescent lamps, batteries, and catalysts as well as from waste ores. South Africa has deposits that are rich in REE but that are currently being discarded as waste. For example, at Namakwa Sands, an REE-rich deposit is currently mined for zircon, ilmenite, rutile, and leucoxene, but most of the REEs are rejected [8]. Another source of REEs is end-of-life nickel metal hydride (NiMH) car batteries [9]. These batteries have a very limited life span and contain valuable components such as nickel, cobalt, and REE.

Given that there is currently very little recycling taking place, it is clear that the future potential for recycling will require a significant amount of research. REE recovery from waste has the potential to contribute significantly to sustainability. Although a significant investment will need to take place, there is the potential for this research to have great benefits, as developing processes that will increase the amount of REE recycling will contribute to the closing of material cycles as well as to the global requirement for circular economies.

Interest in the recovery of REE has increased dramatically in the last few years, with a surge of research and publications [10–16]. These have focussed on recovery using staged precipitation processes carried out in various configurations of stirred tank reactors [14] such as the parallel flow precipitation method [16]. Although it is known that stirred tank reactors are not the most suitable choice for precipitation processes, they are inevitably the default in both industrial processes and research studies. This paper proposes that innovation is needed in the design of both the reactor and solvent system in order to most effectively recover REE and to design efficient, controllable, and sustainable processes. In this work, we focused on a feasibility study for a fluidised bed, antisolvent process for the recovery of REE.

Literature Review

Recently, it has been proposed to use a supercritical antisolvent-fluidized bed process for particle coating in pharmaceutical applications [17]. However, to the best of our knowledge, neither fluidised bed nor antisolvent crystallization has been used to precipitate REE metal salts.

Antisolvent crystallization is used in the pharmaceutical industry [18], but is uncommon in extractive metallurgy. The principle is to decrease the solubility of the solute in the liquid phase through addition of a second solvent, whereupon the solute crystallizes. After filtering off the solid phase, the solvents can be separated and recovered. Antisolvent crystallization opens up new options for metal salts to be precipitated under acidic conditions with high yield, something that is becoming increasingly important if we are to develop processes for recovery of REE.

The problem with antisolvent processes is that, on addition of the antisolvent, high local supersaturation is created, which generally results in uncontrolled primary nucleation and, as a consequence, very small product crystals (<10 micron). Peters et al. [19], when recovering scandium by antisolvent crystallization, found that the crystals generated were extremely small ie. <2 μm. The fine crystals were attributed to a very high supersaturation generated upon adding the ethanol to the strip liquor, resulting in dominance of nucleation over crystal growth. The crystals decreased in size with the increasing ethanol content due to the increasing supersaturation. Such crystals are difficult to dewater and wash, resulting in impurity incorporation from entrained mother liquor. The use of seeding, as well as carefully controlled dosage of the antisolvent, can be used to promote crystal growth, resulting in larger, more pure, product crystals. An even better approach would be to carry out the crystallization in continuously operated fluidized bed reactors. Fluidised bed crystallization processes are highly effective and can be used for selective removal and recovery of components from wastewater, e.g. recovery of fluoride and phosphates and removal of heavy metals [20–24]. By this technique, it is possible to recover salts from dilute streams in the form of pellets, which can be easily handled downstream or included in the final product [20, 25, 26].

The use of fluidized bed crystallizers allows for controlled crystallization and has many advantages over conventional stirred tank crystallizers, like the elimination of significant impellor crystal collisions and the ability to harvest pellets that migrate to lower regions of the reactor once they reach a certain size. Fluidized bed crystallizers use seeds to minimise the generation of fine particles by providing a large surface area for crystal growth or deposition. This lowers the surface energy requirements compared to homogenous nucleation [27]. As the crystallization reaction proceeds, the seeding material becomes covered, either through growth or agglomeration, and gradually migrates to the bottom of the reactor. This allows for the crystallizing compounds to be separated from the treated stream in the form of pellets, which can be harvested. This mechanism has been successfully used in systems such as 150 mg/L Ni [21]; 5–100 mg/L P [28]; 3000 mg/L total metals [29]; 3557 mg/L $NiSO_4$ [30], and 5500 mg/L Ca [31] and in the precipitation of yellow cake in uranium processing

[32]. In these cases, supersaturation is relatively low, thus promoting deposition on the seed surface, which facilitates the solid–liquid separation process and avoids large volumes of sludge streams [21, 33]. While most studies conducted using fluidized bed crystallizers focused on low concentration, inorganic systems, the application of such crystallizers to REE systems has not, to our knowledge, been explored yet. This is a potentially novel contribution of this work.

Therefore, our approach in this work was to combine antisolvent crystallization and a fluidised bed process for the precipitation of REE (See Fig. 1). This novel approach opens up new avenues for future efficient and environmentally sustainable processes that have high recoveries of valuable end products from wastes. It will be vital to study and evaluate the full process, including the recovery of the antisolvent from the product filtrate (Fig. 1).

Aim

As part of developing processes for efficient recovery of REEs, the use of fluidised bed crystallizers is proposed in this work. It is therefore important to carry out the modelling of antisolvent crystallization of REEs and conduct preliminary batch tests in order to establish the potential suitability of the system for this application. Since the aim of this project was to establish the potential feasibility of the combined antisolvent crystallization and fluidised bed reactor process, this was carried out in two parts:

- The first part involved a thermodynamic modelling exercise which simulated the antisolvent crystallization of REE using a model alcohol. The modelling was used to identify the potential products, and to quantify the potential supersaturation;
- The second part involved carrying out batch experiments in order to study the crystals formed and to determine the feasibility of the combined antisolvent crystallization/fluidised bed reactor approach.

Method

Thermodynamic Modelling

Despite the ongoing interest in antisolvent crystallization processes, solubility data of many salt-solvent-antisolvent systems are still not available or easily accessible. This is particularly true for the solubility of REEs in alcohol/water mixtures. The only data that was available was the solubility of $Nd_2(SO_4)_3$ and $Dy_2(SO_4)_3$ in an ethanol/water mixture as part of a newly developed thermodynamic database in OLI Studio 10.0 Stream Analyser [34] and therefore this was used as the model system.

OLI Stream Analyser determines aqueous speciation and phase equilibria for multi-component systems using the Helgeson–Kirkham–Flowers (HKF) Equation of State parameters for the prediction of equilibrium constants. The predictions are executed using the Mixed Solvent Electrolyte (MSE) framework derived from the aqueous electrolyte non-random two-liquid (NRTL) model which uses the Bromley–Meissner equation to extrapolate limited data, the Pitzer model for molecule–molecule and ion–molecule interactions, the Helgeson–Kirkham–Flowers (HFK) equation to approximate standard state properties of species in water, and The Soave Redlich–Kwong Equation of State to determine fugacity coefficients of non-ideal states [34]. The solubility product constants, K_{sp}, of $Nd_2(SO_4)_3$ and $Dy_2(SO_4)_3$ in an ethanol–water mixture of varying ratios from 0 to 1 were modelled.

Batch Tests

Batch tests were carried out using aqueous solutions of $Y_2(SO_4)_3 \cdot 8H_2O$ at 21.96 gL^{-1} of the octahydrate (i.e. $[Y^{3+}]$ = 6.4 g/L) and four different alcohols: ethanol, methanol, 2-propanol, and t-butanol which were added at O/A ratios of 1:1 v/v.

Results and Discussion

Thermodynamic Modelling

The OLI Studio 10.0 Stream Analyser showed that, when ethanol was added to aqueous solutions of $Nd_2(SO_4)_3$ and $Dy_2(SO_4)_3$, the only solid products formed were the REE sulphate salts.

The modelled solubility product constants, K_{SP}, of $Nd_2(SO_4)_3 \cdot 8H_2O$ and $Dy_2(SO_4)_3 \cdot 8H_2O$ in the ethanol–water mixture of varying ratios are illustrated in Fig. 2. The K_{SP} values of other systems that have previously been studied in the fluidised bed reactor are also shown.

In Fig. 2, it can be seen that the solubilities of the two REEs, at a range of O/A ratios, fall into the same range as those of the two metal (II) carbonates, $10^{-6} < K_{SP} < 10^{-9}$, and are much higher than both the calcium phosphate with $K_{sp} < 10^{-27}$ and the copper sulphide ($K_{sp} = 10^{-37}$) (Table 1).

In previous work, it was shown by Guillard and Lewis [21, 22] that $NiCO_3$ is suitable for crystallization in the fluidised bed reactor, as the $NiCO_3$ coated the seed material through heterogeneous nucleation, growth, and agglomeration. Tai [38] showed the same for $CaCO_3$, as did Seckler [37] for $(Ca)_3(PO_4)_2$. Conversely, it was shown by Lewis and van Hille [39] that metal sulphides, such as CuS, are not suitable for crystallization in a fluidised bed reactor due to their extreme insolubility and very

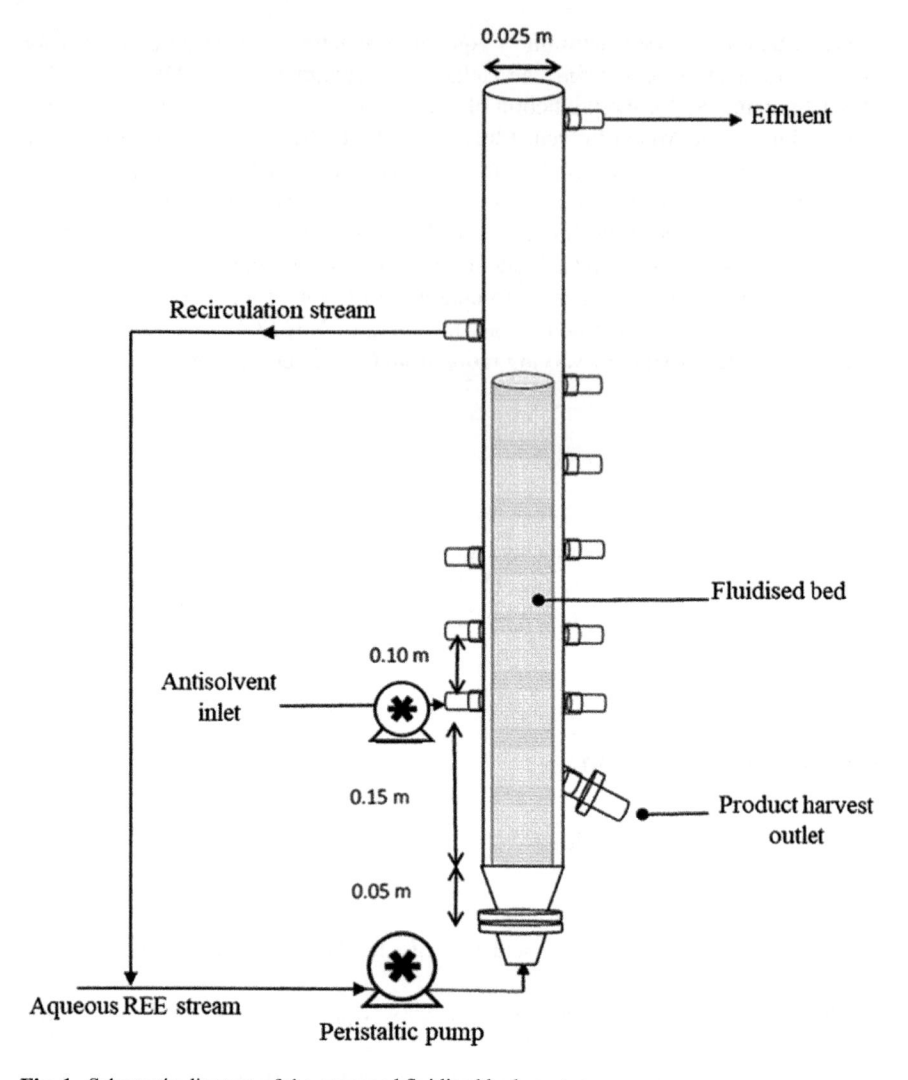

Fig. 1 Schematic diagram of the proposed fluidised bed reactor

high supersaturations. This caused fines to be generated, and these were elutriated out of the fluidised bed reactor.

Therefore, it can be concluded that the solubilities of the REE salts are in the range of acceptable supersaturations and thus are potentially suitable for crystallization in a fluidised bed reactor using an antisolvent crystallization process.

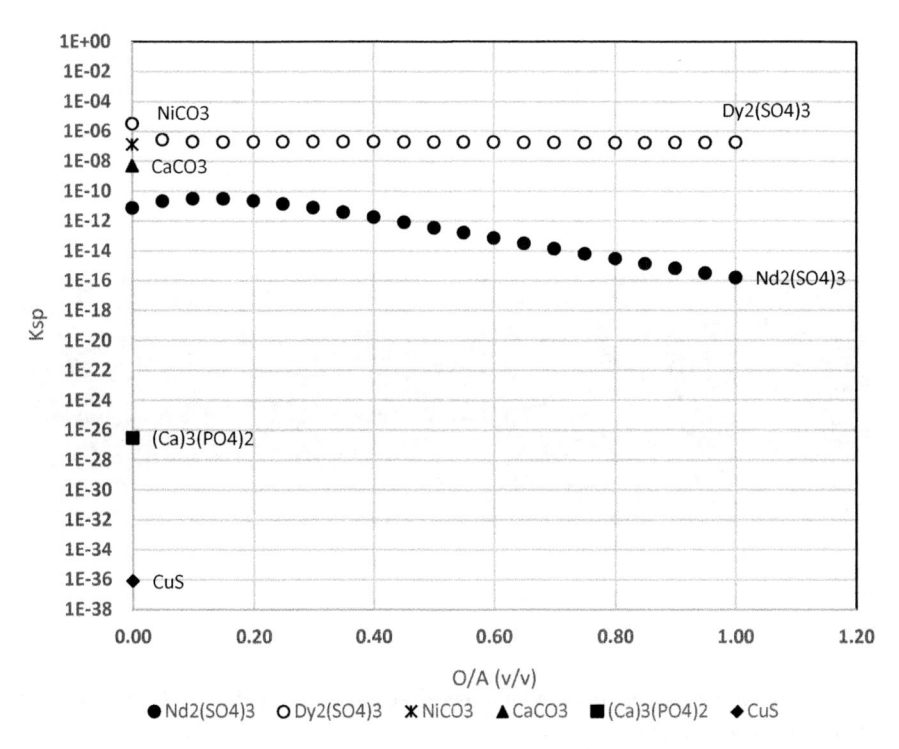

Fig. 2 K_{sp} values of a range of salts, including $Nd_2(SO_4)_3$ and $Dy_2(SO_4)_3$ as a function of O/A ratio where O = ethanol and A = water

Table 1 K_{sp} values of a range of salts

Salt	K_{sp} [35]
$NiCO_3$	1.3E-7 [21]
$CaCO_3$	4.872E-9 [36]
$Nd_2(SO_4)_3$ at O/A (v/v) = 1	2E-16
$Dy_2(SO_4)_3$ at O/A (v/v) = 1	1.84E-7
$(Ca)_3(PO_4)_2$	3E-27 [37]
CuS	7.9E-37 [24]

Batch Tests

The yields for the batch experiments are given in Table 2. From the table, it can be seen that ethanol generates the highest yield, followed by methanol and 2-proponal, with t-butanol giving the lowest yield.

Figure 3 shows the $Y_2(SO_4)_3$ salts produced in the batch tests. As can be seen in these micrographs, the crystals formed in the batch experiments are in general very

Table 2 Yields for batch experiments at initial Y$_2$(SO$_4$)$_3$.8H$_2$O at 21.96 gL^{-1} (of the oxalate) and O/A ratios of 1:1 v/v

Experiment #	REE	Alcohol	Yield (%)
1.1	Y$_2$(SO$_4$)$_3$ · 8H$_2$O	Ethanol	96
1.2	Y$_2$(SO$_4$)$_3$ · 8H$_2$O	Methanol	93
1.3	Y$_2$(SO$_4$)$_3$ · 8H$_2$O	2-propanol	93
1.4	Y$_2$(SO$_4$)$_3$ · 8H$_2$O	t-butanol	86

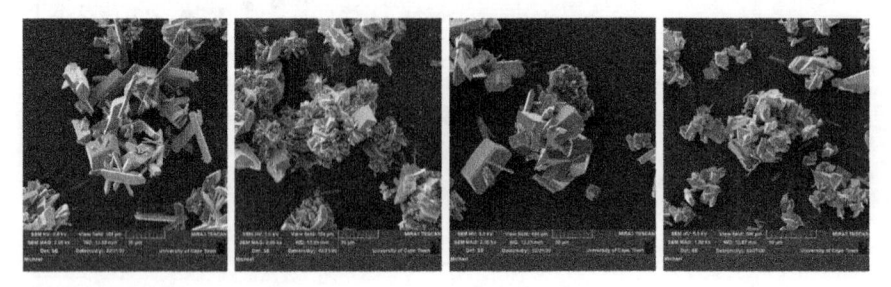

Fig. 3 Y$_2$(SO$_4$)$_3$ salts generated using, from left to right, ethanol, methanol, 2-propanol and t-butanol as antisolvent. Scale bar = 20 micron for all except for t-butanol, where the scale bar = 50 micron

well faceted and isotropic in habit and therefore are good candidates for recovery in the fluidised bed reactor.

For purposes of comparison, Fig. 4 shows the crystals produced in the NiCO$_3$ fluidised bed process, (LHS of Fig. 4) and those produced in the CuS fluidised bed reactor process (RHS of Fig. 4). Whilst the produced NiCO$_3$ forms a uniform coating on the seed particle, the CuS crystals are extremely small and "fluffy" and did not coat the provided seeds.

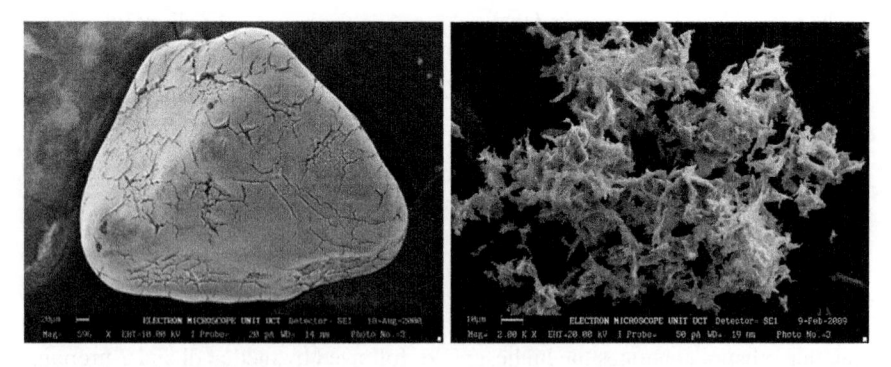

Fig. 4 NiCO$_3$ covered pellet from a fluidised bed reactor (Scale bar = 20 micron) and CuS precipitate (Scale bar = 10 micron)

Conclusions

The thermodynamic modelling showed that, when ethanol is added to a $Nd_2(SO_4)_3$ or $Dy_2(SO_4)_3$ solution as an antisolvent, the only solid products formed were the REE sulphate salts. The modelling also showed that, since the solubilities of the REE sulphate salts at any of the O/A ratios are of similar orders of magnitude to those of salts that have been successfully recovered in a fluidised reactor process, an antisolvent, fluidised reactor process is potentially suitable for REE sulphate salts.

The batch experiments showed that the yields are sufficiently high for a viable process. At the same time, the micrographs show that the nature of the formed crystals are such that they are likely to form a uniform and robust coating on seed particles in a fluidised bed reactor process.

Therefore, our preliminary conclusion is that this REE antisolvent system is well suited for further investigation in a fluidised bed reactor.

References

1. Jha MK, Kumari A, Panda R, Rajesh Kumar J, Yoo K, Lee JY (2016) Review on hydrometallurgical recovery of rare earth metals. Hydrometallurgy 165:2–26
2. Abreu RD, Morais CA (2010) Purification of rare earth elements from monazite sulphuric acid leach liquor and the production of high-purity ceric oxide. Miner Eng 23:536–540
3. Haque N, Hughes A, Lim S, Vernon C (2014) Rare earth elements: overview of mining, mineralogy, uses, sustainability and environmental impact. Resources 3:614
4. Byrne RH, Kim K-H (1993) Rare earth precipitation and coprecipitation behavior: the limiting role of PO43− on dissolved rare earth concentrations in seawater. Geochim Cosmochim Acta 57:519–526
5. Massari S, Ruberti M (2013) Rare earth elements as critical raw materials: focus on international markets and future strategies. Resour Policy 38:36–43
6. Jowitt SM, Werner TT, Weng Z, Mudd GM (2018) Recycling of the rare earth elements. Curr Opin Green Sustain Chem 13:1–7
7. Binnemans K, Jones PT, Blanpain B, Van Gerven T, Yang Y, Walton A, Buchert M (2013) Recycling of rare earths: a critical review. J Cleaner Prod 51:1–22
8. Philander C, Rozendaal A (2012) Rare-earth element and thorium potential of heavy mineral deposits along the west coast of South Africa with special reference to the Namakwa sands deposit, pp 531–539
9. Korkmaz K, Alemrajabi M, Rasmuson Å, Forsberg K (2018) Sustainable hydrometallurgical recovery of valuable elements from spent nickel-metal hydride HEV batteries. Metals 8:1062
10. Lai A, Lai F, Huang L, Qiu J, Zhou X, Xiao Y (2020) Non-ammonia enrichment of rare earth elements from rare earth leaching liquor in a magnesium salt system I: precipitation by calcium oxide. Hydrometallurgy 193:105318
11. Masmoudi-Soussi A, Hammas-Nasri I, Horchani-Naifer K, Férid M (2020) Rare earths recovery by fractional precipitation from a sulfuric leach liquor obtained after phosphogypsum processing. Hydrometallurgy 191:105253
12. Ni S, Chen Q, Gao Y, Guo X, Sun X (2020) Recovery of rare earths from industrial wastewater using extraction-precipitation strategy for resource and environmental concerns. Miner Eng 151:106315
13. Silva RG, Morais CA, Oliveira ÉD (2019) Selective precipitation of rare earth from non-purified and purified sulfate liquors using sodium sulfate and disodium hydrogen phosphate. Miner Eng 134:402–416

14. Vaziri Hassas B, Rezaee M, Pisupati SV (2020) Precipitation of rare earth elements from acid mine drainage by CO_2 mineralization process. Chem Eng J 399:125716
15. Wang Y, Li J, Gao Y, Yang Y, Gao Y, Xu Z (2020) Removal of aluminum from rare-earth leaching solutions via a complexation-precipitation process. Hydrometallurgy 191:105220
16. Yu Z, Wang M, Wang L, Zhao L, Feng Z, Sun X, Huang X (2020) Preparation of crystalline mixed rare earth carbonates by $Mg(HCO_3)_2$ precipitation method. J Rare Earths 38:292–298
17. Li Q, Huang D, Lu T, Seville JP, Xing L, Leeke GA (2017) Supercritical fluid coating of API on excipient enhances drug release. Chem Eng J 313:317–327
18. Lewis AE, Seckler MM, Kramer H, van Rosmalen GM (2015) Industrial crystallization: fundamentals and applications. Cambridge University Press
19. Peters EM, Kaya Ş, Dittrich C, Forsberg K (2019) Recovery of scandium by crystallization techniques. J Sustain Metall 5:48–56
20. Costodes VCT, Lewis AE (2006) Reactive crystallization of nickel hydroxy-carbonate in fluidized-bed reactor: fines production and column design. Chem Eng Sci 61:1377–1385
21. Guillard D, Lewis AE (2001) Nickel carbonate precipitation in a fluidized-bed reactor. Ind Eng Chem Res 40:5564–5569
22. Guillard D, Lewis AE (2002) Optimization of nickel hydroxycarbonate precipitation using a laboratory pellet reactor. Ind Eng Chem Res 41:3110–3114
23. Lewis A, Swartbooi A (2006) Factors affecting metal removal in mixed sulphide precipitation. Chem Eng Technol 29:277–280
24. van Hille RP, Peterson KA, Lewis AE (2005) Copper sulphide precipitation in a fluidised bed reactor. Chem Eng Sci 60:2571–2578
25. Aldaco R, Garea A, Irabien A (2007) Particle growth kinetics of calcium fluoride in a fluidized bed reactor. Chem Eng Sci 62:2958–2966
26. Aldaco R, Irabien A, Luis P (2005) Fluidized bed reactor for fluoride removal. Chem Eng J 107:113–117
27. Mullin JW (2001) Crystallization. Butterworth-Heinemann
28. Seckler MM (1994) Calcium phosphate precipitation in a fluidized bed [Ph.D.]. Delft, Delft University of Technology
29. Zhou P, Huang J-C, Li AWF, Wei S (1999) Heavy metal removal from wastewater in fluidized bed reactor. Water Res 33:1918–1924
30. Wilms DA (1988) Recovery of nickel by crystallization of nickel carbonate in a fluidized-bed reactor. In: Proceedings of the VTT symposium on non-waste technology. Espoo, Finland
31. Tai CY (1999) Crystal growth kinetics of two-step growth process in liquid fluidized-bed crystallizers. J Cryst Growth 206:109–118
32. Planteur S, Bertrand M, Plasari E, Courtaud B, Gaillard JP (2012) Crystal growth kinetics of the uranium peroxide. Procedia Chem 7:725–730
33. Heffels S, Kind M (1999) Seeding technology: an underestimated critical success factor for crystallization. In: Proceedings of the proceedings of the 14th international symposium on industrial crystallization. Cambridge
34. OLI Systems Inc (2020) OLI Studio 10.0.1.24 Stream Analyzer, February. Morris Plains, New Jersey
35. Lewis AE (2017), Precipitation of heavy metals in sustainable heavy metal remediation: volume 1: principles and processes. In: Rene ER, Sahinkaya E, Lewis AE, Lens PNL (eds) Springer International Publishing, pp 101–120
36. Tai CY, Hsu H-P (2001) Crystal growth kinetics of calcite and its comparison with readily soluble salts. Powder Technol 121:60–67
37. Seckler MM, Bruinsma OSL, Van Rosmalen GM (1996) Calcium phosphate precipitation in a fluidized bed in relation to process conditions: a black box approach. Water Res 30:1677–1685
38. Tai CY, Chien WC, Chen CY (1999) Crystal growth kinetics of calcite in a dense fluidized-bed crystallizer. AIChE J 45:1605–1614
39. Lewis A, van Hille R (2006) An exploration into the sulphide precipitation method and its effect on metal sulphide removal. Hydrometallurgy 81:197–204

Recovery of Rare Earth Elements from Recycled Hard Disk Drive Mixed Steel and Magnet Scrap

Tedd E. Lister, Michelle Meagher, Mark L. Strauss, Luis A. Diaz, Harry W. Rollins, Gaurav Das, Malgorzata M. Lencka, Andre Anderko, Richard E. Riman, and Alexandra Navrotsky

Abstract Recycling electronic scrap is a significant source of rare earth metals. Whereas traditional recycling routes for some electronic scrap emphasize the recovery of silver and gold, value can be attained by recovering of rare earth elements from unique feed streams. This paper describes a hydrometallurgical process for the recovery of rare earth elements from hard disk drives using HCl as a re-usable extraction medium. The mixture was selectively leached using HCl to remove the magnet alloy coating from shredded hard disk drives. The dissolved rare earth elements were precipitated using sodium sulfate, recovered as the sodium double salt, and subsequentially converted to hydroxides. The recovery of rare earth elements is consistent with amounts predicted using a thermodynamic model based on the MSE (Mixed-Solvent Electrolyte) framework of precipitated double salts. The effect of HCl concentration was measured upon the magnet dissolution rate. In addition, the leaching rates for steel were evaluated and found to be three orders of magnitude lower than the magnet alloy. An automated system was used to control leachate pH. The magnet and steel dissolution rate were examined for various HCl concentrations. The recovery of rare earth hydroxides was over 80%.

Keywords Rare earth elements · Recycling · Hard drive magnet · Critical materials

---

T. E. Lister (✉) · M. Meagher · M. L. Strauss · L. A. Diaz · H. W. Rollins
Biological and Chemical Processing, Idaho National Laboratory, P.O. Box 1625, Idaho Falls, ID 83415, USA
e-mail: tedd.lister@inl.gov

G. Das · M. M. Lencka · A. Anderko
OLI Systems Inc., 2 Gatehall Dr., Suite 1D, Parsippany, NJ 07054, USA

R. E. Riman
Department of Materials Science and Engineering, Rutgers, The State University of New Jersey, 607 Taylor Road, Piscataway, NJ 08855, USA

A. Navrotsky
School of Molecular Sciences and Center for Materials of the Universe, Arizona State University, Tempe, AZ 85287, USA

© The Minerals, Metals & Minerals Society 2021
G. Azimi et al. (eds.), *Rare Metal Technology 2021*, The Minerals, Metals & Materials Series, https://doi.org/10.1007/978-3-030-65489-4_15

Introduction

Electronic waste (e-waste) or waste electrical and electronic equipment (WEEE) is a growing feed stream of materials which were not recycled until the EU WEEE legislative directive was approved in 2002 [1, 2]. The challenges to e-waste recycling include the fact that manual separation and shredding are labor and equipment intensive, and there are few economically relevant recycling strategies for many types of e-waste [3]. As of 2016, 20% of e-waste was recycled worldwide in documented, proper manner, but rest is either dumped or shipped to other companies where it is recycled in a crude manner. In addition, rare earth elements (REEs) are not recovered to any significant extent [4]. In fact, less than 5% of rare metals were recycled from WEEE as of 2019 [5]. REEs are contained in displays, speakers, cell phones, motors, hard disk drives (HDDs), voice coil actuators, and other materials. Alternatively, according to Adamas Intelligence, the demand for rare earth oxides used in electric cars is predicted to increase by fivefold by 2030 [6]. The United States Department of Energy (DOE) considered neodymium (Nd), dysprosium (Dy), europium (Eu), terbium (Tb), and yttrium (Y) as REE critical materials [7]. The overall crustal abundance of the rare earth is similar to copper (Cu), but due to the technical challenges associated with separation, the enrichment factor required for a rare earth mine must be significantly higher than Cu [8]. For example, the cutoff grade for Cu might be 0.6%, but the cutoff grade for rare earth mine at Mt. Weld Mine is 4% [9–11]. When the most valuable elements are dilute, additional ore must be processed to meet the demand for those dilute metals needed for clean energy. In particular, the magnet REEs (Nd, Dy, Pr) have increasing importance as they are required for clean energy technologies, such as electric motors and wind turbines which justify the criticality. Fortunately, REEs are found in HDD magnets, particularly Nd in $Nd_2Fe_{14}B$ (Nd-Fe-B) magnets [12]. Reviews of REE reserves, suppliers, uses, and potential recycling sources are available [12, 13]. The recovery of REEs from e-waste could meet future criticality and bolster economic growth [14]. If more recycled metals are recovered for value, less mining is required, and waste is minimized.

Because few recycling methods for REEs have reached the commercial scale and most secondary sources of REEs are not being recycled, there is an increased interest in recycling research. The literature has outlined potential REE recycling feedstocks, existing recycling technologies and future needs [15–17]. A review of REE recycling specific to Nd-Fe-B magnets was published, including both pyrometallurgical and hydrometallurgical processes [15]. However, a more recent review outlines more avant-garde methods for treating e-waste for REEs from HDD magnets [18]. Given this recent review, the summary below will focus on the studies most relevant to this paper.

Hydrometallurgical methods for the extraction of REEs are supported by their easy dissolution by mineral acids. The dissolution of scrap as the Nd-Fe-B magnets using sulfuric acid dissolution and subsequent pH adjustment to capture was described in a patent by Lyman in 1992 [19]. Nd-Fe-B magnet alloys react quickly with protons (H^+) with copious evolution of H_2 gas [20]. Due to reactivity, sintered magnets are typically coated with nickel (Ni) or layers of Ni and Cu. Selection of the acid for dissolution (H_2SO_4, HCl, etc.) depends on the downstream recovery processes [21]. For example, downstream solvent extraction methods are supported by hydrochloric and nitric acid leaching and selective precipitation methods prefer sulfuric acid leaching [15].

To recover REEs from solution, selective precipitation reactions have frequently been reported to isolate dissolved REEs. All REEs form sparingly soluble trihydroxides in basic medium [22], which can be used to precipitate REEs. However, a pH shift to approximately pH 6 is higher than precipitation of co-dissolved metals such as iron (Fe), nickel (Ni), and zinc (Zn). The REEs are highly soluble in chloride and nitrate while having low solubility in sulfate which is pH dependent [23]. From moderately concentrated H_2SO_4, a slight increase in pH to a value slightly over 1 will precipitate the REEs as double salt solids (($RE)_2(SO_4)_3 \cdot Na_2SO_4 \cdot xH_2O$ using NaOH). As this pH is below that for precipitation of Fe [22], this method was used to capture REEs from nickel-metal hydride (Ni-MH) batteries [24, 25]. Pietrelli et al. used 2 M H_2SO_4 and obtained over 90% REE release, followed by pH adjustment to 1.5 using NaOH to recover over 70% of REE content. The REE recovery from magnet scrap (swarf) using the H_2SO_4-NaOH route have also been reported [26]. Phosphoric acid dissolution followed by pH increase has been used to precipitate REEs from Ni-MH battery leachate solutions [26].

Recently, this group has focused on a comprehensive recovery process that can produce different value streams from e-waste [20, 27]. An economic evaluation of the comprehensive process has concluded that a cost-efficient recovery of REEs from electronic scrap can only be achieved if other metals are recovered for value [14]. Specifically, for the recovery of REEs, the ferrous fraction (magnetically separated) of mobile phones was treated using H_2SO_4 followed by pH increase to separate REEs [27]. While use of H_2SO_4 is efficient for dissolution and recovery of REEs, the presence of sulfate and sodium (Na) in the extraction media could complicate material separation if the leaching solution was intended to be re-used.

This paper describes a modification to the chemistry used in that work, where an HCl-based system is employed to dissolve REEs from mixed magnet-steel mixtures followed Na_2SO_4 addition to precipitate as the sodium double salt. The resultant double salt is then converted to rare earth hydroxide ($RE(OH)_3$) using NaOH. While previous work included data on the recovery of REEs, this work is dedicated to the REE recovery process using HCl-based dissolution which includes a more thorough assessment of recovery in comparison with the model predictions of recovery, additional purification of the powder product, and an assessment of the possibility of leachate re-use.

Materials and Methods

Materials

Leaching solutions were prepared in deionized water (18 MΩ-cm) using ACS grade or better reagents. To examine the dissolution process, uncoated "Nd" (Nd-Fe-B) disk magnets (1.3 cm diameter, 0.3 cm height) were used (McMaster Carr). Shredded HDD material was obtained from Oak Ridge National Laboratory (ORNL). This material was hand sorted using a coated Nd-Fe-B magnet to remove steel and magnet fragments from aluminum (Al), plastic, and circuit board material. This ferrous material was used for recovery experiments. Magnetic particles and chunks were magnetically attached to steel or tightly bound together onto steel as described previously [15]. Due to the small quantity of non-homogenous scrap attained from Oak Ridge National Laboratory, the team was unable to obtain a statistically relevant analysis of the feed material.

Magnet and Steel Dissolution Studies

Steel coupons (1.6 × 1.6 cm) were cut from C-1100 shim steel. Air or N_2 gas were purged at 0.02 m^3/h into an open beaker at ambient temperature (20–22 °C). Experiments were performed for 6–24 h in 100 mL of 1 M HCl. Uncoated magnets were exposed to a range of diluted HCl solutions at ambient temperature (20–22 °C). A solution volume of 250 mL was chosen to avoid significant concentration change during the test. Stirring was not employed, although bubbles of H_2 produced from the reaction agitated the solution. Magnets were carefully balanced at an angle against the side of the beaker to minimize contact area. Weight of the magnets was recorded three times before and after exposure. Using the weight change, known density (7.4 g/cm^3), surface area (3.88 cm^2), and exposure time (1 h), the uniform dissolution rate was calculated. A uniform rate was used due to the lack of evidence for localized attack.

Leaching System

A diagram of the system used to process material is shown in Fig. 1. The separated HDD material was placed in the reaction column, and a pump was used to continuously flow the acid solution through the vessel containing the material. A pH probe was inserted in-line to inform a pH controller to release acid as needed. Dissolution was performed at normal room temperature (18–22 °C). Solution was pumped through the reaction column at approximately 250 mL/min using a diaphragm pump. While the system was automated, operation occurred under supervision during

Fig. 1 Diagram of the system to perform REE dissolution. (Color figure online)

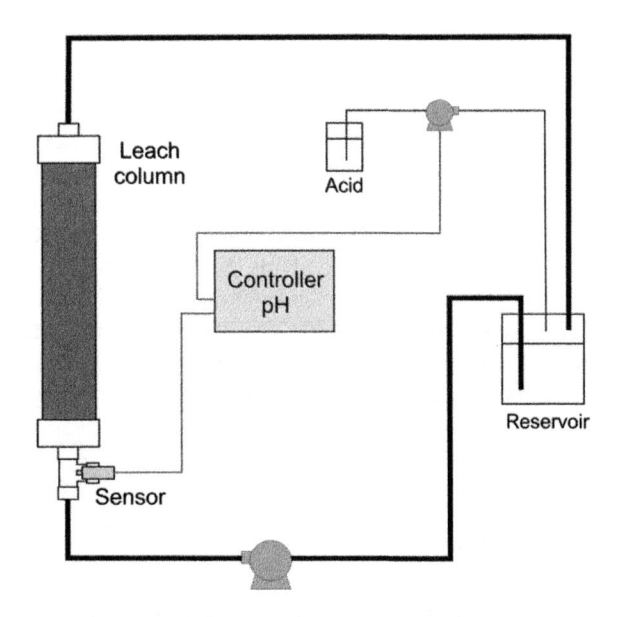

normal work hours as a precaution. The pH controller operated a low flow peristaltic pump to deliver 5 M HCl to the leach reservoir. The pH was controlled to maintain a value below 0. The addition of acid diluted the leach solution over time. Solution was removed periodically and precipitated individually.

Recovery of Dissolved REEs

After dissolution, REEs were precipitated from the leachate as sulfate double salts (NaRE(SO$_4$)$_2$·xH$_2$O) by adding solid Na$_2$SO$_4$. Samples of the leachate, before and after REE precipitation, were taken for chemical analysis. The precipitated double salt was converted to RE(OH)$_3$ (REOH) by reaction in 2 M NaOH for 2 h at 70 °C [26]. The final REOH product was filtered, rinsed, and dried. In some cases, powder was further purified by a second treatment in 10 M NaOH at 70 °C to completely react remaining sulfate double salt and to leach metal impurities.

Analysis

A Bruker (S2 PICOFOX) bench-top total reflection X-ray fluorescence spectrometer (TXRF) was used to analyze solutions and powders. The REOH products were dissolved in HCl before analysis. An aliquot of solution was pipetted onto the instrument sample holders and dried. A selenium (Se) internal standard was used for analysis. The concentrations of Pr, Nd, Dy, Gd in the final powder products were analyzed by an inductive coupled plasma mass spectrometer (ICP-MS) Thermo Scientific iCAP Q. The analysis of Na, S, Fe, and Zn was performed using an iCAP Series 6000 inductively coupled plasma optical emission spectrophotometer (ICP-OES) from Thermo Scientific. Calibration of ICP-MS and ICP-OES was performed using commercially prepared standards (VGH and Spex).

Powders were analyzed using X-ray diffraction (XRD) in a Bruker D8 Advance diffractometer operated at 40 kV and 40 mA, with a cobalt (Co) target (K = 1.78897A) being used for the characterization of the REE deposits formed. The XRD spectra were obtained for the REE sulfate double salts and hydroxides by scanning from 5 to 70° 2θ with a step size of 0.02° 2θ.

Thermodynamic Modelling

Aqueous systems containing REEs were modelled using the previously developed mixed-solvent electrolyte (MSE) model [27, 28]. The model parameters were determined for REE sulfate-containing systems using the procedures that were described in a previous work [30]. These procedures ensure that the model matches the available experimental data for solid–liquid equilibria, vapor–liquid equilibria, and caloric properties. Subsequently, the model was used to calculate the solubility of REE—sodium sulfate double salts and to predict on a thermodynamic basis the amounts of the solids that are expected to precipitate in the experiments.

Results and Discussion

Dissolution Rate Determination

As described in previous work, the magnet alloy was reactive to acid solutions with visible H_2 evolution [20]. Due to challenges of separating magnets from electronic devices, this process examined dissolution from steel-magnet mixtures. As described in Eq. 1, the alloys react directly with H^+ in strong acids to form H_2, as shown in this proposed reaction.

$$Nd_2Fe_{14}B(s) + 3H_2O(aq) + 34H^+(aq) \rightarrow 2Nd^{+3}(aq) + 14Fe^{+2}(aq) + H_3BO_3(aq) + 18.5H_2(g) \tag{1}$$

The H_2 gas produced by the dissolution reaction would need to be dealt with in an industrial process through simple flaring, energy recovery or through dilution with air. Produced H_2 acts as an indicator of reaction progress as does the time between pH doses (H^+ consumption).

To selectively dissolve the magnet alloy, previous work employed a N_2 gas purge to reduce the concentration of dissolved O_2 [25]. This was in attempt to control steel corrosion, as described in the well-known iron corrosion reaction (Eq. 2), by reducing the concentration of dissolved oxygen.

$$2Fe + O_2 + 4H^+ \rightarrow 2Fe^{2+} + 2H_2O \qquad (2)$$

To assess the corrosion of mild steel, coupons of mild steel were exposed to 1 M HCl with either N_2 gas or air bubbling into the solution. The corrosion rate measured over 24 h was 6.82×10^{-4} mm/h in air and decreased to 1.31×10^{-4} mm/h in N_2. Purging with an inert gas to remove O_2 reduces the rate as Eq. 2 suggests.

Uncoated disk magnets were used to measure the dissolution rate of REE magnets as a function of HCl concentration. The rate of dissolution increased almost linearly with HCl concentration (slope = 0.109 mm/h/M), with the rate at 1 M HCl being 0.170 mm/h, three orders of magnitude greater than that of steel. The dissolution rates are rapid demonstrating the reactive nature of the magnet alloy.

However, as will be described below, even in the presence of oxygen, the dissolution rate is three orders of magnitude lower than the rate for the Nd-Fe-B magnet alloy. Thus, it was decided to not purge O_2 from the dissolution reactor, as the O_2 removal may not be critical to control dissolution selectivity. It is anticipated that steel corrosion will be even lower than measured for two reasons: (1) Dissolved hydrogen from magnet dissolution should shift the corrosion potential of the steel to prevent active corrosion and (2) the internally produced H_2 should act to purge O_2.

Recovery of REEs from HDDs

The REE recovery was performed using 200 g of magnetically separated HDD material. Although the dissolution rate was demonstrated to increase with HCl concentration, a pH of 0 (1 M HCl) was selected as the control value. The system was operated over several days for a total of 52 h. A total of 545 mL of 5 M HCl was added during operation. Solutions were pulled periodically (five total solutions) and each processed individually to produce REOH powders. Figure 2 shows the elemental weight, measured with TXRF, for all five solutions (4 pulls and the remains) before and after sodium sulfate precipitation. The data was calculated using concentration and measured volume followed by adding each solution. The inset in Fig. 2 shows the % of mass decrease for the REEs in the solutions after precipitation (recovery). The Nd showed the greatest drop in weight along with Y which was present in very small quantities. The REEs such as Pr and Dy showed intermediate recovery while La showed very poor recovery. The lower recovery was presumably due to the lower starting concentrations close to the solubility limit for the double salt. Note that Fe largely remains in solution while Zn appears to carry over to the precipitate. There

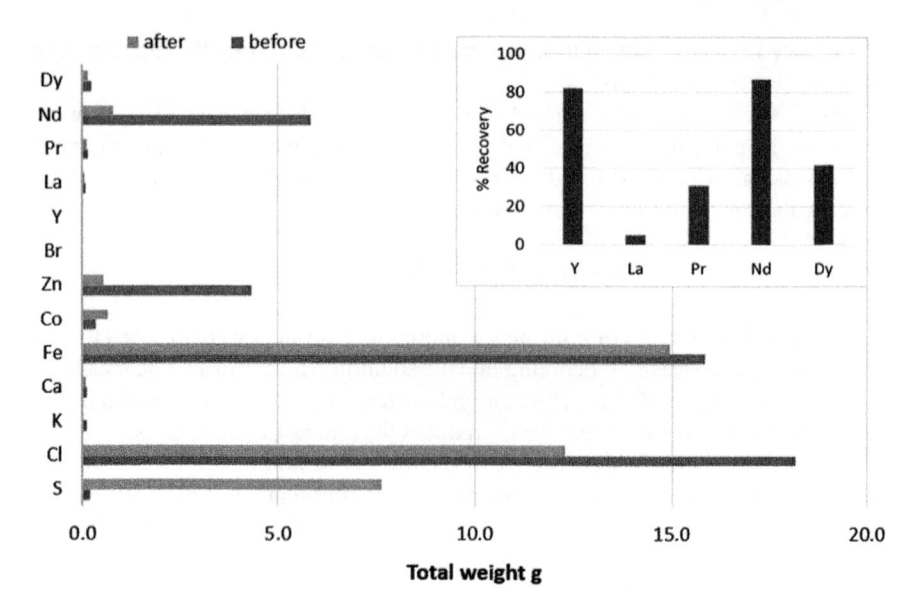

Fig. 2 Total weight of elements before and after precipitation for all five solutions. Inset figure shows the percent change for REEs after precipitation. (Color figure online)

was some carry-over of Cl as well. The S increases due to addition of excess Na_2SO_4. Measurement of Na was not reliable using TXRF and was excluded from Fig. 2.

Table 1 provides recovery data as well as compositions of the recovered REOH powders. The total weight of REOH product was 5.176 g. The % recovery was calculated using the solution analysis before precipitation and compared to the powder analysis and weight of powder recovered. Recoveries of REOH exceeded 80% except for Solution 2. The composition of the powder was determined using various analysis methods described in the table.

In determining the composition %, REEs were assumed to be $RE(OH)_3$, as supported by the XRD data to be discussed below. Using a weighted average (based on fraction of total powder collected), the REOH powder was over 77 wt% $Nd(OH)_3$ and 81 wt% $RE(OH)_3$. Note that the total wt%, which accounts for all analyzed components, averaged 85%. It would be assumed the missing weight would be due to O and H in waters of hydration, hydrous oxides, or sulfates. The adjusted % $RE(OH)_3$ values, obtained by ratio of the total REOH % to the total wt%, show an average of 96%. The presence of significant Na and S is likely carry-over from the Na_2SO_4 in solution 5. The primary transition metal contaminate is Zn which averages 1.3%. Well-known from cementation post-processing, Zn readily dissolves in HCl. As shown in Fig. 2, most Zn reports to the precipitate. The Fe was present at much lower levels, averaging 0.03%. This also agrees with Fig. 3 where most of the Fe remained in solution. Other significant impurities were Ca and Cl. The origin of Ca impurities is not known.

Recovery of Rare Earth Elements ... 147

Table 1 Results for powder products from the five solutions processed. REEs % were reported as hydroxides. Average values were weighted in contribution based on the powder product recovered for that solution

Solution	1	2	3	4	5	Average	Total
Wt REE (g)	1.462	1.155	0.366	0.203	1.99		5.176
% Recovery	86.9	74.2	82.0	86.5	85.6	83.2	
% Na	0.10	0.13	0.16	0.23	3.18	1.30	
% S	0.38	0.38	0.12	0.08	1.32	0.71	
% Cl	0.34	0.17	0.70	1.23	0.87	0.57	
% K	0.08	0.11	0.04	0.05	0.06	0.07	
% Ca	0.22	0.20	0.24	0.12	0.89	0.47	
% Fe	0.02	0.04	0.03	0.03	0.04	0.03	
% Co	0.10	0.09	0.08	0.09	0.11	0.10	
% Zn	1.06	0.92	1.60	1.54	1.62	1.30	
% Br	0.01	0.01	0.01	0.01	0.01	0.01	
% Y(OH)$_3$	0.01	0.00	0.00	0.01	0.01	0.01	
% La(OH)$_3$	0.46	0.41	0.26	0.33	0.42	0.41	
% Pr(OH)$_3$	3.59	3.20	2.95	2.73	2.82	3.13	
% Nd(OH)$_3$	81.23	78.64	75.03	72.84	74.19	77.18	
% Dy(OH)$_3$	0.78	1.25	1.29	1.52	0.87	0.99	
% Gd(OH)$_3$	0.08	0.07	0.07	0.07	0.06	0.07	
Total % RE(OH)$_3$	86.15	83.57	79.60	77.49	78.38	81.78	
Total wt%	88.44	85.61	82.56	80.88	83.30	85.12	
Adjusted % RE(OH)$_3$	97.41	97.62	96.41	95.81	94.09	96.05	

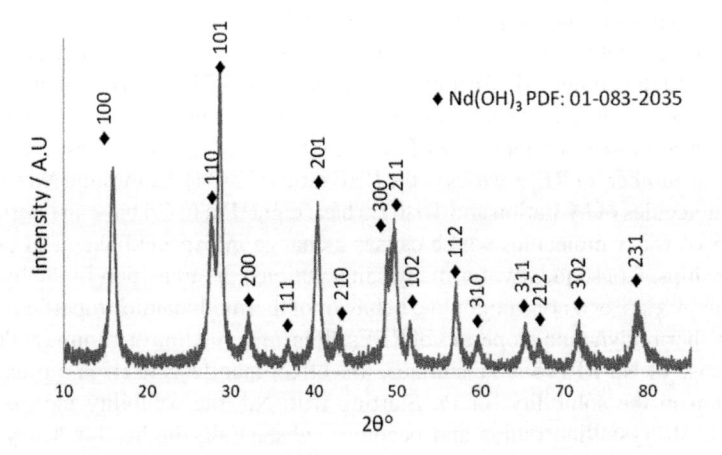

Fig. 3 The XRD spectrum of the solid obtained from solution #5. (Color figure online)

Another way to estimate recovery is through acid consumption. If it is assumed that all the acid consumed (0.545 mol H$^+$) was due to magnet dissolution using Eq. 1, 0.0321 mol of REOH are expected to be in solution. Using the assumption that the powder product was entirely Nd(OH)$_3$ that equates to 0.0265 mol Nd(OH)$_3$. Thus, approximately 0.0056 mol were not recovered (~17%). Considering that the powder product contains some water, this agrees well with the % recovery reported in Table 1.

A sample of the REOH powder recovered from solution 5 was analyzed by XRD as shown in Fig. 3. The powder was identified as a match for Nd(OH)$_3$ with no other phases observed.

Comparison of Recovery to Model

The MSE model predicts the thermodynamic equilibrium state for the reaction of Na, REE, sulfate ion, and water in the following equation:

$$Na_2REE_2(SO_4)_4 \cdot nH_2O(s) = 2Na^+ + 2REE^{3+} + 4SO_4 + nH_2O \qquad (3)$$

Spedding and Jaffe [29] first described the relationship between REE sulfate solubility and ionic radius. Recently, Das et al. [30] developed a comprehensive parameterization of the MSE model for both REE sulfates and Na-REE double sulfate salts by analyzing various experimental thermodynamic data across the REE series. Based on this parameterization, the MSE model was used as the thermodynamic basis for determining the recovery of REEs as sodium double salts from the leaching solution. The analysis of Das et al. [30] revealed a characteristic, non-monotonic trend in the solubilities of the sulfates as a function of the ionic radius. The non-monotonic or "two-series" behavior is in fact a pervasive feature of REE salt solutions. It has been attributed to the combined effect of the change in the REE radius (which decreases smoothly across the REE series) and the variation in the number of water molecules in the first coordination sphere of REE ions. In particular, it was demonstrated that the hydration number for REEs varies in the REE series [32–34]. La through Nd have nine water molecules of hydration and Tb to Lu have eight. Pm to Gd have an intermediate number of water molecules which causes a change in expected chemical property relationships. Consequently, for many salt solutions, the transition in the hydration behavior gives rise to non-monotonic behavior of thermodynamic properties [30]. As a result, thermodynamic properties of REEs often have maxima or minima at Pr or Nd. In the case of Na-REE double sulfates, this effect manifests itself in a pronounced minimum in the solubility for Pr. Starting with Nd, the solubility increases as a function of crystalline radius and becomes substantially higher for heavy REEs. Based on the analysis of Das et al., the solubilities of different Na$_2$RE$_2$(SO$_4$)$_4$ salts were calculated at 25 °C for different Na$_2$SO$_4$ concentrations and are presented in Fig. 4 as a function of crystal cationic radii of REEs [31]. In particular, the solubility of Dy is over an order of magnitude higher. The model also shows a decrease in

Fig. 4 Prediction of $Na_2REE_2(SO_4)_4$ solubility in 0.1 m (blue circles), 0.5 m (pink triangles), and 1 m (green squares) Na_2SO_4 at 25 °C as a function of crystal cationic radii [21] of REEs (listed in reverse atomic radius). (Color figure online)

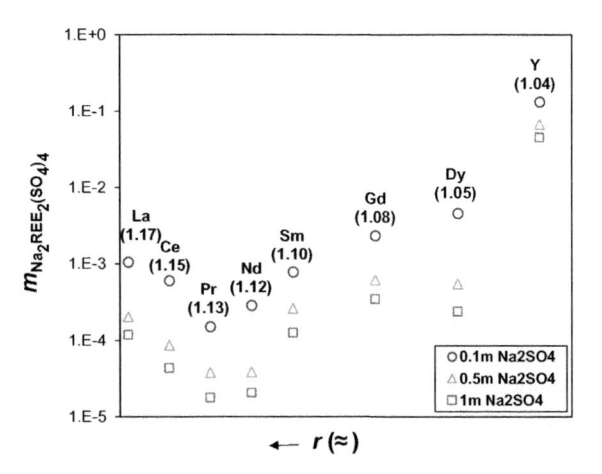

the solubility with the amount of Na_2SO_4 added which is congruent with what Le Chatelier's Principle would predict. This is primarily due to the common ion effect and is in agreement with the experimental data reviewed by Das et al. [30].

Recovery of REEs can be predicted based on the MSE model using the concentration prior to precipitation and the weight of Na_2SO_4 added. Then, the predicted recovery can be compared with the XRF data. A comparison of the model and experimental data is shown in Fig. 5. The actual recovery values for Nd were somewhat lower than predicted by the MSE model, which suggests that nearly all Nd should be removed in the precipitation step. The inset shows the detail for Pr recovery, where

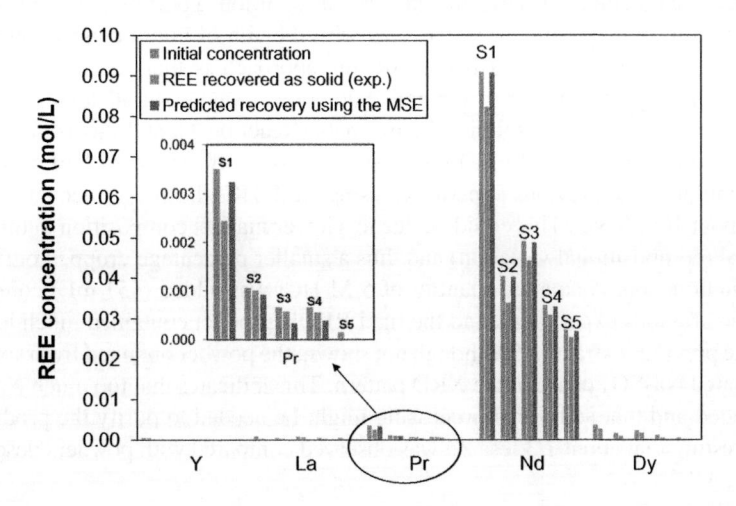

Fig. 5 Comparison of recovery predicted by MSE model versus experimental data by XRF. The green bars represent the initial concentration of REEs in the leaching solution, while the blue and red bars give the experimental and predicted recovery, respectively. The symbols S1–S5 denotes samples 1–5. (Color figure online)

the predicted recovery was somewhat higher than the actual recovery for sample S1 and moderately lower for samples S2–S4. In the case of Pr, the initial concentration was much lower than for Nd and, therefore, near-complete recovery was not predicted because a non-negligible fraction of Pr had to remain in solution according to double salt solubility. Finally, the recovery of Dy was not predicted by the MSE model yet significant recovery was observed. This is due to the very small initial concentrations of Dy, which were lower than the solubility of the double salts. While this observation could not be reconciled based on solubility calculations, the predictions for Nd and Pr were reasonable. The reason for the apparent discrepancy for Dy may lie in the co-precipitation of Dy and Nd due to the overwhelmingly larger amount of Nd in the samples. It is well known that rare earth salts have a strong propensity to form solid solutions containing two or more rare earth elements. This has been observed for multiple classes of compounds, including REE oxides [35–39], chlorides [40, 41], nitrates [40], and cuprates [42]. Although the formation of solid solutions has not been reported for REE—Na double sulfate salts—it can be reasonably presumed that solid solutions are also possible for this class of compounds. If this is the case, then a relatively small amount of Dy may be incorporated into a Nd-dominated double salt, thus leading to the recovery of Dy together with Nd. Also, note that XRF is a semiquantitative measurement method for the rapid measurement of many elements. Future work will include more robust quantitative measurements for the REEs.

Recovery of REEs from HDDs Using Recycled Solution

To assess the potential for recycling the leachate solution, a portion of the remaining solution (after precipitation) was recycled with the aim of minimizing water usage and waste volume. A second benefit of this approach is that REEs not precipitated previously would carry over and thus could reduce overall losses. Acid is the primary reactant in dissolving the magnets (Reaction 1), and this is not significantly improved by recycling. Table 2 shows results of recovered REEs. Results were similar to the previous experiment using fresh HCl; however, recovery values were about 10% lower. This could be due to a lower magnet composition in this feed material (compositional variation) and thus a smaller percentage dropped out in the precipitation step. A smaller quantity of 5 M HCl was added (451 mL) compared with the previous experiment, and the final REOH product contained much less Zn than the previous extraction. Although not shown, the powder obtained from solution 3 indicated Na_2SO_4 phase in the XRD pattern. This indicates that too much Na_2SO_4 was added and that additional processing might be needed to purify the product. It is interesting that about 1/3 less Zn was observed compared with powders described in Sect. 3.2.

Table 2 Results of powder product analysis from the three solutions processed using recycled solution

Solution	1	2	3	Avg.	Total
Total weight (g)	0.31	0.447	2.193		2.95
% Recovery	70.83	76.06	67.68	69.28	
Na	0.63	2.57	7.387	5.95	
S	0.10	1.08	3.190	2.55	
Cl	0.39	0.13	0.142	0.17	
K	0.04	0.05	0.085	0.07	
Ca	1.17	0.93	0.270	0.46	
Fe	0.04	0.04	0.013	0.02	
Co	0.00	0.11	0.120	0.11	
Zn	0.38	0.35	0.242	0.27	
Br	0.01	0.01	0.019	0.02	
Y(OH)$_3$	0.01	0.00	0.004	0.00	
La(OH)$_3$	0.54	0.45	0.466	0.47	
Pr(OH)$_3$	5.82	4.89	2.416	3.15	
Nd(OH)$_3$	70.64	69.00	50.717	55.58	
Dy(OH)$_3$	2.28	4.16	1.404	1.91	
Gd(OH)$_3$	0.13	0.102	0.052	0.07	
Total % RE(OH)$_3$	79.42	78.61	55.059	61.19	
Total wt. %	82.16	83.88	66.526	70.80	
Adjusted % RE(OH)$_3$	96.66	93.72	82.763	85.88	

Elemental results were obtained by TXRF for Ca, Co, Cl, and Br. Na, S, Fe, and Zn analyses were performed using ICP-OES. The remaining elements were analyzed by ICP-MS. The total wt.% includes all analyzed elements. Adjusted % REE uses total wt.% to make an adjusted calculation.

Purification of Recovered Powders

With the aim to improve the purity of the REOH product by removing the unreacted sulfate and other metal impurities, REOH powder 4 (Table 2) was subjected to a second alkaline digestion in 10 M KOH. The purification results shown in Table 3 indicate a 12% increase in the REOH composition after the second alkaline treatment. Elemental analysis performed with XRF did not detect sulfur in the re-processed sample, which indicates that sulfates are no longer present in the refined REOH. Regarding metal impurities, reductions of 50% and 65% were achieved for Fe and Zn, respectively, while Co was completely removed from the REOH product.

Both, Co and Zn can re-dissolve at high pH forming dicobaltite ($HCoO_2^-$) and zincate (ZnO_2^-) [22]. The purified REOH product obtained reached a REE purity of 99.1% metal basis.

Table 3 Elemental composition of REOH product before and after purification in metal basis

Composition (%)	Before	After
Nd	91.35	93.73
Pr	3.40	3.45
Zn	2.58	0.89
Dy	1.96	1.68
La	0.40	0.13
Co	0.15	0.00
Gd	0.09	0.09
Fe	0.05	0.03
Y	0.01	0.00
Total REE	77.5	86.93
Total REE (metal basis)	97.2	99.1

Conclusions

Electronic scrap offers a significant source of REEs and the efficient recovery of REEs could be important to clean energy technologies. This paper was built upon a foundation of previous work regarding the recovery of REEs from mixed steel-Nd-Fe-B alloy material. This mixture is a feed from shredded HDDs as output from data destruction services. Experiments examined HCl concentration upon the rate of dissolution of uncoated Nd-Fe-B magnets. It was determined that the HCl concentration had a linear effect upon leaching rate of magnets. The rate for steel was found to be much lower and was not of significant consequence in the process. A pH-controlled flowing dissolution system was used to dissolve Nd-Fe-B magnet fragments from shredded HDDs. After dissolution, the REEs were precipitated using sodium sulfate and subsequently converted to REOH powder. Analysis demonstrated over 80% of dissolved REEs were recovered. The product showed Na, S, Fe, and Zn were major impurities. The purity level was decreased significantly by a second NaOH treatment. Therefore, a REOH intermediate product with a purity of 99.1% (metal basis) was obtained from processed HDD. Recycling of the acidic dissolution solution resulted in slightly lower recovery of about 70%. It is uncertain if the decreased recovery was due to the recycling of acidic solution or simply a lower Nd-Fe-B composition in the second batch.

Acknowledgements This work is supported by the Critical Materials Institute, an Energy Innovation Hub funded by the U.S. Department of Energy, Office of Energy Efficiency and Renewable Energy, Advanced Manufacturing Office. This manuscript has been authored by Battelle Energy Alliance, LLC under Contract No. DE-AC07-05ID14517. We thank Timothy McIntire (Oak Ridge National Laboratory) for supplying shredded HDD material. We also thank Byron White and Arnold Erickson for providing analytical services that supported this work.

References

1. Mihai FC, Gnoni MG, Meidiana C, Ezeah C, Elia V (2019) Waste electrical and electronic equipment (WEEE): Flows, quantities, and management-a global scenario, no. 2010. Elsevier Inc.
2. Balde CP, Forti V, Gray V, Kuehr R, Stegmann P (2017) The global e-waste monitor 2017
3. Tansel B (2017) From electronic consumer products to e-wastes: global outlook, waste quantities, recycling challenges. Environ Int 98:35–45. https://doi.org/10.1016/j.envint.2016.10.002
4. Yang Y et al (2017) REE Recovery from end-of-life NdFeB permanent magnet scrap: a critical review. J Sustain Metall 3(1):122–149. https://doi.org/10.1007/s40831-016-0090-4
5. Linnenkoper K (2019) Is it now or never for rare earth recycling? Recycling International
6. Adamas Intelligence (2020) Rare earth magnet market outlook to 2030. https://www.adamasintel.com/rare-earth-magnet-market-outlook-to-2030/
7. Mckittrick M, Bauer D, David D, Mckittrick M (2011) U.S. department of energy critical materials strategy. https://doi.org/10.1017/CBO9781107415324.004
8. Haxel GB, Hedrick JB, Orris GJ, Sound S, Of M, Mineral OUR (2002) Rare earth elements—critical resources for high technology. United States Geol Surv Fact Sheet 087:4
9. Dold B (2008) Sustainability in metal mining: from exploration, over processing to mine waste management. Rev Environ Sci Biotechnol, SPEC. ISS7(4):275–285. https://doi.org/10.1007/s11157-008-9142-y
10. Hellman PL, Duncan RK (2014) Evaluation of rare earth element deposits. Trans Institutions Min Metall Sect B Appl Earth Sci 123(2):107–117. https://doi.org/10.1179/1743275814Y.0000000054
11. Buchert D, Manhart M, Bleher A, Pingel D (2012) Recycling critical raw materials from waste electronic equipment Commissioned by the North Rhine- Westphalia State agency for nature, environment and consumer protection authors 49(0):30–40
12. Report F, Greens T, Group EFA, Parliament E (2011) Study on rare earths and their recycling 49:30–40
13. Nguyen RT, Diaz LA, Imholte DD, Lister TE (2017) Economic assessment for recycling critical metals from hard disk drives using a comprehensive recovery process. Jom 69(9):1546–1552. https://doi.org/10.1007/s11837-017-2399-2
14. Binnemans K et al (2013) Recycling of rare earths: a critical review. J Clean Prod 51:1–22. https://doi.org/10.1016/j.jclepro.2012.12.037
15. Anand T, Mishra B, Apelian D, Blanpain B (2011) The case for recycling of rare earth metals-a CR3 communication. Jom 63(6):8–9. https://doi.org/10.1007/s11837-011-0098-y
16. Zhang Y, Gu F, Su Z, Liu S, Anderson C, Jiang T (2020) Hydrometallurgical recovery of rare earth elements from ndfeb permanent magnet scrap: a review. Metals (Basel) 10(6):1–34. https://doi.org/10.3390/met10060841
17. Lyman J, Palmer G (1992) Scrap treatment method for rare earth transition metal alloys. US Patent 5,129,945
18. Lister TE, Wang P, Anderko A (2014) Recovery of critical and value metals from mobile electronics enabled by electrochemical processing. Hydrometallurgy 149(2014):228–237. https://doi.org/10.1016/j.hydromet.2014.08.011
19. Binnemans K et al (2013) Recycling of rare earth: a critical review. J Clean Prod 51:1–22. 10.1016. https://doi.org/10.1016/j.jclepro.2012.12.037
20. Pourbaix M (1974) Atlas of electrochemical equilibria in aqueous solutions, 2nd edn. NACE International, Houston, TX
21. Stevenson PC et al (1961) The radiochemistry of the rare earths, scandium, yttrium, and actinium. https://library.lanl.gov/cgi-bin/getfile?rc000021.pdf
22. Pietrelli L, Bellomo B, Fontana D, Montereali MR (2002) Rare earths recovery from NiMH spent batteries. Hydrometallurgy 66(1–3):135–139. https://doi.org/10.1016/S0304-386X(02)00107-X

23. Bertuol DA, Bernardes AM, Tenório JAS (2009) Spent NiMH batteries-The role of selective precipitation in the recovery of valuable metals. J Power Sources 193(2):914–923. https://doi.org/10.1016/j.jpowsour.2009.05.014
24. Lyman JW, Palmer G (1995) Hydrometallurgical treatment of nickel-metal hydride battery electrodes
25. Diaz LA, Lister TE, Parkman JA, Clark GG (2016) Comprehensive process for the recovery of value and critical materials from electronic waste. J Clean Prod 125:236–244. https://doi.org/10.1016/j.jclepro.2016.03.061
26. Abreu RD, Morais CA (2010) Purification of rare earth elements from monazite sulphuric acid leach liquor and the production of high-purity ceric oxide. Miner Eng 23(6):536–540. https://doi.org/10.1016/j.mineng.2010.03.010
27. Wang P, Anderko A, Young RD (2002) A speciation-based model for mixed-solvent electrolyte systems. Fluid Phase Equilib 203(1–2):141–176. https://doi.org/10.1016/S0378-3812(02)00178-4
28. Das G, Lencka MM, Eslamimanesh A, Anderko A, Riman RE (2017) Rare-earth elements in aqueous chloride systems: thermodynamic modeling of binary and multicomponent systems in wide concentration ranges. Fluid Phase Equilib 452:16–57. https://doi.org/10.1016/j.fluid.2017.08.014
29. Spedding FH, Jaffe S (1954) Conductances, solubilities and ionization constants of some rare earth sulfates in aqueous solutions at 25 °C. J Am Chem Soc 76(3):882–884
30. Das G et al (2019) Rare earth sulfates in aqueous systems: thermodynamic modeling of binary and multicomponent systems over wide concentration and temperature ranges. J Chem Thermodyn 131:49–79. https://doi.org/10.1016/j.jct.2018.10.020
31. Shannon RD, Prewitt CT (1969) Effective ionic radii in oxides and fluorides . Acta Crystallogr Sect. B Struct Crystallogr Cryst Chem 25(5):925–946. https://doi.org/10.1107/s0567740869003220
32. Habenschuss A, Spedding FH (1979) The coordination (hydration) of rare earth ions in aqueous chloride solutions from x-ray diffraction. II. $LaCl_3$, $PrCl_3$, and $NdCl_3$. J Chem Phys 70(8):3758–3763. https://doi.org/10.1063/1.437928
33. Habenschuss A, Spedding FH (1979) The coordination (hydration) of rare earth ions in aqueous chloride solutions from x-ray diffraction. I. $TbCl_3$, $DyCl_3$, $ErCl_3$, $TmCl_3$, and $LuCl_3$. J Chem Phys 70(6):2797–2806. https://doi.org/10.1063/1.437866
34. Habenschuss A, Spedding FH (1980) The coordination (hydration) of rare earth ions in aqueous chloride solutions from x-ray diffraction. III. $SmCl_3$, $EuCl_3$, and series behavior. J Chem Phys 73(1):442–450. https://doi.org/10.1063/1.439895
35. McCullough JD (1950) An X-ray study of the rare-earth oxide systems: Ce^{IV}—Nd^{III}, Cr^{IV}—Pr^{III}, Ce^{IV}—Pr^{IV} and Pr^{IV}—Nd^{III}. J Am Chem Soc 72(3):1386–1390
36. Schneider SJ, Roth RS (1960) Phase equilibria in systems involving the rare-earth oxides. Part II. Solid state reactions in trivalent rare-earth oxide systems. J Res Natl Bureau of Standards. Sect A, Phys Chem 64(4):317–332
37. Reddy BM, Katta L, Thrimurthulu G (2010) Novel nanocrystalline $Ce_{1-x} La_x O_{2-\delta}$ (x= 0.2) Solid solutions: structural characteristics and catalytic performance. Chem Mater 22(2):467–475
38. Reddy BM, Vinodkumar T, Durgasri DN, Rangaswamy A (2017) Synthesis and characterization of nanostructured $Ce_{0.8}M_{0.2}O_{2-\delta}$ (M= Sm, Eu, and Gd) solid solutions for catalytic CO oxidation. Proc Natl Acad Sci, India, Sect A 87(1):155–161
39. Han X, Wang Y, Hao H, Guo R, Hu Y, Jiang W (2016) $Ce_{1-x}La_xO_y$ solid solution prepared from mixed rare earth chloride for soot oxidation. J Rare Earths 34(6):590–596
40. Nikolaev AV, Sorokina AA, Vilenskaya AY, Tsubanov VG (1967) Solubility in ternary mixtures containing lanthanide salts and water. Izv Sibir Otd Akad Nauk SSSR Ser Khim Nauk 14:5–21
41. Sokolova NP, Bagryantseva LL, Komissarova LN (1983) Solid solutions in the systems $NdCl_3$-$LnCl_3$-H_2O at low temperatures. Izv Sibir Otd Akad Nauk SSSR Ser Khim Nauk 4:89–92
42. Trusova EM, Popov VP, Tikhonov PA, Glushkova VB (2002) Solid solutions of rare-earth cuprates and their electrical properties. Glass Phys Chem 28(4):264–267

Extraction Chromatography for Separation of Rare Earth Elements

Meher Sanku, Kerstin Forsberg, and Michael Svärd

Abstract Developing efficient and viable processes for separation of critical metals is essential to meet the increasing demand. Rare earth elements (REEs) are identified by the EU as critical resources, and moreover, they are difficult to separate due to their similar properties. Extraction chromatography is a powerful method suitable for difficult, high-purity separations, which could form part of a separation process for recovery of REEs from various sources. In the present work, separation of REEs from synthetic apatite leach solutions is investigated using physically immobilized extractants. By means of reverse-phase columns, reversibly functionalized by acidic organophosphorus compounds, the metals are separated by elution with nitric acid solution.

Keywords Chromatography · Rare Earth elements · Separation

Introduction

Rare earth elements (REEs) are difficult to separate from each other by conventional methods owing to their similarity in chemical properties. They are also found in very low concentrations both in natural ores as well as in urban waste. With traditionally employed solvent extraction processes, large quantities of organic solvents are usually needed to separate the REEs. Therefore, developing alternative separation processes, focusing on efficiency, environmental sustainability and economy, is necessary. Extraction chromatography is a promising "green" candidate component of such an alternative process, with the potential to separate REEs into individual pure fractions. Benefits include the potential to achieve higher purity levels than solvent extraction, using lower amounts of extractants and solvents, increased potential to recycle chemicals, and a vastly reduced number of process steps required. In extraction chromatography, extractants which have shown applicability in solvent extraction processes are reversibly adsorbed (physisorption) to the non-polar chain groups

M. Sanku · K. Forsberg · M. Svärd (✉)
Department of Chemical Engineering, KTH Royal Institute of Technology, Teknikringen 42, 11428 Stockholm, Sweden
e-mail: micsva@kth.se

of the particle surfaces of a reverse-phase (RP) chromatographic column, which can subsequently be used for separation by elution with a suitable solvent. The method of reversible physisorption allows significant flexibility in terms of extractant and coverage. Ideally, the extractant used should offer sufficient selectivity to allow resolution of the target elements; should be present at sufficiently high coverage to allow sufficiently high sample loads to be processed; and the physisorption should be stable under the elution conditions to allow the column to be used repeatedly without regeneration. Extraction chromatography for separation of metals from solution, including REEs, has been investigated in several studies [1–5].

In the current project, preparative extraction chromatography is being evaluated as the final stage of a hydrometallurgical process of recycling of REEs from aqueous waste solutions. The point of departure for the chromatography process is the leaching of REE-containing material such as ore, permanent magnets, or battery electrodes [6]. The present contribution is an account of experiments using the extractant bis (2-ethylhexyl) phosphate (HDEPH) together with elution by nitric acid gradients, for the separation of six REEs from synthetic leach solutions, as part of a series of experiments to establish a proof of concept.

Materials and Methods

Materials

The different solutions used in this study are described below. These solutions were prepared by using the individual components as received.

Column conditioner: A solution of ethanol and water, the concentration of which matches the concentration of ethanol and water in the acidic organophosphorus feed solution. In the present account, a solution composition of 55.5 wt% ethanol with the balance being milli-Q grade water is used.

Acidic organophosphorus feed solution: 71.8 mg Bis (2-ethylhexyl) phosphate (HDEPH) per g solvent mixture was dissolved in a solution of ethanol and water of similar composition as the conditioner. The resulting solution was verified to be a homogenous one-phase liquid.

REE solution: A solution of six REEs predominant in apatite ore, prepared from standard solutions mixed in equal amounts. The different REEs used in this mixture were La, Ce, Nd, Y, Pr, and Dy, and the final concentrations of each REE are 167 mg/L (1000 mg/L with respect to total REE concentration).

HNO_3 solution: Concentrated HNO_3 (69%) diluted in water to obtain a 4 M HNO_3 solution.

Arsenazo III solution: Arsenazo III, urea and acetic acid dissolved in water to form a solution used for post-column reaction.

NaOH solution: A 0.25 M NaOH solution.

HPLC Setup

A Thermo Scientific Dionex ICS-5000+ Ion Chromatography System was used. All tubing and connections are made of PEEK. The solutions to the HPLC column were degassed under vacuum, and the flow rates controlled with a 4-channel gradient pump. An injection valve fitted with an internal coil of exact volume (50 μL) was used to inject REE solutions. The temperature of the solutions, the flow path, and column has been maintained at a set temperature using a cryostatic water bath and a thermal compartment. The tubing between the solution bottles and the column was thermally insulated. A dedicated isocratic pump was used to deliver Arsenazo III solution to the eluate in a T-connector downstream of the column. A knitted braid downstream of this connection was used to provide the necessary time to allow complex formation between Arsenazo III and the REEs; a preprequisite for REEs to be detectable by a UV detector at 658 nm. An automatic fraction collection module was used to allow the REE concentrations to be resolved individually using inductively coupled plasma optical emission spectrometry (ICP-OES). The chromatographic column used was 150 mm in length with an internal diameter of 4.6 mm. The column was packed with Kromasil (Nouryon) C18-functionalized mesoporous spherical particles, with a diameter of 10 μm and a pore size of 100 Å.

Column Preparation and Performance

The retention of ligands on the column is a result of interactions between the hydrophobic C18 chains and the hydrophobic long chains on the HDEPH molecules. Before column impregnation with acidic organophosphorus compounds, any residual acid from previous runs were eluted with ethanol. The column conditioner was then run through the column until steady state, followed by feeding the acidic organophosphorus solution to the column. Finally, the column was washed with water to remove any loosely bound HDEPH. To optimize the column functionalization step and to evaluate the coverage for a given feed solution, the amount of adsorbed organophosphorus compound was determined through elution with ethanol and subsequent titration with NaOH solution.

For separation experiments, 50 μL of REE solution was injected. The retained REEs were eluted using various combinations of isocratic steps and gradients ranging from 0 to 4 M HNO_3. Experiments were conducted at two temperatures: 25 and 40 °C.

Results and Discussion

Column Preparation

It was found that an equilibrium surface coverage was established after elution with 40 column volumes of feed solution, applied at room temperature, with no further improvement obtained with longer preparation runs. The resulting coverage was 1.05 mmol HDEHP adsorbed on the column particles, which translates into a ligand density of 0.64 mmol/g C18-functionalized Kromasil material.

Separation Experiments

Preliminary experiments of REE separation carried out under conditions of linear gradient elution from 0 to 4 M HNO_3 established that the column performance is maintained and does not deteriorate over repeated runs under these conditions. Moreover, as expected, increasing the eluent flow rate from 0.5 to 2.0 mL/min resulted in some deterioration of the peak resolution, at a slight gain of process time. Finally, increasing the temperature from ambient to 40 °C led to a significant improvement in the peak resolution. The flow rate and the temperature for further experiments were thus fixed at 1.0 mL/min and 40 °C.

However, the chromatograms obtained under these conditions were insufficiently resolved with respect to the light REEs (eluted early, under low acid concentrations) while it would be inefficient to extend the total process time required to elute all six REEs by elongating the gradient. A combination of isocratic elution at low acid concentration to separate the four lighter REEs (La, Ce, Pr, and Nd) followed by a steeper gradient to separate and elute the heavier REEs (Dy, Y) was therefore evaluated. Figure 1 shows the resulting chromatogram obtained. The element-resolved chromatogram obtained by ICP analysis of the collected fractions are shown in Fig. 2 for the isocratic part of the process. It can be seen that this column has the potential to separate all the REEs evaluated in this study, notably providing an almost perfect separation even for the notoriously hard-to-separate pair Nd/Pr.

Significance and Outlook

This work shows that, for loads just exceeding the analytical range, complete separation into individual fractions of REEs—even for the hard-to-separate pair Nd/Pr—can be achieved with extraction chromatography in a single stage from synthetic acidic solutions. To attain industrial significance, however, the challenge will be to

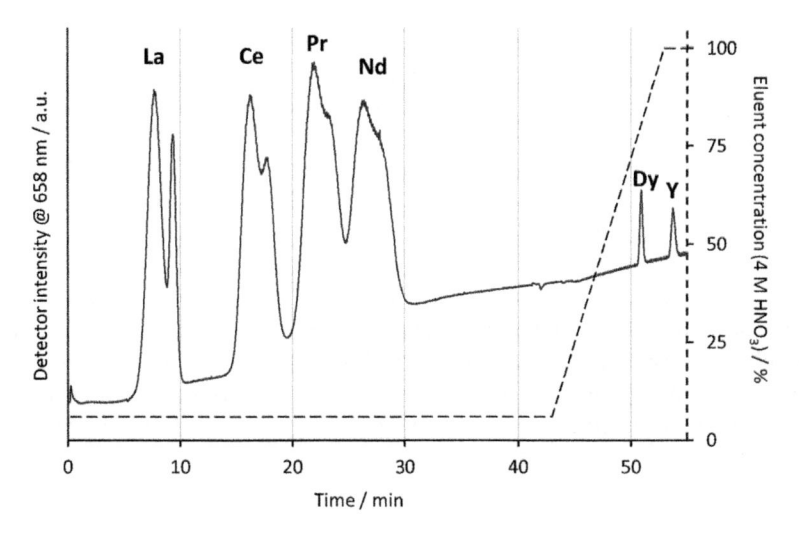

Fig. 1 The eluent concentration (dashed black line) and the elution profile of REEs obtained (red solid line) at 40 °C using a combined isocratic–gradient elution. (Color figure online)

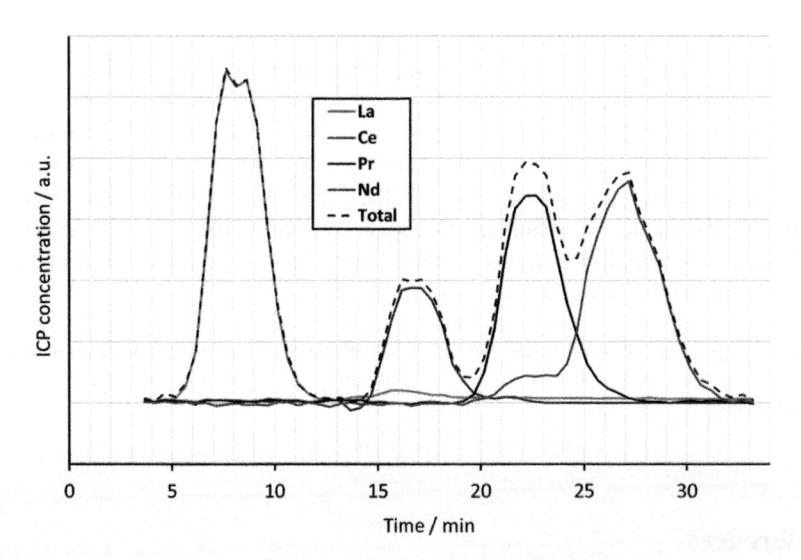

Fig. 2 ICP-OES resolved chromatogram from the isocratic part of the run shown in Fig. 1. (Color figure online)

increase the limited productivity, which is the main drawback of chromatography as a separation method. The total mass load, given by the volume and concentration of REE-containing solution, has to be increased at a sufficiently maintained resolution. Therefore, as a natural next step, the performance of the column at mass loads in the overloaded regime must be evaluated, and the selectivity and the ligand surface coverage optimized for this purpose. Further areas of study include the influence of other components and impurities present in real leach solutions on the column performance and stability, and the optimization of the productivity with respect to all variables of importance.

If an economically viable chromatographic method of separation of REEs into individual pure fractions could be developed, the benefits would include a lower amount of extractants required, a reduced carbon footprint, and a decreased amount of waste to be handled in the process, since the amounts of organic solvents involved would be drastically reduced. An extraction chromatography-based process would be flexible and enable a high product purity in a single stage, without the need for multiple mixer-settlers. Such a process could be a green, safe, and efficient alternative to established techniques available for recovery of these important metals from several different sources, as well as opening up new avenues of recycling that have hitherto not been commercially feasible.

Conclusions

In this work, functionalization of a hydrophobic C18 column with HDEPH has been investigated, and the resulting column evaluated for separation of REEs. The column under low sample loads has been verified to be suitable for separation of all six evaluated REEs in a total process time of below 1 h, using a combination of isocratic (0.24 M) and gradient (0.24–4.0 M) elution with nitric acid at a flow rate of 1 mL/min and a temperature of 40 °C. The performance of the column has been verified to be stable and maintained over several repeat runs. Overall, these tests indicate the potential of extraction chromatography for separation of REEs from acidic leach solutions as part of a recycling process.

Acknowledgements This work was carried out within the REEform project financed by Formas (grant no. 2019-01150.) Nouryon Pulp and Performance Chemicals, Bohus, Sweden, are gratefully acknowledged for supplying Kromasil C18 columns.

References

1. Max-Hansen M (2014) Modeling and optimization of rare earth element chromatography. Lund University
2. Kifle D (2013) Separation of rare earth elements and other precious metals by high performance liquid chromatography and solid phase extraction: University of Oslo
3. Knutson HK (2016) Robust multi-objective optimization of rare earth element chromatography. Lund University
4. Schaeffer N, Grimes SM, Cheeseman CR (2017) Use of extraction chromatography in the recycling of critical metals from thin film leach solutions. Inorganica Chim Acta 457:53–58

5. Schmidt A, Mestmäcker F, Brückner L, Elwert T, Strube J (2019) Liquid-liquid extraction and chromatography process routes for the purification of lithium. E-Mobil Circul Econ 959:79–99
6. Korkmaz K, Alemrajabi M, Rasmuson ÅC, Forsberg K (2018) Recoveries of valuable metals from spent nickel metal hydride vehicle batteries via sulfation, selective roasting, and water leaching. J Sustain Metall 4:313–325

Tool and Workflow for Systematic Design of Reactive Extraction for Separation and Purification of Valuable Components

Hana Benkoussas, David Leleu, Swagatika Satpathy, Zaheer Ahmed Shariff, and Andreas Pfennig

Abstract Rare-earth metals, such as La(III), Nd(III), Eu(III), or Y(III), which are recycled from electronic waste in urban mining, can be separated and purified utilizing reactive extraction. Process development and equipment design then aim to determine the optimal reactive extractant, diluent, any additional components, equipment type, structure of the equipment internals, as well as all process parameters. This requires a deep understanding of the chemistry of the underlying complexing reactions as well as engineering expertise on extraction-process development as well as equipment design. To aid this design task, a tool was developed based on cascaded option trees, which combines the expertise from both sciences. Process design is supported by a prototypic workflow and by systematically structured and quantitative information on the underlying thermodynamics. The method is also applicable for extraction of diluted components from aqueous solutions, as encountered in fermentation broth in the context of bioeconomy. The method will be presented and applied to examples from urban mining.

Keywords Urban mining · Process design · Option trees · Reactive extraction · Optimal process · Separation

Challenges in Extraction Design

In hydrometallurgy, reactive-extraction processes have been designed, which allow efficient separation and purification of valuable metal components like precious metals and rare-earth metals. These processes have been developed over several decades, and a major body of experience has been built up that allows optimization of the various processes [1]. In contrast, separating valuable metal components in the context of urban mining is a relatively new field of research, complicated by the fact that metals are combined in different mixtures and ratios as compared with the natural minerals. Depending on the specific source of the urban raw material, very

---

H. Benkoussas · D. Leleu · S. Satpathy · Z. A. Shariff · A. Pfennig (✉)
Department of Chemical Engineering, University of Liège, Liège, Belgium
e-mail: andreas.pfennig@uliege.be

different separation steps are required, which may be quite variable. This creates new challenges for separation-process and equipment design.

An additional challenge in process design is the need to combine expertise from chemistry and chemical engineering to arrive at an optimal process. For example, it is not sufficient to optimize for equilibrium separation, if under process conditions, the extract and raffinate cannot be easily separated. This in turn may strongly depend on the matrix from which the feed has been obtained as well as the leaching conditions. Thus, physical concerns such as coalescence and phase separation as well as mass-transfer kinetics need to be considered early on in process design. Thus, engineering expertise is essential. On the other hand, the concepts that help the chemist to optimize the chemistry of the extraction, like the hardness of metal ions and extractants, may not be as commonly known within the chemical engineering field.

To facilitate linking the expertise from both areas – chemistry and chemical engineering – a methodology has been designed, which allows systematic process development and eases communication between chemists and chemical engineers involved in process design.

Option Trees as Support for Process Design

Based on previous experience in cooperation projects with industry, the cascaded option-tree methodology had been proposed previously to support design of new processes [2]. As starting point for an option tree for a specific separation-process design task, the options for that separation process on a high level are collected as well as the basic criteria that have to be fulfilled. Each option is then evaluated to judge to which degree the individual criteria are fulfilled, which is recorded in a corresponding matrix. The first criteria evaluated should be those that are most restrictive. The result of this evaluation can be coded with colors or symbols. For example, green or a ' + ' indicate that the option meets the criterion, while yellow or '0' show that the option would work for the criterion, but some challenges like higher cost or more process effort are foreseeable. If an option does not fulfill a criterion, this is indicated in red or '–'. As soon as for an option a criterion is evaluated as red, that option does not need to be considered further. Once all criteria are evaluated, the optimal process option as well as the next-best options on the chosen level of refinement become obvious. If no option should remain, either the criteria have to be relaxed – if that is acceptable – or other options must be considered. Barring that, the separation considered is not possible.

For the best option(s) the option-tree procedure can then be repeated at a finer level of detail. While at a high level different principal separation steps may be considered, at the next finer level of detail in case of extraction different extractants may be evaluated. This procedure can be repeatedly refined to include equipment options and operating conditions. Thus, in a systematic way, the options available at each level of detail are evaluated and the evaluation recorded in a transparent manner. The methodology also has the advantage that the criteria evaluated as well as the underlying thermodynamic data for the metal components remain identical for different separation tasks. As a consequence, it is possible to develop prototype option trees, a systematic workflow, and a data basis that can be generally utilized for varying applications.

Option Tree for Reactive Extraction

The cascading option-tree methodology was originally developed for the separation of bio-based components from fermentation broth, and has since also been applied to the separation of phosphorous from sewage sludge, from which heavy metals should be removed [3], and is currently applied to rare-earth and transition metal separation. As a first example for the purpose of illustrating the method, the separation of cobalt and nickel shall be considered because of the simplicity of the system, which nevertheless poses a separation challenge due to the similar behavior of the metals. The methodology can directly be applied to other mixtures of metals and more complex systems involving more components.

It is assumed that in a general process the metals are first leached from a solid starting material, possibly after dismantling the urban waste, crushing, and grinding. The choice of these preliminary steps as well as the leaching acid can already be based on the option-tree method. Since here the focus shall be on the extraction step, it is assumed that a suitable acid has been found. After the extraction, typical steps are re-extraction and precipitation. Re-extraction, i.e., a second extraction step, in which the loaded organic phase is extracted with an aqueous phase e.g. at different pH, is used to regenerate the extractant and to access one separated component in an aqueous phase. The desired products are typically precipitated either from the aqueous raffinate leaving the primary extraction step and/or from the aqueous extract leaving re-extraction.

For extraction, as a first step different extractants can be considered. From a chemical perspective, the hardness or softness of the ions to be separated and the hardness or softness of the extractant can be used as a framework for systematically sorting and interpreting the extraction results. The hardness of a metal ion is defined as the arithmetic mean of the electron affinity and the ionization potential, typically expressed in eV [4]. The extractants which are O-, N-, or, P-donors are called hard extractants and those which are S-, F-, or Cl-donors are called soft extractants. Additionally, the extractants are sorted according to their decreasing pK values. As a result, a matrix is obtained which systematically characterizes the equilibrium constants as shown for some selected extractants and ions in Table 1. The source of these data is a large collection of literature references, where care has been taken that the pK describes the true equilibrium constant, i.e., the different concentrations of the extractant in the respective experiments have been corrected for [5–10]. Since in many publications, only cases are reported which aim at optimizing a separation, i.e., where degree of extraction is close to 0 or 1, this only allows to define a lower or upper bound, respectively. In general, for any new separation task, these pK values can

Table 1 Matrix with pK values for different combinations of extractants and ionic species to be separated, evaluated from [5–10]

Metal ion	hardness eV	Extractant Cyanex 272 pKa 6.37 pK	Cyanex 302 5.63	Versatic 10 5.17	DEHPA 3.24
Fe^{3+}	15.3		0.9	1.75	0.5
Cu^{2+}	9.5	1.75–3.6	<0	4	2.8
Zn^{2+}	9.0	1.8–2	1.5	5.5	0.75–1.5
Ni^{2+}	8.5	5.75–6.2	6.5	6.4	5
Co^{2+}	8.2	4	3.9	6.6	4

be obtained, for example from the literature. If the required data are not available, dedicated experiments have to be performed and evaluated to fill in the gaps in the table.

Table 1 is the basis for evaluating the first criteria of the option tree for extractant selection. In principle, it is assumed that hard metal ions are best extracted by hard extractants and soft metals by soft extractants. Since metal ions need to be separated, in the illustrative example cobalt and nickel, the difference in their pK should be considerable, if a separation with few equilibrium stages is desired. For the examples shown in Table 2, both Cyanex extractants have a similarly high difference in pK so that separation is possible. For DEHPA, the difference is smaller, which means that more stages are required for an identical degree of separation. Versatic 10 apparently does not allow to separate cobalt and nickel, so it is indicated as red and no further criteria are evaluated for this extractant.

As the next criterion in Table 2, the ease of extraction can be considered. This includes the pH range at which extraction and re-extraction are to be performed, which for all the remaining extractants is in a technically feasible range. Finally, the re-extraction efficiency is considered, which refers to an agent for re-extraction being available as well as the pH difference required between extraction and re-extraction. If that difference is large, large amounts of more acid and base are required, leading to excessive salt production. For this criterion, other alternatives can also be considered. If the equilibrium for the desired separation is sufficiently temperature dependent, instead of a pH shift a temperature shift may be an option, since it produces less salt in the waste streams. In some cases, it may be possible to use ammonia to increase the pH which can then be stripped from the aqueous solution by desorption. Such options highly depend on the specific separation considered.

As shown in Table 2, it is then possible to obtain an overall rating for each option, which is indicated to the left of the extractants. This overall evaluation should be considered with care, since simply taking the arithmetic average of the evaluations implies an equal weighting for all criteria that are included. In principle, it is possible

Table 2 Option tree for the extractive separation of cobalt and nickel. Only a selection of options and criteria is shown. (Color figure online)

overall evaluation	extractant	separation possible	easy extraction	re-extraction efficiently
1	Cyanex 272	1	1	1
0.67	Cyanex 302	1	1	0
-1	Versatic 10	-1		
0.67	DEHPA	0	1	1

to introduce individual weights for each criterion, but this can introduce new uncertainties such that the overall benefit may be limited compared with the effort. Also, it should be mentioned that a red entry should only be indicated if the option is completely infeasible, as this leads to that option not being considered any further.

Next, the diluent can be selected from a variety of options using the same general procedure, as shown in Table 3. While a sufficient solubility of the extractant is of course the basic requirement, the subsequent criteria on the physical properties of the diluent are essential for proper operation in a commercial extractor. The density difference between the organic and aqueous phases should be high, and the viscosity should be sufficiently low to avoid any impairment in phase separation after extraction. The interfacial tension should be in an intermediate range; if too low, emulsions are formed at low levels of energy input and if too high, too much energy is required for dispersing into sufficiently small drops to allow fast mass transfer. The last criteria refer to the economy of the process. If the solubility of the diluent in the aqueous phase is too high, this will result in excessive effort to remove it from the raffinate stream. A high toxicity to environment or humans means excessive effort for safety measures that need to be considered in the process design. Here, it is assumed that carcinogenic and potentially carcinogenic compounds shall be excluded. If the diluent is prone to degradation, the purge and makeup streams would be large, also affecting cost efficiency. The last criterion on the ease of phase separation has obviously not yet been evaluated in Table 3. It will be evaluated in the near future with a standardized settling cell, which has been developed in previous research. Quantitative evaluation of the results obtained with the standardized settling cell allows the design of

Table 3 Option tree for the diluent selection for exemplary diluents and criteria. (Color figure online)

overall evaluation	diluent	sufficient solubility in extractant	density difference between phases	low viscosity	intermediate surface tension	low solubility in water	non toxic	resistance to degradation	cost; economy	ease of phase separation
0.86	kerosene	1	1	1	1	1	0	1	1	
-1.00	carbon tetrachloride	1	1	1	1	0	-1			
-1.00	benzene	1	1	1	1	0	-1			
0.57	toluene	1	1	1	1	0	0	1	0	
0.71	xylene	1	1	1	1	1	0	1	0	
0.86	biodiesel	1	1	0	1	1	1	1	1	
-1.00	chloroform	1	1	1	1	-1				

technical-scale settlers and quantifies coalescence properties for extraction-column design [11, 12].

The cost criteria are not necessarily critical for the process operation itself. As a consequence, in such a case, it may be beneficial if the criteria are split into two groups. One group containing the essential criteria, where a negative entry means that this option is completely infeasible. In a second group of criteria that are evaluated separately, 'nice to have' criteria can be collected, where a red means some significantly higher level of effort and cost, but does not render the option infeasible in principle. For example in the case of toxicity, certain levels of toxicity mean just significantly more effort for safety measures, but have no influence on the process itself. Thus, an option with a red entry in the 'nice to have' criteria would not be realized unless it is the very last option that remains available after evaluation of all other criteria.

Another aspect demonstrated in Table 3 is that the criterion 'cost, economy' is at least partly redundant to the criteria on aqueous solubility, toxicity, and degradation. This again shows that care needs to be taken in using the overall evaluation since the redundancy may not be as obvious as in this illustrative example. As indicated above, weighting can be attributed to the individual criteria. This example also shows that this may not eliminate the redundancy, since the cost criterion would include the previous cost-related subcriteria but may also refer to other cost effects not captured by previous criteria.

In the workflow proposed, corresponding option trees are then used to select possibly required complexing agents, synergistic components, and modifiers. For these choices, additional basic data are collected in corresponding tables comparable in structure to Table 1. Also, the equipment selection and choice of operating conditions are structured as option trees.

The information used to fill the option trees can obviously be drawn from rather different sources. The easiest way is to refer to literature and databases as well as expert knowledge which may be available, especially if certain options are to be ruled out. Of course, sufficient care needs to be taken to assess the quality of the data available. Own experiments may be considered as a reliable source of information, but is the most time and resource-intensive option. The advantage of experiments is that the original matrix of feedstock can be used; this is especially important for phase separation, as it is known that matrix effects can have an extremely strong influence [12]. The experiments do not include just the equilibrium data, which are a starting point. The experiments also allow to determine parameters that are relevant for equipment design like the settling time obtained from the standardized settling cell already mentioned [11, 12] or experimental data on drop sedimentation and mass transfer, which can also be obtained from dedicated lab-scale equipment [13, 14]. Also, simulation results can be utilized in evaluating the criteria for the different options, where it has been shown that especially the drop-based simulations allow a reliable design not only of the processes but of the individual equipment [15]. This also allows to include equipment options in the option trees at the corresponding level of refinement.

Discussion

Regarding the overall procedure, the method of cascading option trees is easy to apply in general and leads to a very transparent overview of the options considered and criteria evaluated. At each point in the process development, it is very clear which options still exist and which of those are the best options available and why. The method does not only allow to choose the best option but also second-best options are directly visible. This is especially beneficial if an option at a later stage of evaluation turns out to be infeasible for other reasons, since then the next best options are directly accessible. Due to the transparency, the option-trees method has turned out to be very helpful in delivering such evaluations to project partners.

The basic underlying data shown as an example in Table 1 translate the more intuitive concepts on soft and hard ions and extractants into quantifiable molecular and thermodynamic properties, which allows easier access for engineers. This structure, which meanwhile has been extended to include a variety of rare-earth and transition metals and several more reactive extractants, allows to get a quick overview of the available options for new separation tasks. If a new combination of metal ions to be separated is already contained in the table from other mixtures, this information can directly be utilized to set up the option tree for extractant selection, among others. It is also obvious that most entries in the option tree for diluent selection are independent of the extractant. Of course, this has to be verified for each individual case, but the available information is already a rather good first basis. In addition, the other components to be added to the system as a synergistic component or modifier, even though their effect is very specific to the metal and extractant, nevertheless are usually chosen from a limited collection of options. Thus, not only the data tables but also the option trees can easily be reused for new separation tasks. Finally, the overall workflow is rather generic so that it can be applied directly in designing extraction processes for any new separation task. As already mentioned, beside the application to rare-earth separation and purification, it is currently also applied to bio-process design as well as to the recovery and purification of phosphorous from sewage sludge. Since it allows to include the chemistry as well as the chemical-engineering perspective, it has proven to be useful on all levels including for equipment design, which has been validated for the phosphorous recovery where this methodology has been the basis for the design of a demonstrator on pilot-plant scale for a specifically optimized process [16].

This overall workflow, the prototype option trees, and the basic tables on molecular and thermodynamic properties form a sound basis for optimized extraction-process and equipment design. Meanwhile, this methodology has been extended to rare-earth and transition metals, with the separation from waste neodymium magnets and fluorescent lamps as illustrative examples. The methodology and the data are currently validated experimentally, the underlying documents being extended detail of information and continually refined.

References

1. Bart H-J (2000) Reactive extraction. Springer, Berlin
2. Bednarz A, Rüngeler B, Pfennig A (2014) Use of cascaded option trees in chemical-engineering processdevelopment. Chem Ing Tec 86(5):611–620. https://doi.org/10.1002/cite.201300115
3. Shariff ZA, Fraikin L, LéonardA, Pfennig A (2019). Recovery of phosphorus from sewage sludgeand subsequent purification using reactive extraction. Paper presented at The 12th Europen Congress of Chemical Eniginering (ECCE12), Florence, Italy, 15–19 September 2019
4. https://www.knowledgedoor.com/. Accessed 8 September 2020
5. Silvia Y, Nurqomariah A, Fajaryanto R (2019). Extraction of Co and Ni metals using emulsion liquid membrane and liquid-liquid extraction with cyanex 272 as extractant. AIP Conference Proceedings 2085(March):020071–1–020071–7. https://doi.org/10.1063/1.5095049
6. Menoyo B, Elizalde MP (1997) Extraction of Cobalt(II) by Cyanex 302. Solvent Extr Ion Exch 15(1):97–113. https://doi.org/10.1080/07366299708934468
7. Ayanda OS, Adekola FA, Baba AA, Ximba BJ, Fatoki OS (2013) Application of Cyanex® extractant in Cobalt/Nickel separation process by solvent extraction. Int J Phy Sci 8(3):89–97. https://doi.org/10.5897/IJPS12.135

8. Ghonem EA, Hamed MM, Mohamad YT, El-Reefy SA, Hegazy M, Abd El-Wahed A (2015) Extraction of cobalt (II) from sulphate medium and possible separation from nickel (II). Arab J Nucl Sci Applic 48(3):17–21
9. Tait BK (1993) Cobalt-nickel separation: the extraction of cobalt (II) and nickel (II) by Cyanex 301, Cyanex 302 and Cyanex 272. Hydrometallurgy 32:365–372. https://doi.org/10.1016/0304-386X(93)90047-H
10. Torkaman R, Asadollahzadeh M, Torab-Mostaedi M, Maragheh MG (2017) Recovery of cobalt from spent lithium ion batteries by using acidic and basic extractants in solvent extraction process. Sep Purif Technol 186(October):318–325. https://doi.org/10.1016/j.seppur.2017.06.023
11. Leleu D, Pfennig A (2019) Drop-based modelling of extraction equipment. In: Moyer BA. (ed) Ion exchange and solvent extraction, vol 23. Changing the Landscape in Solvent Extraction. CRC Press, Boca Raton, pp 253–285
12. Kopriwa N, Pfennig A (2016) Characterization of coalescence in extraction equipment based on lab-scale experiments. Solvent Extr Ion Exch 34(7):622–642. https://doi.org/10.1080/07366299.2016.1244392
13. Ayesterán J, Kopriwa N, Buchbender F, Kalem M, Pfennig A (2015) ReDrop—a simulation tool for the design of extraction columns based on single-drop experiments. Chem Eng Technol 38(10):1894–1900. https://doi.org/10.1002/ceat.201500097
14. Altunok MY, Kalem M, Pfennig A (2012) Investigation of mass transfer on single droplets for the reactiveextraction of zinc with D2EHPA. AIChE J 58(5):1346–1355. https://doi.org/10.1002/aic.12680
15. Pfennig A, Pilhofer T, Schröter J (2006) Flüssig-Flüssig-Extraktion. In: Goedecke R (ed) Fluid-Verfahrenstechnik, vol 2. Wiley-VCH, Weinheim, pp 907–992
16. https://www.nweurope.eu/projects/project-search/phos4you-phosphorus-recovery-from-waste-water-for-your-life/. Accessed 8 September 2020

Rethinking Mineral Processing and Extractive Metallurgy Approaches to Ensure a Sustainable Supply of High-tech and Critical Raw Materials

Yousef Ghorbani, Glen T. Nwaila, Steven E. Zhang, and Jan Rosenkranz

Abstract Raw materials (RM) are crucial for maintaining our standard of living internationally. The fourth industrial revolution and the energy transition are reliant on access to various RMs. High-tech RMs are usually extracted as by-products from ore deposits. To increase the production of rare high-tech RM, it is essential to modify the existing bulk RM production processes and utilize partial, secondary, or waste streams. This study aims to present and discuss the necessities of redefining the concept and scope in mineral processing and extractive metallurgy approaches in order to secure a sustainable supply of high-tech and critical raw material (CRM) for the economy in modern society. We introduce a list of paths and trends for developing future concepts and methods in mineral processing and extractive metallurgy in pursuit of the sustainability of high-tech CRMs from all resources.

Keywords Critical raw materials · Mineral processing · High-tech raw materials · Extractive metallurgy · Process development

Introduction

Modern societies require many raw materials (RM), which are primary commodities that are used to produce derived goods and energy, such as crude oil, coal, water, lumber, minerals and metals. Critical raw materials (CRMs) are currently non-substitutable RMs, which most consumer countries are dependent on importing, and whose supply is dominated by one or a few producers [1]. The definition and assessment of criticality differ by country and change over time (e.g. [2]). Various

---

Y. Ghorbani (✉) · J. Rosenkranz
Department of Civil, Environmental and Natural Resources Engineering, Luleå University of Technology, 97187 Luleå, Sweden
e-mail: yousef.ghorbani@ltu.se

G. T. Nwaila
School of Geosciences, University of the Witwatersrand, Private Bag 3, Wits 2050, South Africa

S. E. Zhang
PG Techno Wox, 43 Patrys Avenue, Helikon Park, Randfontein 1759, South Africa

factors affect the criticality of an RM, such as: production ceilings; decreases in reserves; changes in the production ratio of bigger to smaller deposits; inefficient price systems and; increases in extraction costs [1]. Hence, the sustained provision of CRMs is contingent on a complex system of multiscaled socioeconomic and environmental policies, technological and scientific progress, and international trade and geopolitical stability. As the list of CRMs grows, there are serious challenges to the sustainability of CRMs. The EU, Australia, the USA, UK, China, and Canada have launched several campaigns, including the Raw Materials Initiative [3]. Their main goal is to ensure the sustainable supply of RMs, while increasing benefits for society as a whole [3]. Even as a major supplier of RMs, China has also established its first policy and catalogue of "strategic minerals" in November 2016, which is to be updated every 5 years. Solutions to the sustainability of RMs are likely to be multisectoral and multidisciplinary and could include: increasing extraction and processing efficiency and security; finding suitable substitutions; and increasing recycling [3]. As many CRMs are dominantly minor constituents of ore deposits, they are generally extracted as by-products and in limited quantities. Therefore, it is indispensable to amend the existing production processes and to incorporate partial, secondary, or waste streams. In this paper, we focus on various future-oriented ideas, technology, and solutions to improve extraction and processing efficiency aligned with a sustainable supply of high-tech CRMs for the economy in modern society. In addition, supply potential from both primary and secondary resources are presented.

High-Tech CRMs

Definition and Changes in Assessment Over Time (2011–2017)

Changes in external factors such as consumption patterns and the proliferation of new technology renders the list of CRMs dynamic. In 2011, the first list of 14 CRMs was published by the EU, and reassessments would occur every 3 years. In 2014, the list included 20 CRMs, which expanded to 27 in 2017 [2]. In 2014, one additional CRM was silicon, which is the basis for current computing technology and remains on the list in 2017, for which China is the world's major producer at 61% [2]. In contrast, chromium was added onto the list in 2014, but subsequently removed in 2017 as the EU refined its assessment methodology [2]. This is reflective of the fact that criticality of a RM depends not only on its definition and changes in technology but also the socioeconomic context. For example, the EU assessment methodology uses the economic importance and supply risk to calculate criticality and adopts a fixed baseline based on a threshold model [1]. The limiting factor varies and could be monopolistic suppliers (e.g. REEs, [2]) or dwindling supply in known deposits (e.g. platinum-group metals (PGMs), [2]). Although many CRMs are or have been historically supply-constrained, different conjectures exist on the extent of supply risk of CRMs. In 2010, due to China's embargo on rare earth elements (REEs) stemming from a China–Japan territorial dispute, many feared that production–consumption imbalances in the CRM trade would become a foreign policy tool [1, and references therein]. However, the implications are mixed since most REEs are abundant in the earth's crust. China merely hosts one of the largest insitu concentrations of economic REEs and also beneficiate various CRMs as by-products from various streams. Although companies such as Molycorp Inc. have in the past dominated the extraction of REEs, China currently leads the production of many CRMs because of production scale advantages, cheap labour, and relaxed environmental policies. Eventually, the abundance of CRMs inspired innovations outside of mining and effectively weakened China's grip on the market. In a similar manner, supply–risk mitigation consists of recycling and technological innovations in exploration, extraction, and processing [1].

Concept of Sustainability

The concept of sustainability predates resource extraction and its application is pervasive. In 1987, the World Commission on Environment and Development (the "Brundtland Commission") coined the term "sustainable development" as "development that meets the needs of the present without compromising the ability of future generations to meet their own needs." The current human experience spans three main areas: environmental; economic; and social, with the overlap of the three, a sustainable combination. In practice, sustainability is challenging to define, monitor, and implement. It is therefore mainly a theoretically motivated but empirically operationalized goal to derive relatively more sustainable outcomes. For example, the United Nations dissects the concept of sustainability into 17 distinct goals [4], of which, many are relevant to CRMs, such as: affordable and clean energy; industry, innovation and infrastructure; sustainable cities and communities; responsible consumption and production; climate action; and decent work and economic growth. To approach sustainability for some CRMs, a general solution is increased circularity and extraction/processing efficiency [1], while for others that are not conducive to recycling, only enhanced extraction/processing efficiency is applicable outside of the market and consumer considerations.

Innovative Techniques and Solutions to Sustainable Access to High-tech CRM

Influence of Mineral Crystallography and Fluid Inclusions on CRM Extraction Route

The knowledge of the formation of a particular ore deposit (thermodynamics, chemistry, structures, emplacement conditions, spatial relationships) provides constraints and guidance for mining and processing (MnP) and is sometimes useful for metallurgists to improve MnP. To date, most observations occur at the macroscale regarding ore processes, although it is possible that microscale (e.g., crystalline/molecular) innovations are on the horizon to support sustainable MnP through more selective reagents and processes. Microscale observations can be achieved through the use of technologies such as: quantum chemical modelling; photoemission electron microscopy (PEEM)/scanning photoelectron microscopy (SPEM); nano-X-ray fluorescence (nano-XRF); transmission electron microscopy (TEM); atomic force microscopy (AFM); X-ray crystallography (XRC); and three-dimensional electron microscopy (3DEM). Most of these are available at institutions (e.g., MAX IV Laboratory, CERN European research laboratory, and Diamond Light Source). Microscopic inclusions of fluid (1–20 μm) inside a host mineral during its formation contains information of its parent hydrothermal fluid (chemistry and thermodynamics) and are abundant (e.g., milky quartz containing ~1 billion inclusions per gram) [5, 6]. To our knowledge, this type of information is currently unused, but could be applied in a reverse-engineering approach to develop novel MnP methods for high-tech CRMs. In addition, this information can be used to grow synthetic crystals or to concentrate desirable CRMs, both of which aid CRM production and recovery.

Smart Multifiltering System to Identify Specific Reagents

Chemicals are utilized in MnP from grinding, flotation, hydrometallurgy, solid–liquid separation, tailings treatment, to materials handling [7]. Reagent design and complexity is likely to increase to satisfy CRM demands and to cope with more challenging ores and impure water. Additionally, increasing environmental and health concerns likely imply chemicals with more specificity and less toxicity. Artificial intelligence (AI) can significantly increase automation and augment decision making and along with big data are referred to as the "fourth paradigm of science" and the "fourth industrial revolution." Data/AI-driven approaches (e.g. high-throughput sequencing in "-omics" fields) combined with predictive computational chemistry have the potential to transform reagent design. In essence, computational modelling of chemicals and reactions can generate a wealth of data, which combined with AI can rapidly identify suitable reagents or predict ideal reagent design. A smart tool can be built around these technologies to automate, to the extent possible, the discovery and design of reagents. This approach could identify mineral structure- and process-specific reagents and would certainly assist the mining industry to maximise the performance of their operations through optimised reagents and constant innovation. Such tools are already employed in other industries such as pharmacology, which evaluates huge number of chemical reagents and effective parameters, in order to optimise pharmaceutical research and design. In comparison to traditional test work, our proposed approach could evaluate many possible scenarios rapidly. In addition, it can identify target conditions and reagent types, in order to conduct more effective and detailed physical test work [8].

Downstream Extraction of CRMs from Primary and Secondary Sources

Most CRMs are found in a variety of mineral systems, such as ore deposits and in either complex ores or in low concentrations that are uneconomical for use. In primary ore systems, economic recovery of CRMs is often limited by our ability to isolate marketable concentrates. This calls for novel hydrometallurgical processing techniques that isolate CRMs better and therefore also minimize waste. Hydrometallurgical techniques have shown the greatest potential for metal extraction from both primary and secondary raw material resources. Hydrometallurgical processing of complex low-grade ores and concentrates is becoming increasingly important as the mining and metallurgical industry seeks to exploit mineral deposits that are difficult to treat conventionally. The metal extraction steps are typically characterized by approaches that range from metal leaching by chemical reagents or bacterial action under suitable thermodynamic conditions, and in reactors to vats or heaps (both chemical and biological) to in situ recovery. Certain environmentally benign and selective hydrometallurgical processes that require low capital cost for an equivalent metal production rate may offer a solution to the current CRMs extraction challenge. The selection of an optimal solution tends to distract engineers from realizing value through integration or modification of existing flowsheets. For example, solution-based processes such as Solvent Extraction (SX) and Ion Exchange (IX) provide a starting point to explore for amenable solutions [9]. Both SX and IX are well-known techniques and have been applied in a commercial scale for over 40 years. If applied individually, the shortcomings may limit usage in a number of commodities as illustrated in Table 1.

Integrating SX with IX can eliminate the negative characteristics of both processes. The benefits of developing an integrated IX/SX flowsheet include, but are not limited to: (a) recovery of metals from Pregnant Leach Solution (PLS) with a broad concentration range, (b) extraction of metals from very low-graded PLS, (c) avoids the construction of large ponds for prefiltration of the PLS, (d) effectiveness

Table 1 Characteristics, advantages and disadvantages of using IX and SX techniques [10, and references therein]

Technique	Advantages	Disadvantages	Application
Solvent extraction: *Used for purification of valuable metals such as the concentration of valuable metal(s) with concentrations that range from under 1 g/L up to 100 g/L and conversion from one anion to another: chloride to sulphate or ammonia to acid sulphate*	• Operates in low temperature (<50 °C) • Complete recycling of barren solutions in many cases • Very high selectivity at removal and purification of metals • Metal conversion to a form suitable for easy subsequent recovery	• Employs large quantities of volatile organic substances, which lead to large footprint of SX circuits • Very sensitive to low temperatures and quantity of total suspended solids in feed solutions • Flammable • Organic losses can be deleterious to the environment	Copper, uranium, nickel, cobalt, molybdenum, rare earth and other metals.
Ion exchange *Selective recovery of metals from Pregnant Leach Solution (PLS) sourced from leaching operations coming through IX columns, where PLS is passed through the resin laid in the columns and valuable metal is adsorbed on the resin. The column is then regenerated to obtain highly purified metals*	• Good selectivity with respect to metal ions • Long life of resins • Low maintenance cost • Ease of operation • No disengagement of phases, • Economic feasibility for use in low concentrations of metal ions • Environmental safety • Relatively immune to the presence of suspended solids	• Success depends largely on the structure of the resin and specifically on the ionic radius and ionic charge density of the metal ions • Calcium sulfate fouling, iron fouling, adsorption of organic matter.	Platinum group metals (PGMs), gold, silver, vanadium, mercury, copper, zinc, cobalt, uranium, vanadium, waste/process water, spent sulphuric acid catalysts, fly-ash, petrochemical residues, AMD and tailings dumps

in a variety of conditions including turbid solutions and solutions containing large quantities of suspended solids, (e) reduction in organic losses, and (f) reduction in plant size. An added benefit is that preconditioning of feed material may be unnecessary, hence a mixture of solution streams that share synergies from spent heaps and dumps, acid mine drainage and other industrial streams can be processed without modifying the flowsheet. This shows the possibilities that exist from combining existing hydrometallurgical processes, which expands the possibility for recovery of CRMs as by-products from secondary streams.

Accelerated Selective Percolation Leaching

Percolation leaching processes can be defined as the selective removal of metal from a mineral by allowing an appropriate solvent or leaching agent to trickle into and through a mass or pile of material containing the mineral. Subject to the characteristics of the deposit and the ore, commercial percolation leaching is commonly categorized into different groups [11, and references therein], of which, heap Leaching (HL) and in situ leaching (ISL) are the two main percolation-leaching methods that have made a significant contribution to mineral extraction (Fig. 1). Considering their substantial technoeconomic and environmental advantages, they could play a key role in the processing of high-tech CRMs.

Heap Leaching

Heap leaching from low-grade ores has become a key contributor to the overall global production of Cu, Au, Ag, and U [10]. It is also sometimes applied to small higher grade deposits in remote or politically high-risk locations to lessen capital cost. Heap leaching also has been considered for Zn, Ni, and for ores bearing PGMs and electronic scrap. Some types of heap leaching are commercially successful (U, Cu, and Au) and some remain novel ideas or under commercialized (Zn, Ni, Mn, Co, and primary-sulphide copper). Lack of success of heap leaching of some commodities is less attributable to problems around mobilizing them through the heap leaching, but more due to the ability to attain high recovery of the metal from low-concentration solutions without destroying the lixiviant. SX reagents, IX resins, and activated carbon adsorption were the important enablers for successful Cu, U, and Au/Ag heap leaching. On the other hand, correspondingly effective reagents have not yet been identified for the recovery of Zn, Co, Mn, and Ni without significant pH adjustment to the PLS through neutralization of the sulphuric acid lixiviant [11, and

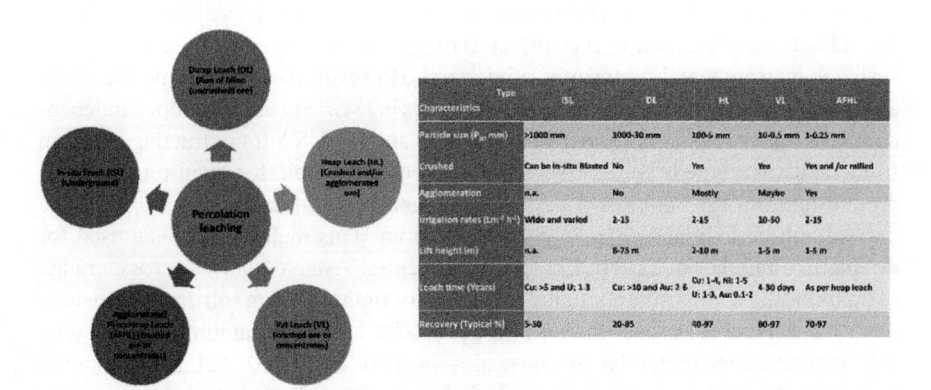

Fig. 1 The percolation leaching types and a summary of typical criteria (adapted from [10]). *Note* n.a. denotes not applicable. (Color figure online)

references therein]. The feasibility of heap leaching also depends on other factors such as economic, environmental, and the long-term operational plan of the mine that supplies the feed ore, as well as social concerns. When the disadvantages of long leach times and slow extraction rates can be overcome, heap leaching offers substantial prospective for further innovation. Many new ideas for heap leach processes, whether based on different commodities (Zn, Ni, PGMs), or novel chemistry (use of thiosulphate for Au, ammonia for base metals, chloride or thermophile micro-organisms for chalcopyrite) seem viable at the laboratory scale, but are possibly hindered by limited heap permeability at the industrial scale. Environmental concerns around heap leaching necessitate critical discussion. Incidentally, heap leach technology could also play a role in the low-cost recovery of metals from secondary resources to reduce the need for primary mining, as has already been shown in the context of certain e-wastes. Care must be taken not to treat heap leaching as some sort of 'primitive' technology. For its optimal operation, appropriately detailed knowledge of the ore and the mechanisms of heap leaching is vital to understand and evaluate the impact of any specific interference to advance performance. The long-term success of heap leaching involves an intensive high-level cross-disciplinary engineering approach from research and development from design to construction, operation and closure. A comprehensive review on the theoretical background of numerous heap leach processes, including a scientific and patent literature overview on technology developments in commercial heap leaching operations around the world was provide by Ghorbani et al. [10]. In their review paper, they have also identified factors that affect the selection of heap leaching as a processing technology, defined challenges to exploiting these innovations, and concludes with a discussion on the future of heap leaching.

In Situ Leaching

In situ leaching (ISL, also solution mining, or in situ recovery (ISR)) passes a leaching solution using injection wells through an ore deposit to solubilize metal(s) of interest. The PLS is then returned to the surface through recovery wells. Consequently, the surficial environmental footprint is minimized. However, the ore body permeability and groundwater contamination are key concerns in ISL. ISL was developed independently in the 1970s in the former Soviet Union and the USA for extracting uranium from sandstone-type uranium deposits that were unsuitable for open-pit or underground mining. ISL requires water-saturated and permeable sands hosting the ore deposit, which is bounded by impermeable strata. This method has been used for mining in a number of eastern European and central Asian countries across a range of commodities, most notably in the extraction of uranium from roll-front sandstone deposits [12, and references therein]. In 2019, 57% of world uranium was mined by ISL methods, a share that has risen steadily from 16% in 2000 [12]. Most uranium mining in the USA, Kazakhstan, and Uzbekistan is now by ISL and the method is becoming more popular in Australia, China, and Russia. In the USA, ISL is seen as the most cost-effective and environmentally acceptable method of mining. Copper is the second-most popular commodity that is mined by ISL. There are references to primitive forms of ISL of copper in Roman times, and perhaps even earlier in China. ISL has been used for supplemental recovery of copper from established open-pit and underground mines using both sulfuric acid and acidic ferric sulfate solutions. In addition to ISL of intact material, there have also been numerous commercial and experimental projects recovering copper after blasting, fracking, block caving, and partial underground mining. ISL has also been used for the recovery of evaporites such as soda ash, potash, and salt. For example, ISL was used in 1994 on the Gagarskoye Gold Deposit in the Ural Mountains region (Russia) for leaching of gold-bearing regolith, which prompted the application of ISL to other gold deposits in weathered regolith and deep placers. Pilot tests of ISL for nickel were successful for some silicate nickel deposits. The ISL processing of tailings and pyrite ash is also useful for remediation, not only for extraction. Scandium, rhenium, rare earth elements, yttrium, selenium, molybdenum, vanadium were leached in situ as by-products in pilot tests at the uranium deposits in the 1970s–1980s. More recent developments in ISL of nonuranium elements suggests a more widespread adoption of ISL should be considered.

The major economic challenge of ISL recovery of copper is limited exposure of minerals to the leach solution in complex subsurface flow paths, which has generally resulted in lower recoveries (20–70%) compared with established processing methods (65–90%). The major environmental challenge for ISL is containment integrity in the target ore zone to protect adjacent groundwater. Containment is unreliable in some places, due to the existence of natural low-permeability zones and/or by the hydrogeological management. Where ISL is possible, major environmental benefits compared with conventional mining include reduced energy consumption, near-elimination of waste rock and tailings, lessened land disturbance, reduced dust and noise, and possibly lower water consumption. The variability of ISL recovery is associated with minimal control of geometallurgical variables such as fragmentation, aeration, and temperature. To overcome these deficiencies, an in situ mining (ISM) method has been proposed, in which leaching and conventional underground mining methods are integrated in an innovative way [13], such that ISL is applied to blasted rock in a large stope developed using the sublevel stoping method. It is projected that ISM will improve ISL recovery by providing better control of metallurgical variables.

Farming-Based Strategies of High-tech and CRMs

The interest in metal recovery from secondary resources (e.g., metal-containing soils) has increased following the increase in global demand for metals. In this context, the research of phytomining, now called agromining, which is a feasible superficial-contaminated remediation technique based on hyperaccumulator plants, has been growing [14, 15]. Agromining is an integrated concept that aims to optimize ecosystem services of ultramafic regions at the landscape level after mining. Hyperaccumulator plants have been extensively investigated and are capable of significant bioaccumulation of heavy metals in their aerial parts, which are then harvested and burnt to produce metal(loid)-enriched ash or "bio-ore." The synergistic activity between plants and microorganisms may contribute to soil management strategies in natural metal-enriched soils. In a Ni-agromining framework, more than 400 Ni-hyperaccumulators were identified and new tropical Ni hyperaccumulators were discovered recently [16, and references therein]. It is considered a commercially viable technique in the case of high-value elements such as Ni, Co, or Au [86]. Phytomining has been advanced and optimized from field agronomy in both North America and Albania to full metallurgical processes and carried to the pilot scale. Investigations are occurring for other metals, especially those with high added value (Ni, Mg, P, K, Fe, and Mn), as well as REEs [17, 18]. The growing interest in this emerging technology is echoed through the recent funding of two EU projects: Agronickel and LIFE-Agromine, which aim to implement agromining at large scales, following the "Agromine" project funded by the ANR (Agence Nationale de la Recherche) in France.

A Move Toward Synchronized Process and Extractive Metallurgy Dry Laboratories

Big data and artificial intelligence (AI) have the capacity to transform business decision-making, by providing simultaneously more reliable and in-depth insights, automating routine decisions and avoiding human decision fatigue. MnP is rich with data, which is unfortunately siloed, ill-managed and ungoverned presently. However, a proliferation of initiatives that produce complex data to improve plant performance and metal recovery [19] means that the adoption of AI in MnP is inevitable. We proposed the establishment of the dry lab, which in the MnP context is a laboratory that predominantly utilizes computational mathematical analyses through some combination of forward- and inverse-modelling, including AI. Its focus is insight generation and data analytics. Dry labs can provide the interdisciplinary innovation space, data management and monetization opportunities for the development of AI, such as smart tools to guide metallurgy. Actionable insights can be extracted at various timescales and spatial scales and feedback to the MnP can occur in the form of periodic optimization, synchronous data acquisition and real-time analytics, predictive maintenance and long-term research and development [20].

Future Potential Resources

Primary Resources

Advances in Deep Seabed Mining of CRMs

Traditional marine mining occurs in shallow waters, where heavy mineral sands (e.g., Richards Bay mineral sand deposits, South Africa), phosphorite mining (e.g., New Zealand waters), and diamonds (e.g., marine diamond deposits, Namibia) are dominant deposit types [21]. Explorations of nontraditional sources of CRMs in the deep sea are mainly driven by the need for more CRMs [22]. Deep sea mining is a process of extracting mineral deposits from the deep sea (~50% of the Earth's surface is ocean that is deeper than 200 m and could provide CRMs [23]. Some oceanic geological features host important mineral deposits, such as massive (polymetallic) sulfides around hydrothermal vents, cobalt-rich crusts on the flanks of seamounts and fields of manganese (polymetallic) nodules on the abyssal plains. There is a growing interest in for deep sea mining, especially in under-regulated international waters. However, the target areas for future seabed mining are ecologically fragile. The International Seabed Authority (ISA) was formed in 1982 and is the current body that is regulating human activities on the deep-sea floor beyond the continental shelf. The ISA was established under Part XI of the United Nations Convention on the Law of the Sea (UNCLOS) and it is an independent organisation based in Kingston, Jamaica with a total of 167 member states including the European Union (International Seabed Authority, 2012, 2013). In addition to ISA, the European Union has developed an MIN-Guide initiative which is an online repository for information on minerals and related policies for Member States. To date, the ISA has issued 27 contracts for mineral exploration, each valid for 15 years from the date of issue, encompassing a combined area of >1.4 million km^2, and continues to develop rules for mining. A total of 18 of these contracts are for exploration for polymetallic nodules in the Clarion-Clipperton Fracture Zone (16 contracts) and Central Indian Ocean Basin and Western Pacific Ocean (one contract each). There are seven contracts for exploration of polymetallic sulphides in the South West Indian Ridge, Central Indian Ridge and the Mid-Atlantic Ridge and five contracts for exploration for cobalt-rich crusts in the Western Pacific Ocean.

Remote Asteroid Mining of CRMs

In recent years, there has been a rising interest in mining resources on the Moon and asteroids since our solar system is full of potentially economic CRMs [23]. Although asteroid mining is an old idea, it is only recently that the technology and logistics for such a mission have become borderline possible, particularly for the near-Earth asteroids [24]. This is a unique opportunity for enhancing and developing a space economy, contributing also to the economic and social development of future generations. The current situation in space mining of CRMs and its forecast for the year 2025 is discussed by Garcia-del-Real et al. [25]. As a first step toward remote asteroid mining, some universities have started graduate programs. For example, the space resources program at the Colorado School of Mines is a multidisciplinary program and was launched in 2018. This program focuses on developing core knowledge and gaining design practices in systems for exploration, extraction, and use of resources in the solar system.

Secondary Resources and Urban Mining

Problems associated with the primary extraction of various CRMs have made secondary resources more attractive, which supports CRM sustainability. Secondary streams including waste are some of the potential sources for the future supply of CRM. To maximize the value from extracting CRMs from secondary streams, hydrometallurgical processes should be designed to tolerate environmental contamination, such as lead, cadmium, beryllium, brominated flame-retardants and even hazardous materials. Solution-based processes such as SX and IX are unrefined initial solutions and biological treatment of secondary streams might provide a low-cost alternative for recovery of multiple CRMs from a single source. It is important that early stage process design should consider the possibility of selective sorting of CRM by-products to avoid further dilution. Contamination of primary products should be minimized through incorporating sufficient intermediary stream-material isolation/discharge stages. The USGS [26] estimates that almost 60% of annual supply of vanadium is recovered from slags; about 20% from mining as a primary product; and the remainder from secondary sources, including oil residues and fly ash. The existence of ample vanadium-rich tailings and slags has made it possible to selectively recover vanadium using a hybrid IX technology. The laboratory and pilot scales have been proven by IONEX, a member of X Group Technologies, whom demonstrated selective recovery of V_2O_5 at the 16.5 million metric tonnes EVRAZ Highveld calcine residue stockpile in South Africa (X Group technologies, 2020; Chris Grobler—March 2020, personal communication).

Summary and Outlook

Although it is impossible in general to predict the requirement of CRMs in the future and particularly in the long term, the need for CRMs is unlikely to evaporate as natural resource consumption has always been, and will continue to be, a part of human culture, which evolves continuously but overall gradually. Moreover, more RMs are likely to become CRMs internationally and geopolitical competition would continue to intensify. Sustainability of CRMs has many facets that require critical examination, from political, cultural, socioeconomic, scientific, engineering and technological, to environmental. Maximizing both primary production diversity and efficiency, and the development of a circular economy via secondary streams and recycling are unavoidable to address our needs. Innovations such as biological leaching, in situ leaching; nontraditional sources such as deep sea and off-Earth mining; secondary/waste streams and recycling from historical waste sources and urban sources are some of the immediately foreseeable solutions to implementing any notion of sustainability.

References

1. Overland I (2019) The geopolitics of renewable energy: debunking four emerging myths. Energy Res Social Sci 49:36–40
2. European Commission (2017) On the 2017 list of critical raw materials for the EU, 2017. Communication from the commission to the European parliament, the council, the European economic and social committee and the committee of the regions
3. Ghorbani Y (2018) Gaining access to high-tech and critical raw materials: a driving factor in developing future concept and methods in mineral processing and extractive metallurgy. In: Third congress of industrial engineering of the North of Chile, INDUNOR 2018, Antofagasta-Chile
4. United Nations (2020) Sustainability development goals. https://www.un.org/sustainabledevelopment/sustainable-development-goals/. Accessed 29 Aug 2020
5. Faramarzi NS, Jamshidibadr M, Heuss-Assbichlerb S, Borg G (2019) Mineral chemistry and fluid inclusion composition as petrogenetic tracers of iron oxide-apatite ores from Hormuz Island, Iran. J Afr Earth Sci 155:90–108
6. Klyukin YI, Steele-MacInnis M, Lecumberri-Sanchez P, Bodnar RJ (2019) Fluid inclusion phase ratios, compositions and densities from ambient temperature to homogenization, based on PVTX properties of H2O-NaCl. Earth Sci Rev 198:102924
7. Hutton-Ashkenny M, Ibana D, Barnard KR (2015) Reagent selection for recovery of nickel and cobalt from nitric acid nickel laterite leach solutions by solvent extraction. Miner Eng 77:42–51
8. Ames Laboratory (2020) Selective REE recovery from magnet leachate. https://www.ameslab.gov/cmi/research-highlights/selective-ree-recovery-from-magnet-leachate
9. Mohebbi A, Abolghasemi Mahani A, Izadi A (2019) Ion exchange resin technology in recovery of precious and noble metals. In: Rangreez TA, Asiri AM (eds) Applications of ion exchange materials in chemical and food industries. Springer, Cham
10. Ghorbani Y, Franzidis J-P, Petersen J (2016) Heap leaching technology-current state innovations and future directions: a review. Miner Process Extr Metall Rev 37:73–119
11. Seredkin M, Zabolotsky A, Jeffress G (2016) In-situ recovery an alternative to conventional methods of mining: Exploration resource estimation environmental issues project evaluation and economics. Ore Geol Rev 79:500–514
12. WNA (World Nuclear Association) (2020) In-Situ Leach (ISL) mining of uranium. https://www.world-nuclear.org/information-library/nuclear-fuel-cycle/mining-of-uranium/in-situ-leach-mining-of-uranium.aspx
13. Bahamondez C, Castro R, Vargas T, Arancibia E (2016) In-situ mining through leaching: experimental methodology for evaluating its implementation and economic considerations. J South Afr Inst Min Metall 116:689–698
14. Nti Nkrumah P, Echevarria G, Erskine PD, Chaney RL, Sumail S, van der Ent A (2019) Growth effects in tropical nickel-agromining 'metal crops' in response to nutrient dosing. J Plant Nutr Soil Sci 182:715–728

15. Bouman R, van Welzen P, Sumail S, Echevarria G, Erskine PD, van der Ent A (2018) Phyllanthus rufuschaneyi: a new nickel hyperaccumulator from Sabah (Borneo Island) with potential for tropical agromining. Bot Stud 59
16. Simonnot MO, Vaughan J, Laubie B (2018) Processing of bio-ore to products. In: van der Ent A, Echevarria G, Morel JL, Baker A (eds) Agromining: farming for metals. Springer, Cham, pp 39–52
17. Houzelot V, Ranc B, Laubie B, Simonnot M-O (2018) Agromining of hyperaccumulator biomass: study of leaching kinetics of extraction of nickel, magnesium, potassium, phosphorus, iron, and manganese from Alyssum murale ashes by sulphuric acid. Chem Eng Res Des 129:1–11
18. Cecchi L, Bettarini I, Colzi I, Coppi A, Echevarria G, Pazzagli L (2018) The genus Odontarrhena (Brassicaceae) in Albania: taxonomy and Nickel accumulation in a critical group of metallophytes from a major serpentine hot-spot. Phytotaxa 351:1–28
19. McCoy JT, Auret L (2019) Machine learning applications in minerals processing: a review. Miner Eng 132:95–109
20. Ghorbani Y, Nwaila GT, Zhang SE, Hay MP, Bam LC (2020) Guntoro PI (2020) Repurposing legacy metallurgical data Part I: a move toward dry laboratories and data bank. Miner Eng 159(1):106646
21. Miller KA, Thompson KF, Johnston P, Santillo D (2018) An overview of seabed mining including the current state of development, environmental impacts, and knowledge gaps. Front Mar Sci 4:418
22. Aldred J (2019) The future of deep seabed mining. China Dialogue Ocean. https://chinadialogueocean.net/6682-future-deep-seabed-mining/
23. Toro N, Robles P (2020) Jeldres RI (2020) Seabed mineral resources, an alternative for the future of renewable energy: a critical review. Ore Geol Rev 126:103699
24. Garcia-del-Real J, Barakos G, Mischo H (2020) Space mining is the industry of the future…or maybe the present? In: SME annual meeting February 23–26, 2020, Phoenix, AZ
25. Changbai L, Deepika LR, Mika SÄ, Eveliina R (2021) Separation and concentration of rare earth elements from wastewater using electrodialysis technology. Sep Purif Technol 254:117442
26. U.S. Geological Survey (USGS) (2020) Mineral commodity summaries—vanadium. https://pubs.usgs.gov/periodicals/mcs2020/mcs2020-vanadium.pdf

Extraction of Rare Earth Metals: The New Thermodynamic Considerations Toward Process Hydrometallurgy

Ajay B. Patil, Rudolf P. W. J. Struis, Andrea Testino, and Christian Ludwig

Abstract Successful management of secondary waste resources is essential for the viable circular economy. E-waste could serve as the potential urban mining source for the alternative supply chain of critical metals such as rare earth elements (REEs). The hydrometallurgical processes for REEs are mainly designed for primary mining. Conventional approaches lack sustainability and the economic- and value chain-based aspects that are significant in the current era with its increasing focus and pressure to reduce environmental impact. We have performed thermodynamic calculations to simulate the solution chemistry behaviour of REEs, such as Neodymium (Nd), Dysprosium (Dy), and Praseodymium (Pr) present in NdFeB magnets. The results suggested that one could exploit the different solubility of these REE hydroxides by controlling the pH value and separating the REEs further using extractive processes. In contrast to primary mining, the use of appropriate wet chemistry, extractive conditions with selective ligands and supported liquid membrane methods with secondary (urban) mining could open up more sustainable and economic recycling of rare earth magnets with reduced environmental impact and direct scalability.

Keywords Circular economy · Sustainability · Rare earths · Recycling · E-waste · Resource management · Critical raw materials · Magnets

---

A. B. Patil (✉) · R. P. W. J. Struis · A. Testino · C. Ludwig (✉)
Chemical Processes and Materials Research Group, Energy and Environment Division, Paul Scherrer Institut, Forschungsstrasse 111, 5232 Villigen PSI, Switzerland
e-mail: ajay.patil@psi.ch

C. Ludwig
e-mail: christian.ludwig@psi.ch

A. B. Patil · R. P. W. J. Struis · C. Ludwig
École Polytechnique Fédérale de Lausanne (EPFL), ENAC IIE GR-LUD, 1015 Lausanne, Switzerland

Introduction

With the advent of efficient element detection and separation technologies, humankind has gained access to smart minerals and raw materials. Their remarkable and astonishing properties altered the technological landscape around the world. One of the key raw materials behind this transformation is 15 lanthanides, along with Sc and Y, rendering in a total of 17 rare earth elements (REEs) [1–5]. The applications and thus the demand for REEs are continuously increasing. However, the primary REE supply chain is associated with an enormous environmental impact due to harsh mining, processing, and radioactive by-products [1, 2]. As the future involvement of these raw materials in our day-to-day lives will increase, there will be increasing piling up of these materials in the back-end waste streams [6–9]. E-wastes have significantly limited recycling rates at the moment for several reasons (Fig. 1). With the >10% annual growth rate, e-waste is potentially harmful to the environment and making the cities and ecosystem unsustainable. Recent strategies view e-wastes as a potential source of the required raw materials. This will also help to keep the REEs in the ecosystem and can decrease the part of impact by decreasing the burden on the front-end value chain [10, 11]. Currently, REEs-containing secondary products are being assessed for their recycling potential.

Recent literature focuses on the recycling process developments for the end-of-life fluorescent lamp powder and magnets e-wastes [10–12]. The attempts in new recycling and process technology developments are often challenged by the complex chemistry of lanthanides and similarities in physical and chemical behaviours. However, the recycling sector has limited resources and capabilities to invest in such developments. Therefore, it is necessary to develop economically viable recycling approaches for REEs rapidly [10–12]. Recent reports have used simulation-based approaches to either screen or develop new process steps/loops based on solution chemistry or chromatographic separations of REEs. Such simulations minimize the efforts and costs in process developments and can provide insights useful for the overall process flow sheet developments [10, 13].

Fig. 1 Complexity and reasons for the low recycling rates of REEs. (Color figure online)

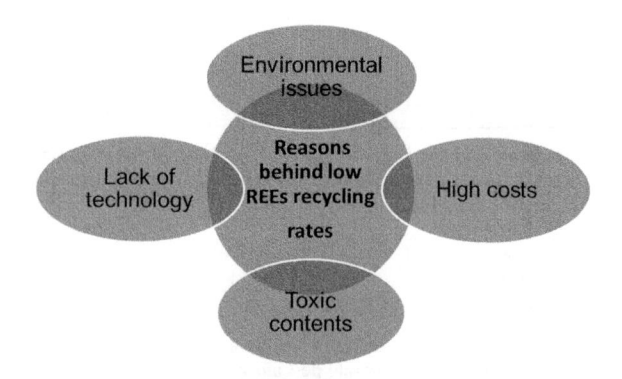

Process hydrometallurgy enables access to the rare metals from the original or secondary resources such as mined ores or wastes, respectively. The hydrometallurgy of REEs is complex and requires highly acidic conditions. To remove associated metallic and non-metallic impurities, it is necessary to dissolve such matrices in mineral acids, such as nitric acid or hydrochloric acid. The thus obtained complex feed-in solution is processed further using wet and extractive treatments to concentrate pure REEs [12, 14].

In e-waste recycling, also solution chemistry is the key to achieve pure recycled products. Recently, we reported a process to recycle REEs from end-of-life lamp phosphor to produce single, >99% pure, Yttrium (Y), Europium (Eu), and Terbium (Tb) [12, 15]. The process involved sequential digestion and selective extraction of the REEs with a minimum number of purification steps and low cost for achieving these recycled REEs. Such solution chemistry approach can also be adopted for the purification of Nd, Dy, and Pr from magnet e-waste feeds. However, cost-effective material processing and chemical treatment can be guided and assessed utilizing aqueous speciation simulation of the processing conditions.

With the recycling of NdFeB magnets in mind, the present study reports and discusses simulation results for adjustable conditions in aqueous multi-REE solutions. We have tested the capabilities of the simulations using the tool involving the electrolytes thermodynamics, process calculations in aqueous and multicomponent solution conditions. The findings reaffirmed that the slightly different solution chemistry behaviours of different REEs could play a vital role to separate them from each other in waste (or any other complex feed) solution with the use of relatively cheap chemicals, hence, low costs. In particular, it is inferred here that with mineral acid-leached magnet solutions, Neodymium (Nd), Dysprosium (Dy), and Praseodymium (Pr) could be separated from each other to a large extent as they precipitate as hydroxide $REE(OH)_3$ species at very different pH values. This was simulated for surrogate REE mixture solutions with a mixture composition similar to that of real-magnet waste.

Experimental

Simulation Method and Software

The solution chemistry study presented in the current article was done with the OLI software (version 10.0, Studio and ESP packages [13]). The proportionate amount of Nd, Dy, and Pr with real-magnet e-waste of 85% Nd, 14% Dy, and 1% Pr [11] was defined in the software with a total REE concentration of 3 mol per litre as input (feed) solution. In the simulation, the anion form with the input solutions was either nitrate or chloride. The concentration of the NaOH added to the input solutions is the variable quantity in the calculation, ranging between 0 and 12 mol/L NaOH. The output results as a function of the NaOH concentration were the pH value, the

REE-specific speciation in terms of trivalent and other ionic species in the aqueous phase ("Aq"), and that of solid REE hydroxides (which are the targets of interest). Unfortunately, the software database system does not have comprehensive data of the Dy(OH)$_3$ solid phase. Noteworthy, Nd has the atomic number (AN) of 60 with the periodic system, and Pr has AN = 59. They are both considered as "lighter lanthanides," whereas Dy (AN = 66) is a "heavier" lanthanide [15]. Therefore, the fate of the "Dy(OH)$_3$ solid" phase is deduced here by focusing on the "Er(OH)$_3$ solid" with Erbium (Er; AN = 68) (due to database availability), which is a close and higher lanthanide neighbour to Dy. In another magnet-representing simulation, the run was performed with Pr and Nd, but with Er instead of Dy. In future work, the solubility behaviour of the "Dy(OH)$_3$ solid" will be determined experimentally and will be implemented with the software database.

Results and Discussion

Different Solution Systems Containing REEs

We have simulated the solution conditions of the REEs for a mixture ratio resembling the REE composition of spent NdFeB magnets with an Nd:Dy:Pr = 85:14:1 mol ratio [11]. In another simulation run, Er was included as heavy lanthanide element instead of Dy. As input, we defined a wholly dissolved solution, with either nitrate or chloride as the anion. The speciation runs were varied as a function of the molarity (mol/L) of NaOH thought to be added in stepwise increased amounts to the surrogate magnet solutions. The speciation results in the nitrate and chloride medium without the addition of NaOH differed in that with nitrate (Fig. 2a) the REE(NO$_3$)$^{2+}$ form prevailed in the aqueous phase but with chloride (Fig. 3a) the trivalent REE^{3+} cations. Because of this, the amount of trivalent REE cations available in the chloride medium is somewhat larger than with the nitrate medium.

In both media, the identified zones (I–IV) can be exploited for the preconcentration of the REEs of interest as follows: With Fig. 2, Nd could be precipitated preferentially as "Nd(OH)$_3$ solid" in Zone I. The precipitation of Dy could follow it as second metal after adding only 4.5 mol/L of NaOH in nitrate conditions, or, 4 mol/L of NaOH in chloride conditions to the feed solution. Therefore, a mixture of solid Nd and Dy hydroxides without Pr content could be achieved preferentially in Zone II, and a mixture of solid Nd, Pr, and Dy hydroxides could co-exist together in Zone III. Focusing with Figs. 2b and 3b on Er, it shows an "Er(OH)$_3$ solid" phase builds up to nearly 100% with increasing alkalinity in Zone III, but also that it decreases after that due to re-dissolution in the solid into the anionic form of erbium tetrahydroxide, Er(OH)$_4^-$. Also noteworthy is that Dy and Er both show very similar courses for the REE(OH)$_4^-$ species as a function of the NaOH concentration, except that they are shifted from each other on the alkalinity scale. With Dy, the incomplete set of species (because "Dy(OH)$_3$ solid" is not available), leads to "second best" results in

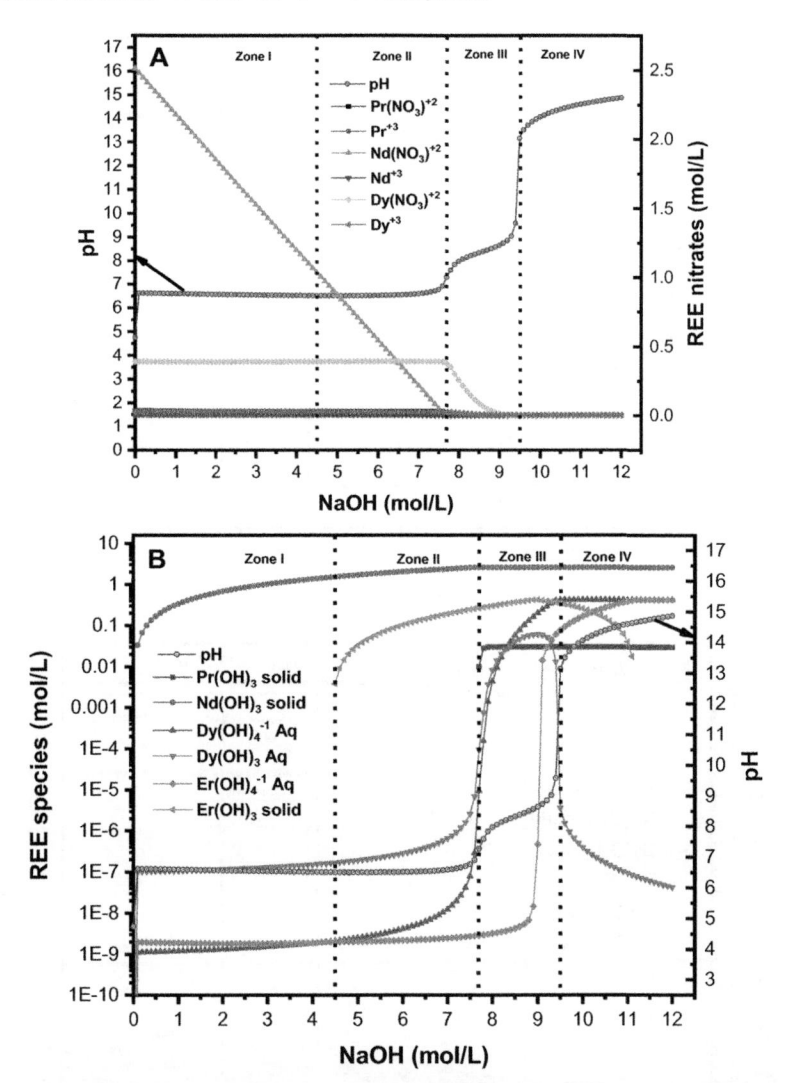

Fig. 2 Precipitation of REE hydroxides (A: Initial feed = REE nitrates; B: Hydroxide product speciation). (Color figure online)

the calculated amount of one (or more) species is overestimated to comply with the total mass balance condition set by the input data. The most likely candidate here is that of "$Dy(OH)_3$ Aq," in which Figs. 2b and 3b both show a very rapid decline when going from Zone III to Zone IV. In analogy to the fate of the "$Er(OH)_3$ solid," the declining amount of the "$Dy(OH)_3$ Aq" species may thus portray a partwise or complete re-dissolution of the "$Dy(OH)_3$ solid." From the coincidence of results with Dy and Er, it is anticipated here that in Zone IV (Figs. 2b and 3b), "$Dy(OH)_4^-$ Aq"

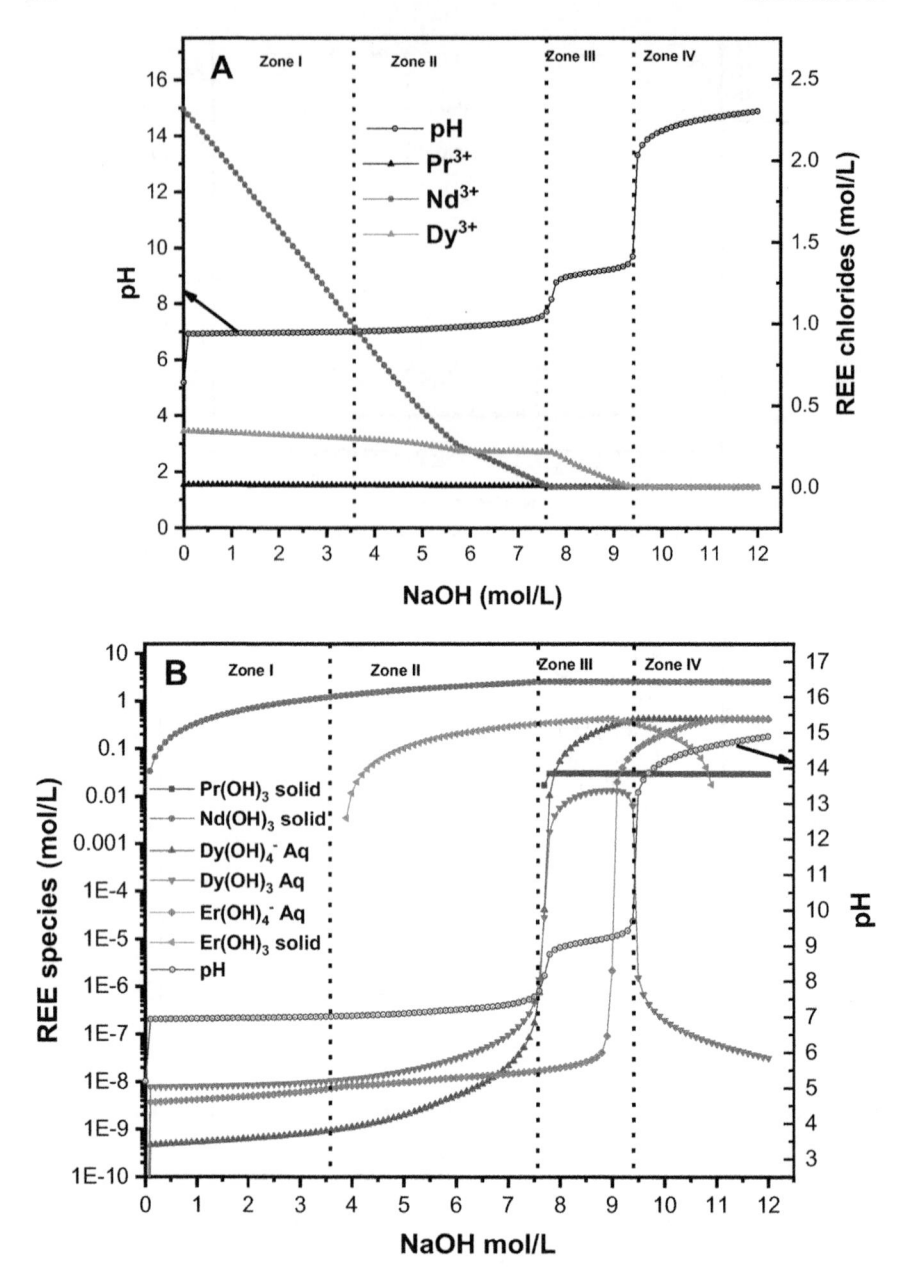

Fig. 3 Precipitation of REE hydroxides (A: Initial feed = REE chlorides; B: Hydroxide product speciation). (Color figure online)

could also be separated as a soluble species, leaving the stable Nd and Pr hydroxide behind.

Significance of the Simulated Chemistry to the REE Recycling from Magnets

The simulation results clearly demonstrate the usefulness of the fine-tuning of solution conditions to achieve the pre-concentration of the valuable and strategic REEs, such as Nd, Dy, and Pr. Such information will also help to avoid excess use of alkaline or acidic solutions and lower the costs and environmental impact. The preconcentrated and less complex mixtures could be separated in a relatively affordable manner and process conditions after that, e.g. by employing extractive, chromatographic, or membrane techniques to make the overall recycling strategy realistic and viable.

Conclusions

We have applied electrolyte thermodynamic-based simulation software to the complex solution chemistry of REEs present in NdFeB magnet waste. The solution conditions chosen in the present study are essential to feed streams in primary mining as well as in recycling processes. The simulation results showed that there is an opportunity to separate Nd, Pr, and Dy from each other with the help of a carefully tuned precipitation approach based on the addition of sodium hydroxide to the magnet e-waste feed solution. In contrast to conventional approaches, it would allow avoiding excessive use of chemicals for achieving an improved pre-concentration (by precipitations) and pre-purification of the targeted REEs with few costs and reducing the environmental impacts.

Acknowledgements The Swiss Federal Office for the Environment (FOEN) is acknowledged for co-funding the present work (project no.: UTF-1011-05300).

References

1. Zepf V, Reller A, Rennie C, Ashfield M, Simmons J (2014) Materials critical to the energy industry—an introduction, 2nd edn, BP
2. De Lima IB, Filho WL (eds) (2015) Rare Earths industry; technological, economic, and environmental implications. 978-0-12-802328-02016, 269
3. Rademaker J, Kleijn R, Yang Y (2013) Environ Sci Technol 47:10129–10136
4. Patil AB, Struis RPWJ, Schuler AJ, Tarik M, Krebs A, Larsen W, Ludwig C (2019) Rare earth metals recycling from e-wastes: strategy and perspective. In: Progress towards the resource

revolution Prof. Chr. Ludwig, Valdivia S (eds), Subchapter 23, 162–164. ISBN 978-3-9521409-8-7
5. Kumar R, Thenepalli T, Whan J, Kumar P, Woo K, Lee J (2020) J Clean Prod 267:122048
6. Tunsu C, Petranikova M, Ekberg C, Retegan T (2016) Sep Purif Technol 161:172–186
7. Mohammadi M, Forsberg K, Kloo L, De La Cruz JM, Rasmuson Å (2015) Hydrometallurgy 156:215–224
8. Hatanaka T, Matsugami A, Nonaka T, Takagi H, Hayashi F, Tani T, Ishidak N (2017) Nature Comm. 8:15670
9. Legaria EP, Rocha J, Tai C, Kessler VG, Seisenbaeva GA (2017) Sci. Rep. 7:43740
10. Ding Y, Harvey D, Wang NL (2020) Green Chem 22:3769–3783
11. Dupont D, Binnemans K (2015) Green Chem 17:2150–2163
12. Patil AB, Struis RPWJ, Schuler A, Ludwig C (2019) Method for individual rare earth metals recycling from fluorescent powder e-wastes. WO2019/201582A1
13. https://www.olisystems.com/
14. Patil AB, Pathak P, Shinde VS, Godbole S, Mohapatra PK (2013) Dalton Trans 42:1519–1529
15. Patil AB, Tarik M, Struis RPWJ, Ludwig C (2021) Resour Conserv Recycl 164:105153

Part IV
REEs and Sc

Developing Feasible Processes for the Total Recycling of WEEE to Recover Rare Metals

Jae-chun Lee, Manis Kumar Jha, Rekha Panda, Pankaj Kumar Choubey, Archana Kumari, and Tai Gyun Kim

Abstract The present paper reports several application-oriented processes developed for the recovery of various non-ferrous (Cu, Ni, Al, Pb, Sn), rare (Li, Co, In), precious (Au, Ag, Pt, and Pd), and rare earth metals (Nd, Ce, La, Y, Eu) from various urban ores, i.e., waste electrical and electronic equipments (WEEE), liquid crystal displays (LCD), batteries, magnets, fluorescent tubes, etc. Initially, the WEEE and various wastes were classified and dismantled. Further, the materials were pretreated to separate plastics, epoxy, ceramics, rubber, iron cover, and metallic concentrates. Based on their properties, plastic, epoxy and rubber could be either pyrolysed for production of marketable low-density oil and saleable activated carbon or directly recycled. The pre-treated metallic concentrates were processed by hydrometallurgical techniques, i.e., leaching, solvent extraction, ion-exchange, electro-winning for maximum recovery of metals. Various flow sheets discussed for rare metal extraction and processing strictly comply with environmental regulations.

Keywords Rare metals · E-waste · Recycling · Hydrometallurgy

Introduction

Nowadays, lots of electrical and electronic wastes (e-waste) are being generated with alarming rate worldwide and around 80% of the valuables present in these wastes end up in landfills. Although, strict regulations are being imposed by government in many countries for proper treatment of these electronic scraps but only 20% of that

---

J. Lee (✉)
Mineral Resources Research Division, Korea Institute of Geoscience and Mineral Resources (KIGAM), Daejeon 34113, South Korea
e-mail: jclee@kigam.re.kr

M. K. Jha · R. Panda · P. K. Choubey · A. Kumari
Metal Extraction and Recycling Division, CSIR-National Metallurgical Laboratory, Jamshedpur 831007, India

T. G. Kim
Research & Product Development, TAE-HYUNG Recycling, Gimcheon-si, Gyoengsangbuk-do 39066, Republic of Korea

is being recycled in an organized manner. In developing countries, lack of proper collection system, illegal recycling by unorganized sector as well as lack of cost-effective simple technology for processing e-waste, adversely affect the environment. Therefore, a feasible technology with minimum environmental impact is required for the recovery of metals from such scraps. Recycling is the imperative solution to the growing e-waste problem and it has become a significant economic activity. Major fraction of e-waste constitutes personal computers, mobile phones, batteries, magnets, fluorescent tubes, etc. PCBs of personal computers contain Cu, Ni, Sn, Pb, and precious metals; however, mobile phone PCBs/connectors contain metals of economic interest mainly the precious metals(Au, Ag, Pt, Pd) and Cu. The magnets present in various electronic goods contain Nd (rare earth) whereas the lithium-ion batteries (LIBs) is composed of rare and strategic metals such as Co, Li, Ni, etc., which should be recovered in an eco-friendly manner.

The conventional technology for the recycling of e-waste has been based on the pyrometallurgical process, which has the advantage of accepting any physical form of e-waste. However, it is associated with certain drawbacks such as a long-term recovery options, air pollution, and loss of the noble metals with low recovery of critical metals. Accordingly, in recent years, we have made concerted effort to develop the hydrometallurgical routes for the total recycling of metallic and other components from a variety of e-wastes. The total recycling of e-waste particularly requires elaborate mechanical pretreatment including the pyrolysis as well as the chemical processing because of the complicated structures and compositions, whereas for the selective recovery of metals, chemical leaching followed by advanced separation techniques viz., solvent extraction/ion-exchange techniques are considered appropriate. Thus the present research reports various novel, feasible, and scientifically validated hybrid pyro-, chemical-, and hydro-metallurgical processes developed for the total recycling of metal values such as rare, rare earths, and precious and non-ferrous metals from a variety of e-wastes.

Pre-treatment of E-waste

E-waste is mixture of plastics, epoxy resins, ceramics, and metals requires pre-treatment prior to the hydrometallurgical processing. Initially, the e-wastes are dismantled, classified, and pre-treated to collect/remove metals present in it. Direct leaching of e-waste coated or encapsulated with plastics and ceramics rarely accomplishes effective extraction of valuable metals. Thus, metallic fractions have to be liberated from the non-metallic parts (plastics and ceramic, epoxy, etc.) before hydrometallurgical treatment. Methods of mechanical pre-treatment/chemical processes, i.e., organic swelling [1], pyrolysis [2, 3], etc. used for pre-treatment of e-waste have been described below.

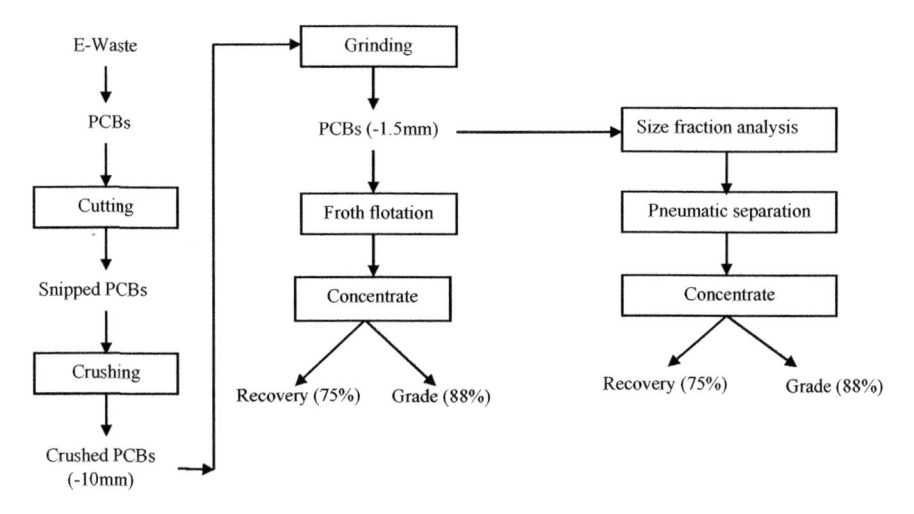

Fig. 1 Mechanical pre-treatment of e-waste to get enriched metallic values [4]

Mechanical Pre-treatment of E-waste

All e-waste contains PCBs, which are made up of various materials and metals. Kumar et al. (2013) developed a mechanical beneficiation route to process the PCBs present in e-waste in eco-friendly manner to get metallic concentrates for further processing (Fig. 1). The separation of material mainly depends on the distribution of metallic and non-metallic fractions and their liberation size due to their physical properties. The experimental data showed the enrichment of metals in coarser particles ($-1000 + 150$ μm) and non-metals in the finer particles <150 μm size following pneumatic separation and froth flotation process. In this case, recovery achieved was 75% with a grade of 88% on applying froth flotation. Simultaneously, lower grade of ~75% with ~65% recovery was obtained using pneumatic separation with <1500 μm powder. Finally, the grade of metals in average obtained was ~88% by controlling the feed charge and rate of air flow during pneumatic separation. The concentration of rare metals presents in the different fractions varying from 1.88% to 4.18% was enriched up to 9% using this beneficiation technique [4]. The developed process contributes to enrichment and the recovery of various metals like Cu, Al, Fe, Zn, Pb, Sn, rare metals, etc. Advantage of this process is the high metal recovery and directed towards the alternative metal resources from e-waste.

Chemical Pre-treatment of E-waste

The mechanical pre-treatment of PCBs is not effective in some cases, due to the lack of feasible milling devices. In order to address such bottle-necks, the feasible milling

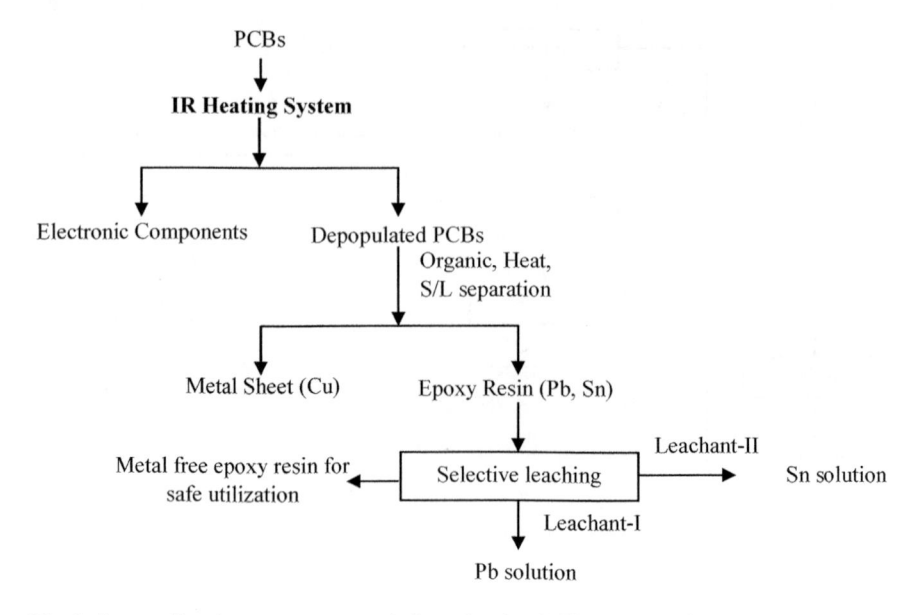

```
                    PCBs
                      ↓
              IR Heating System
                      ↓
        ┌─────────────┴─────────────┐
        ↓                           ↓
Electronic Components        Depopulated PCBs
                             Organic, Heat,
                             S/L separation
                      ┌─────────────┴─────────────┐
                      ↓                           ↓
               Metal Sheet (Cu)          Epoxy Resin (Pb, Sn)
                                                  ↓
                                        Leachant-II
Metal free epoxy resin for   ← Selective leaching →  Sn solution
   safe utilization                       ↓
                                    Leachant-I
                                          ↓
                                     Pb solution
```

Fig. 2 Process flowsheet to extract metals from the chemically pre-treated e-waste [1]

technique for PCBs was explored. Jha et al. [1] reported a bench-scale laboratory test using an organic to liberate the layer of metals from the PCBs without using any mechanical pre-treatment. Several thin layers of metals from the swelled PCBs were effectively separated leaving solder metals on the surface of the liberated epoxy resin [1]. The samples were processed as per the novel organic-treatment technique developed in our group. In this pre-treatment, the depopulated PCBs were sized into 7 cm × 7 cm and then dipped in a glass container containing the swelling organic. The container having cut PCBs and organic was heated and save for overnight. The outermost layer of PCBs containing soldering metals was removed after separating the swollen material, as presented in Fig. 2, and the thin layer of metal sheets encapsulated by resins was removed one by one. The sheets were dried after washing with hot water to remove the entrapped organic. Sheets were then sized further to ~0.5–0.7 cm and 0.5 mm thick, which contained 74.76% Cu, 0.48% Pb, and 0.52% Sn.

Pyrolysis of E-waste

Pyrolysis is a thermal treatment process, where the organic materials are decomposed to low molecular weight products of liquids or gases that can be used as fuel while the inorganic components containing metals and glass fibers are separated. Kumari et al. [2] and [3] developed processes for pyrolysis of the depopulated PCBs of e-waste in a vacuum pyrolysis device, where the evolved gases were collected and condensed to be reused as a fuel. The pyrolysed PCBs were further beneficiated to separate the metallic content from the burnt epoxies (Fig. 3). The separated metallic content was further screened to get low metal concentrate of size <2 mm size, which was generally lost with the carbon ash. This low metal concentrate (small fraction) is focused here for the recovery of base metals.

The typical chemical analysis of low metal concentrate obtained in our studies showed the presence 34.26% Cu, 2.45% Fe, 0.43% Ni, 1.11% Pb along with negligible amount of several metals and ~45% of burned non-metal carbon. In order to recover metals from the pyrolysed sample, it was physically beneficiated to separate the metallic's from the non-metallic fraction. The sample received was screened and it was observed that the concentrate possessing >2 mm size contained ~65% metals along with polymers/epoxies, while ~45% carbonic ash along with ~38% metallic

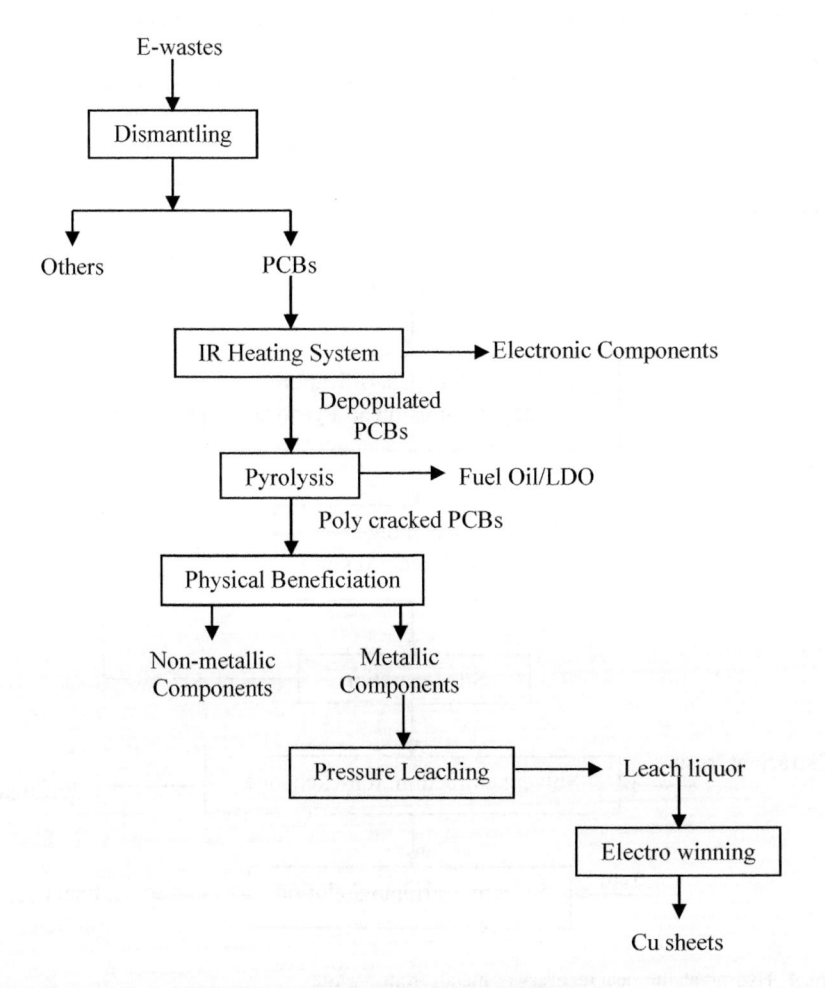

Fig. 3 Process flow-sheet to extract metals from pyrolysed PCBs [2, 3]

content and some plastic parts were separated in <2 mm size. Hydrometallurgical technique was further used to recover/separate metals from the low metal concentrate of pyrolysed PCBs [2, 3].

Hydrometallurgical Processes Developed for the Recovery of Metals from E-waste

Leaching of Pre-treated E-waste

Leaching process is employed for the selective extraction of metals after pre-treatment with application of different leachants viz. acidic, alkaline, and neutral type depending on the nature of the material to be leached, and the presence of other impurities which may affect the leaching process (Fig. 4). Mixed leachants of acidic/alkaline or acidic/salt may also be employed to investigate the leaching behaviour of metals. Jha et al. [5] carried out leaching studies to study the dissolution behaviour of various metals from PCBs with different acids viz. H_2SO_4, HCl, and HNO_3.

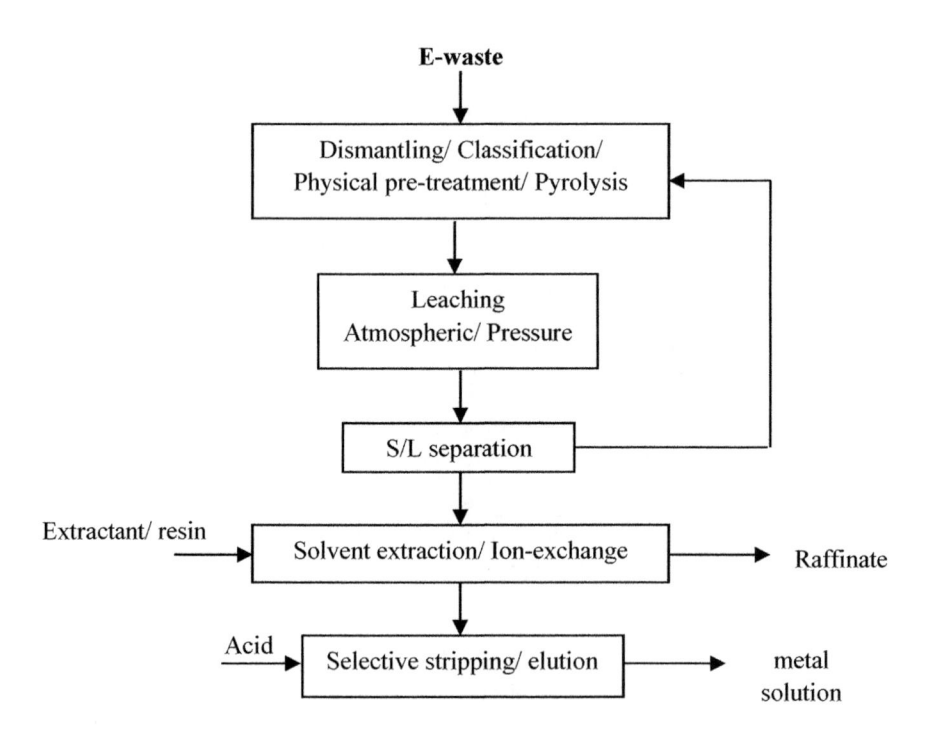

Fig. 4 Hydrometallurgical recovery of metals from e-waste

Sulfuric acid was not found to be a suitable reagent for the leaching of metals from PCBs even at elevated temperature and high concentration. Hydrochloric acid was also not suitable for the dissolution of copper; however, the dissolution of Sn was found to be satisfactory. Nitric acid was found to be a suitable reagent for the dissolution of most of the metals. The Fe and Ni present in the PCBs dissolved easily within 10 min of contact time, whereas 38% Pb leached out in 40 min. The dissolution of Sn was very poor with nitric acid even at high temperature with 6.0 M HNO_3 at S/L ratio 100 g/L and 90 °C. Around 99.99% Cu, Fe, and Ni were found to be leached along with 36.66% Pb. The NOx gas generated during leaching was absorbed in a suitable scrubbing solution. Further leach liquor was purified by solvent extraction, and from the purified solutions, salt/metals were obtained using suitable hydrometallurgical techniques [5].

Another hydrometallurgical recycling process for waste PCBs was also developed, which involved the novel pre-treatment consisting of organic swelling of PCBs to liberate thin layers of metals followed by sulfuric acid leaching of the metals so liberated. Leaching studies were performed for the recovery of Cu from the crushed and organic swelled liberated metals using sulfuric acid in the presence of hydrogen peroxide under atmospheric and pressure conditions [1]. The percentage recovery of copper was found to be 97.01% with addition of 15% (v/v) hydrogen peroxide keeping solid to liquid constant at 30 g/L. Jha et al. [6] and [7] also reported selective leaching of Pb (99.99%) from the liberated epoxy resin of waste PCBs using 0.2 M HNO_3 at 90 °C whereas 97.79% Sn was leached out from swelled and liberated epoxy resin using 4.5 M HCl at 90 °C in 60 min mixing time [6, 7].

Recovery of Metals from Leach Liquor of Metallic Concentrate

The solvent extraction (SX) processes were developed by Jha et al. [19] for the removal/recovery of hazardous metals from the complex leach liquor of electronic scraps following by recovery of valuable metals. As the leach solution of e-scraps contains various metallic constituents such as Fe, Cu, Zn, Cd, and Ni, different organic extractants viz. LIX 84, DEHPA, Ionquest 801, Cyanex 272, Cyanex 923, Cyanex 302, etc. diluted in kerosene have been used for metal separation and recovery. The solvent was modified with isodecanol to improve the phase separation. In order to extract Cu from the sulfate solution in continuous mode in MSU, basic studies have been made for the copper extraction at different pH, extraction isotherms, A/O ratio etc. using 5.0% LIX84 diluted in kerosene. Results were validated in mixer settler unit (MSU—having 620 mL mixing and 860 mL settling capacity) maintaining specific leach solution flow rate and A/O ratio. The results showed almost complete extraction (~97%) of Cu in three stages leaving minor metals Zn, Cd, and Ni in raffinate. The loaded Cu was completely stripped after scrubbing of minor metals using water with diluted sulfuric acid in two stages.

After extracting Cu, the SX study was carried out for the separation of Zn, Cd, and Ni from sulfate solution using different extractants. DEHPA (v/v) was found to be suitable for extraction of metals in comparison to Cyanex 923 and Cyanex 272 and the order of extraction with pH was found to be in the following sequence Zn > Cd > Ni. The kinetics of extraction showed that the equilibrium extraction of both Zn and Cd was attained in 2 min. The results showed the increase in Zn extraction at above 2.5pH. Cd was totally extracted even at lower pH without extracting Ni. The stripping of loaded Zn was carried out with 1.0% sulfuric acid at A/O ratio 1/1 maintaining a contact time of 5 min at room temperature whereas loaded Cd was effectively stripped with 10% hydrochloric acid at A/O ratio 1/1 [8].

Recovery of Metals from LIBs

Mobile phones constitute one of the major fractions of e-waste. These mobile phones consist of lithium ion batteries (LIBs), which are made up of rare and strategic metals such as Co, Li, Mn, Cu, Ni, Al, etc. These LIBs composed of an anode, a cathode, a separator, and an electrolyte. The black cathodic material present in the LIBs contains higher amount of Co (~20 wt% of LIBs) along with Mn, Li, Cu, Al, etc. The presence of ample metals in these discarded batteries has made their recycling very essential to not only cope up with the metals supply but also to comply with the regulations for the disposal of the used batteries. In view of the above, a hydrometallurgical flowsheet has been developed by Jha et al. [9, 10], and Dutta et al. [11] for recycling of spent LIBs to recover rare and strategic metals. Initially, the LIBs were crushed in a scutter crusher followed by physical beneficiation to separate the metallic fractions, plastics, and black cathodic powder [9–11].

Leaching of metals was carried out from the black cathodic material using suitable lixiviant at optimized conditions to dissolve maximum metals. The obtained leach liquor was further purified using precipitation followed by solvent extraction technique. Initially, the pH of the solution was maintained at 1.5 using sodium hypochlorite with constant stirring for 1 h in order to precipitate out Mn. This procedure was repeated thrice for complete removal of Mn. The leach liquor free from Mn was then mixed with 10% LIX 84IC maintaining O/A ratio 1:1 for 5 min to make the solution free from Cu, Fe, and Ni. It was found that ~99% Cu, Ni, and Fe were extracted in two stages at pH 4.5. Further, the loaded LIX 84IC was stripped using 10% H_2SO_4 to get the respective metal solution. Now, the leach liquor containing Co and Li was evaporated to increase the concentration of the metal ions in the solution. Reduced leach liquor containing Co and Li was again treated with 20% Cyanex 272 diluted in kerosene for 10 min maintaining O/A ratio 1:1. It was observed that at pH ~ 5, 98% Co was extracted leaving Li in the raffinate in two stages. The loaded Cyanex 272 was stripped thrice using 10% H_2SO_4 for complete back-extraction of Co from the loaded organic. Thus, 99% pure cobalt solution was obtained at above-described condition which can be further evaporated or electro-won to get Co salt or metal sheet, respectively. The raffinate containing Li was evaporated to get Li salt. The developed process flowsheet for recycling of spent LIBs to recover rare and strategic metals is presented in Fig. 5.

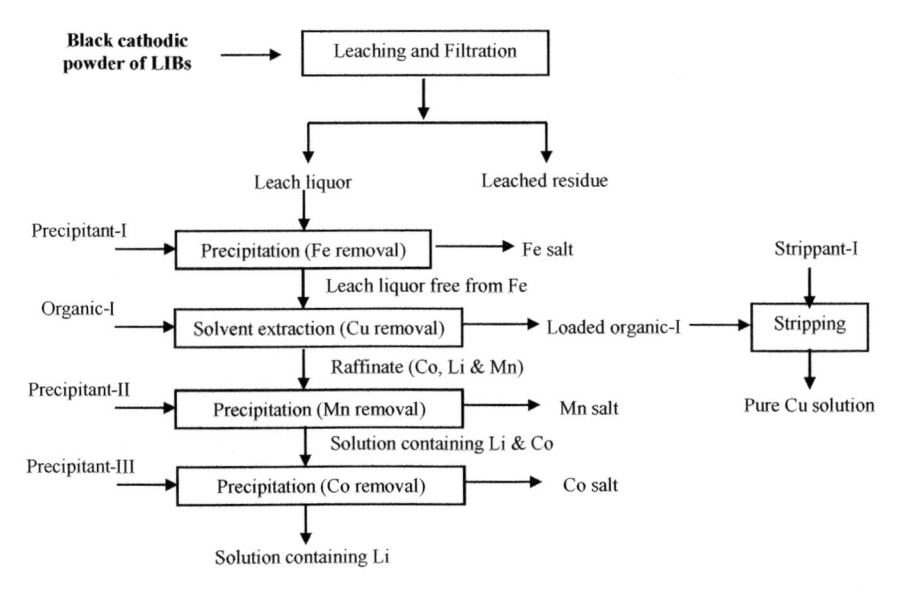

Fig. 5 Process flowsheet to extract metals from the black cathodic material of Li-ion batteries [9–11]

Recovery of Precious Metals from E-waste

A hydrometallurgical process has also been developed by Panda et al. [12] for the extraction of precious metals (Au, Ag, Pd, and Pt) from the e-waste. Initially, the e-waste was dismantled and depopulated to liberate all parts. Further, classification of parts containing precious metals was carried out. The parts containing precious metals were pretreated and physically beneficiated to get enriched metallic concentrate. Leaching of the metallic concentrate was carried out in a suitable lixiviant to get the precious metals into solution. The leach liquor obtained was further processed using standard separation techniques (solvent extraction/ion-exchange, precipitation, etc.) to get purified solutions of individual precious metals [12]. From the purified solutions, value-added marketable products of precious metals could be obtained using the method of evaporation/electrowinning techniques. The flowsheet to extract precious metals from e-waste is presented in Fig. 6.

Recovery of Metals from Scrap LCD Panels

Indium (In) is being widely used as indium tin oxide (ITO) on the glass substrate of the LCD (liquid crystal display) panel [13–15]. About 70–80% In is used in the LCD panel of different electronic goods but due to technological advancement, the lifespan of such electronic goods is decreasing day by day, which has resulted in the generation

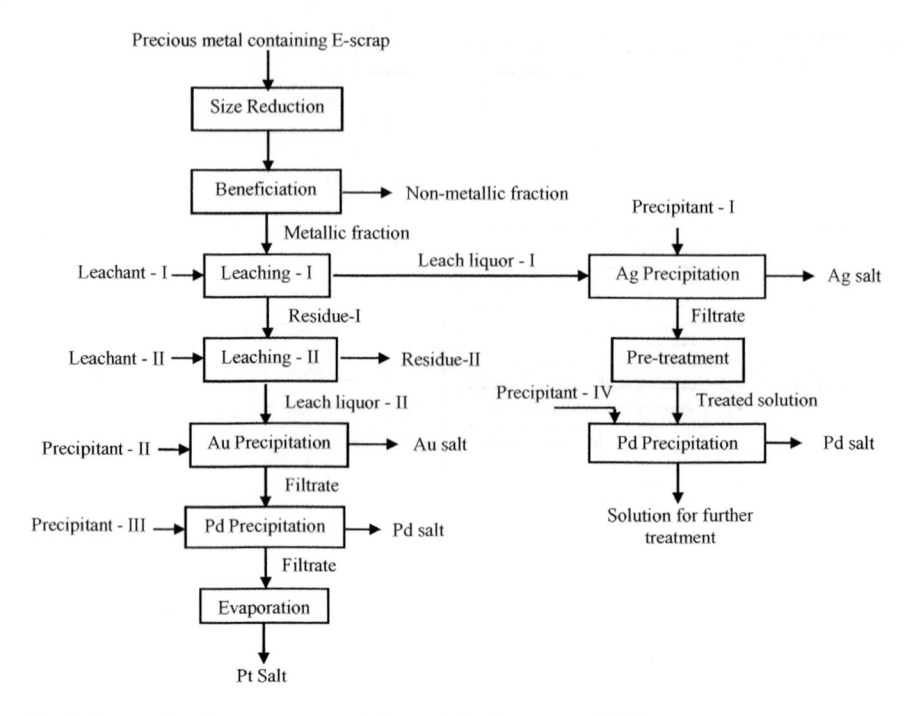

Fig. 6 Process flowsheet to extract precious metals from e-waste [12]

of massive amount of LCD panels [16]. In this regard, a hydrometallurgical flowsheet has been developed by Choubey et al. [17] to recover the indium from discarded LCD panels. Initially, the monitors of personal computers were dismantled to separate the LCD panels. Further, the LCD panels were pulverized and physically beneficiated to get enriched metallic fraction. Selective leaching of In was carried out from the metallic concentrate using different minerals acids [17]. Almost ~94.1, 93.8, and 93.2% In were leached using 1.0 M solution of HCl, H_2SO_4, and HNO_3, respectively at 60 °C in a reaction time of 75 min. Further, ion-exchange studies were carried out to selectively adsorb In from the leach liquor using Indion BSR resin. About 99.2% In got adsorbed onto the resin at pH ~ 1.7 in contact time of 60 min. Elution of the loaded resin was achieved using diluted H_2SO_4 in five consecutive contacts of 10 min each. From the purified solution of In, salt or metal could be produced by evaporation/electrowinning method. The complete process flowsheet is presented in Fig. 7.

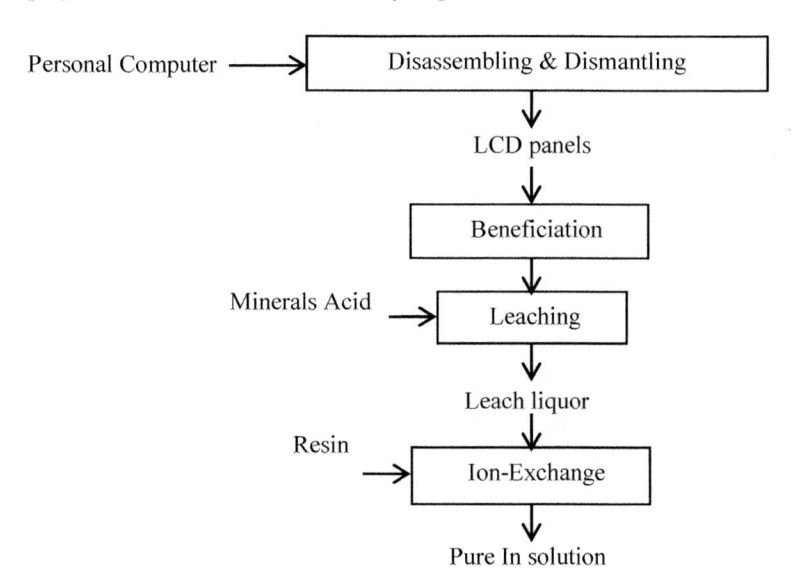

```
Personal Computer ──→ Disassembling & Dismantling
                              │
                              ↓
                         LCD panels
                              ↓
                        Beneficiation
                              ↓
       Minerals Acid ──→ Leaching
                              ↓
                         Leach liquor
            Resin ──→ Ion-Exchange
                              ↓
                        Pure In solution
```

Fig. 7 Process flowsheet for the recovery of In from scrap LCD panels of personal computers [17]

Recovery of Metals from Magnets and Tube Lights

E-waste can serve to be a potential source for REMs along with other valuable metals. About 25–30 wt% of Nd is found to be present in permanent magnets of computer hard disk whereas significant amount of Eu and Y is present in the phosphor powder of tube lights. Hydrometallurgical process has been developed by Kumari et al. [18] for the recycling of scrap magnets to extract Nd, the rare earth metal. Initially, the Nd–Fe–B magnets were demagnetized, crushed, and charged to a chemical leaching reactor. About ~99.99% Nd was leached along with Fe using 1.0 M H_2SO_4 at room temperature in a reaction time of 90 min and 50 g/L pulp density. The leach liquor obtained was subjected to acid extraction using organic extractant TEHA diluted in kerosene for 10 min and O/A ratio 2:1. Precipitation of Nd was carried out from the acid-free leach at pH 1.65 using ammonia solution. The Fe present in the solution was also recovered at pH above 3 [18]. Another flowsheet has also been worked out for the selective separation of Europium (Eu) and Yttrium (Y) from phosphor powder of obsolete tubelights leaving other REMs and impurities in the residue. Phosphor powder obtained from the scrap tubelights was analyzed and found to contain five REMs (2–8% La, 2–10% Ce, 0.5–5% Tb, 0.5–5% Eu, and 10–30% Y) with other metals like Calcium, Aluminum, etc. Depending on the composition of phosphor powder leaching was carried out using suitable lixiviant followed by recovery of REMs from the leach liquor using solvent extraction technique [19].

Conclusion

Varieties of E-wastes are the potential alternative resources to recover rare, rare earth, precious and valuable non-ferrous metals by using the developed hydrometallurgical processes flowsheets, which will conserve the natural resources, save the energy as well as make the environment cleaner and pollution free. Most of these resources can be treated by such approaches while designing treatment steps appropriately to achieve the optimum results with respect to process efficiency with high metal recovery, products in desired forms, and purity.

Acknowledgements The processes are developed under the joint Indo-Korean collaborations, Korean Internship programmes, support from Indian recycling industries and CSIR-New Delhi initiatives. The permission of Director, CSIR-NML, Jamshedpur, India to publish this paper is also highly acknowledged by authors.

References

1. Jha MK, Lee JC, Kumari A, Choubey PK, Kumar V, Jeong J (2011) Pressure leaching of metals from waste printed circuit boards using sulphuric acid. J Metals 63(8):29–32
2. Kumari A, Jha MK, Singh RP (2016) Recovery of metals from pyrolysed PCBs by hydrometallurgical techniques. Hydrometallurgy 165:97–105
3. Kumari A, Jha MK, Lee JC, Singh RP (2016) Clean process for recovery of metals and recycling of acid from the leach liquor of PCBs. J Clean Prod 112:4826–4834
4. Kumar V, Lee JC, Jeong J, Jha MK, Kim BS, Singh R (2013) Novel physical separation process for eco-friendly recycling of rare and valuable metals from end-of-life DVD-PCBs. Sep Purif Technol 111:145–154
5. Jha M K, Shivendra, Kumar V, Pandey B D, Kumar R, Lee J C (2010) Leaching studies for the recovery of metals from the waste printed circuit boards (PCBs). 139th Annual meeting and exhibition (TMS-2010). In: EPD Congress 2010, The Minerals, Metals and Materials Society 2010, pp 945–952
6. Jha MK, Choubey PK, Jha AK, Kumari A, Lee JC, Kumar V, Jeong J (2012) Leaching studies for tin recovery from waste e-scrap. Waste Manag 32(10):1919–1925
7. Jha MK, Kumari A, Choubey PK, Lee JC, Kumar V, Jeong J (2012) Leaching of lead from solder material of waste printed circuit boards (PCBs). Hydrometallurgy 121–124:28–34
8. Jha MK, Gupta D, Choubey PK, Kumar V, Jeong J, Lee JC (2014) Solvent extraction of copper, zinc, cadmium and nickel from sulfate solution in mixer settler unit (MSU). Sep Purif Technol 122:119–127
9. Jha MK, Kumari A, Jha AK, Kumar V, Hait J, Pandey BD (2013) Recovery of lithium and cobalt from waste lithium ion batteries of mobile phone. Waste Manag 33(9):1890–1897
10. Jha AK, Jha MK, Kumari A, Sahu SK, Kumar V, Pandey BD (2013) Selective separation and recovery of cobalt from leach liquor of discarded Li-ion batteries using thiophosphinic extractant. Sep Purif Technol 104:160–166
11. Dutta D, Kumari A, Panda R, Jha S, Gupta D, Goel S, Jha MK (2018) Close loop separation process for the recovery of Co, Cu, Mn, Fe and Li from spent lithium-ion batteries. Sep Purif Technol 200:327–334
12. Panda R, Jha M K, Dinkar O S, Pathak D D (2020) Reclamation of precious metals from small electronic components of computer hard disks. In: Azimi G, Ouchi T, Kim H, Alam S, Forsberg K, Baba A (eds) Rare metal technology 2020, pp 243–250

13. Amato A, Beolchini F (2018) End of life liquid crystal displays recycling: a patent review. J Environ Manag 225:1–9
14. Li RD, Yuan TC, Fan WB, Qiu ZL, Su WJ, Zhong NQ (2014) Recovery of indium by acid leaching waste ITO target based on neutral network. Trans Nonferr Metals Soc China 24:257–262
15. Rocchetti L, Amato A, Fonti V, Ubaldini S, Michelis ID, Kopacek B, Veglio F, Beolchini F (2015) Cross-current leaching of indium from end-of-life LCD panels. Waste Manag 42:180–187
16. Gotze R, Rotter V S (2012) Challenges for the recovery of critical metals from waste electronic equipment—a case study of indium in LCD panels. In: Proceedings of the conference Electronics Goes Green (EGG), IEEE, Berlin, 2012
17. Choubey P K, Jha M K, Gupta D, Jeong J, Lee J C (2014) Recovery of rare metal indium (In) from discarded LCD monitors. In: EPD Congress 2014, San Diego, California. USA, p 39
18. Kumari A, Jha MK, Pathak DD (2020) An innovative environmental process for the treatment of scrap Nd-Fe-B magnets. J Environ Manage 273:111063
19. Jha M K, Jha A K, Hait J, Kumar V (2014) A process for the recovery of rare earth metals from scrap fluorescent tubes. Patent No 0280DEL2014, India

Rare Earth Elements Extraction from Coal Waste Using a Biooxidation Approach

Prashant K. Sarswat, Zongliang Zhang, and Michael L. Free

Abstract Rare earth elements (REE) are moderately abundant and are presently extracted from limited resources of monazite, bastnaesite, and loparite minerals and ionic clays. Other potential sources include large coal resources in active coal mines and existing coal waste dumps. A technology has been proposed and demonstrated that can be used to deliver clean coal for the market as well as REE-bearing non-coal material that is concentrated in REE content and suitably sized for heap leaching. Additionally, a separate stream of concentrated sulfide minerals can be produced from mid to high sulfur coals suitable for the bio-oxidation production of a lixiviant suitable for leaching REE from the non-coal rock in a heap leach setting. The removal of the sulfide minerals cleans the coal, accelerates subsequent REE extraction, and eliminates the future potential for acid-rock drainage. For cost-effective enhanced leaching, bio-oxidation has been used that has been applied to coal-based materials. During bioleaching, Fe^{3+} ions generated from bioleaching oxidize sulfide minerals such as pyrite, and subsequent production of acid. These two species (acid and Fe^{3+} ions) are key drivers for REEs dissolution, as well as residual sulfides removal, thereby controlling future acid mine drainage and related liabilities. This paper discusses some associated results acquired for the proposed process.

Keywords Rare earths · Coal refuse · Bioleaching

Introduction

The REE resources in coal-related materials are vast. Assuming a coal production rate of 544 Mt/year with an average REE content of 200 ppm, the potential REE production rate is 109,000 t/year, which is a large fraction of the annual global production of REE. Most of those coal-related REE resources are found in association with the gangue or ash-based content of the coal ore. In the current research, the authors have identified at least four separate coal resource samples from various coal

P. K. Sarswat (✉) · Z. Zhang · M. L. Free
Department of Materials Science and Engineering, University of Utah, Salt Lake City, UT 84112, USA
e-mail: saraswatp@gmail.com

Table 1 Mass mean size of coal refuse samples

	CR-A	CR-B	CR-C1	CR-C2	CR-D1	CR-D2
Mass mean size (mm)	14.1	11.1	12.9	12.9	12.4	13.6

basins that are targeted to meet a 300 ppm REE minimum feedstock level. These four sites, and other sites operating in similar coal seams, were prioritized in an earlier site identification phase and are located in Virginia, Pennsylvania, and West Virginia. The coal waste samples were hydrometallurgically tested in order to recover rare earth elements.

Coal Ore Samples Crushing, Splitting, and Screening Process

The acquisition of coal refuse samples from the four prioritized sites was completed using automated cross belt samplers following appropriate ASTM protocols [1]. After receiving the coal waste samples, all samples were prepared using a standardized crushing, splitting, and screening process. Samples occupying multiple barrels were first homogenized, and, each sample was next crushed to −1 in. [1]. Following the initial crushing, several representative splits of each sample were obtained for preliminary testing and characterization, while the remaining −1-in. material was sealed and stored. The representative splits were obtained by repeatedly coning and quartering each barrel sized sample. One representative split of each sample (approximately 3–4 kg) was used for size and ash-by-size analysis. Ash was determined using a LECO 701, and the sieve sizes used in the size analysis included: 1", ½", ¼", and 6 meshes. The coarse sizing distribution data (−1 in.) were further assessed using the Gates–Gaudin–Schuhmann equation [2]. Table 1 summarizes the measurement of sizes for different samples.

Leaching

Leaching [3] tests (see details in next paragraph) and digestion tests were conducted for the six samples (see Table 1). As well as leaching tests, digestion tests [4] were conducted using aqua regia followed by ICP analysis after dilution [5] to determine the total REE content. It can be seen that the content of Pr was highest among all REEs in each of the coal waste samples (see Fig. 1). Ce was found in some of the samples and the content was highest for CR-C1 (~6 ppm of coal waste).

For leaching these coal waste samples, the authors used 10 g/l ferric sulfate solution in water using 30 wt% coal waste in the leach slurry. Most of the experiments were conducted at a temperature of ~30 °C. The slurry was continuously stirred for

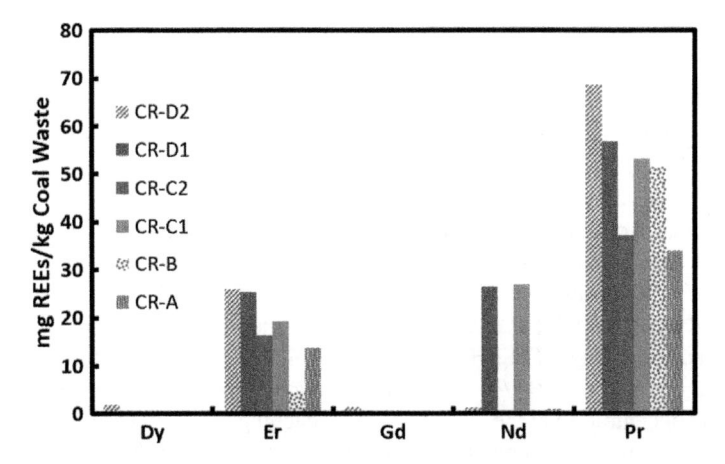

Fig. 1 Rare earth elemental analyses of coal waste sample acquired (see Table 1 for sample details). (Color figure online)

105 h. The pH was adjusted to ~1.5 after every 8 h using sulfuric acid. It was observed that acid consumption (over 25 h duration) was highest for the CR-A sample(~31 kg/t of coal waste), although during the initial leaching period of 12 h the consumption was similar for each sample. Acid consumption was lowest (over 25 h duration) for CR-B sample (16 kg/t of coal waste). It was also observed that final pH was highest for CR-A sample (pH = 1.98). Note that the mass mean particle size was highest for CR-A coal waste sample, which could explain the relatively high acid consumption. The sulfur content for CR-A sample was also lowest (~0.12%). After leaching, elemental analysis was performed on the solutions using ICP that suggested that the recovery of Pr was higher than other REEs.

Bioleaching

Bioleaching or bio-oxidation is one of the less expensive methods among several methods of leaching [1]. In case of metal extraction, most of the microbial oxidations are based on chemical degradation that is enhanced and supported by activity of microorganism. A representative example in present context is the bacterial oxidation of Fe^{2+} ions to Fe^{3+} ions, followed by the Fe^{3+} ions oxidizing minerals to dissolve metals. Hence, bio-oxidation is often indirectly conducted by the supply of oxidant provided by the microorganisms. In case of indirect bioleaching of sulfide minerals, dissolution can be controlled to favor the production of elemental sulfur or sulfate/sulfuric acid. Although the most common method of accelerated leaching using bacteria involves ferrous ion oxidation, there is significant evidence that microorganisms also enhance leaching by oxidizing elemental sulfur. Pyrite leaching is limited by a combination of microbial activity, cell population density, mass transport, and electrochemical kinetics. A common reaction for pyrite leaching where ferric ions formed during bio-oxidation are instrumental in oxidizing pyrite (which produces acids), can be written as [1]:

$$8H_2O + FeS_2 + 14Fe^{3+} = 15Fe^{2+} + 2SO_4^{2-} + 16H^+$$

Both Fe^{3+} ions and acidic media are useful for REEs dissolution, just as for evacuating lingering sulfides, consequently forestalling future corrosive mine waste and related concern. The most microorganisms associated with mineral bioleaching are Acidithiobacillus acidophilus, Acidithiobacillus thiooxidans, Leptospirillum ferrooxidans, Acidithiobacillus ferrooxidans, and Sulfolobus-like bacteria. In our case, for leaching operation, a 9K medium was supplied in every 3 h with a flow rate equivalent to a 7.5-day reactor (1000 ml) residence time. Additional tests have been conducted using daily input of pure pyrite is 15 g and the resident time of solution in the bioreactor is 5 days. The pH of the effluent solution is about 1.3 with a $Fe_2(SO_4)_3$ concentration of about 10 g/L. In this case, we achieve ~4000 ppb of total rare earth elements concentration at the end of ~40 h of leaching. Note that bio-leaching to get concentrated rare earth elements in solution is underway—in the present paper, we just show the extraction feasibility from a ferric sulfate leached solution where the initial concentration of rare earth elements was in the range of 5–50 ppm.

Solvent Extraction of Pregnant Solution from Sample CR-A

For solvent extraction, the authors used di-(2-ethylhexyl) phosphoric acid (D2EHPA) [6]. A typical solvent extractant (SX) contains 10% D2EHPA in kerosene. The authors used 10 mL of 10% DEHPA organic to contact 30 mL of leach solution with mixing performed until a condition of equilibrium using a shaker (300 rpm).

The University of Utah team performed precipitation tests on strip solution from the CR-A sample. The pH value was adjusted to 3.75 using 1.0 M NaOH solution and was maintained for ~12 h with continuous stirring to remove iron. Once precipitation was completed, filtration was performed. It can be seen (see Fig. 2) that precipitate color is yellow-orange, that indicates a significant presence of the iron in sample.

Precipitation

Precipitation tests have been performed for the solutions that were collected from each column after appropriate time of leaching. In one case, precipitation test was conducted using magnesium carbonate while in different test sodium carbonate was used. It was noted that the key relevant species for rare earth elements and iron in the acidic sulfate-based solutions used for leaching are the soluble sulfate complexes such as $FeSO_4^+$, $DySO_4^+$, etc. and the dissolved metal ions such as Fe^{3+}, Dy^{3+}, etc. Visual Minteq assisted speciation diagram (fraction of total element vs. pH) indicates that most of the iron get precipitated for pH greater than 4.0. These precipitation events can be helpful in selectively removing iron compounds such as goethite, ferrihydrite, jarosite, hematite, and magnetite. The saturation index-based calculations show that hematite begins to precipitate at about pH 2 when there is an equal ratio of ferrous and ferric iron in solution, whereas the pH decreases to about 1.8 when the ratio of ferric to ferrous iron is increased to 94% ferric iron and 6% ferrous iron. Our precipitation tests

Fig. 2 Image of leached CR-A sample. **a** After precipitation and **b** filter cake. (Color figure online)

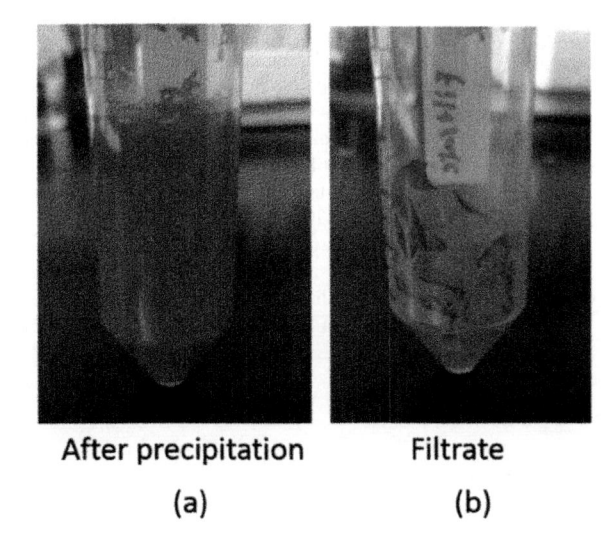

After precipitation
(a)

Filtrate
(b)

(conducted using technique of pH adjustment by adding magnesium carbonate or calcium hydroxide) indicate that most of the precipitation occurs at up to pH ~3.82 or pH ~5.50 for the case of magnesium carbonate or calcium hydroxide, respectively, in room temperature condition. However, when temperature increases, the precipitation occurs even at lower pH for both cases. The precipitate that formed was analyzed to determine if rare earth elements were precipitated and to determine the compound that formed. Inductively Coupled Plasma Mass Spectroscopy (ICP-MS) analysis of the dissolved precipitate indicated that the rare earth content was ~2.9 ppm, hence additional tests have been conducted to recover most of the REEs in solution.

Summary

The authors have examined coal-based resources to determine their potential to yield rare earth elements. The coal waste samples were prepared after appropriate crushing, grinding, sizing, and splitting steps then analyzed for REE content and leached in a ferric sulphate solution. The results suggest that REEs, most notably praseodymium (Pr), can be extracted from selected coal waste samples. The REEs were recovered after leaching, solvent extraction, and precipitation steps.

References

1. Sarswat PK, Leake M, Allen L, Free ML, Hu X, Kim D, Noble A, Luttrell GH (2020) Efficient recovery of rare earth elements from coal based resources: a bioleaching approach. Mater Today

Chem 16:100246. https://doi.org/10.1016/j.mtchem.2020.100246
2. Petrakis E, Stamboliadis E, Komnitsas K (2017) Evaluation of the relationship between energy input and particle size distribution in comminution with the use of piecewise regression analysis. Part Sci Technol 35(4):479–489. https://doi.org/10.1080/02726351.2016.1168894
3. Free ML (2013) Hydrometallurgy: fundamentals and applications. Wiley
4. Rodchanarowan A, Sarswat PK, Bhide R, Free ML (2014) Production of copper from minerals through controlled and sustainable electrochemistry. Electrochim Acta 140:447–456. https://doi.org/10.1016/j.electacta.2014.07.015
5. Balcerzak M (2002) Sample digestion methods for the determination of traces of precious metals by spectrometric techniques. Anal Sci 18(7):737–750
6. Barnes JE, Setchfield JH, Williams G (1976) Solvent extraction with di (2-ethylhexyl) phosphoric acid; a correlation between selectivity and the structure of the complex. J Inorg Nucl Chem 38(5):1065–1067

Scandium Extraction from Bauxite Residue Using Sulfuric Acid and a Composite Extractant-Enhanced Ion-Exchange Polymer Resin

Efthymios Balomenos, Ghazaleh Nazari, Panagiotis Davris, Gomer Abrenica, Anastasia Pilihou, Eleni Mikeli, Dimitrios Panias, Shailesh Patkar, and Wen-Qing Xu

Abstract This work presents the results of scandium extraction from Greek Bauxite Residue (BR) using sulfuric acid as the leaching agent and a composite extractant-enhanced ion-exchange resin for a new novel, selective-ion recovery (SIR) process developed by II–VI. The BR produced in Mytilineos's plant contains approximately 75–130 mg/kg of Sc and given the plant's current production capacity, more than 100 t of Sc are discarded each year within the BR stream. The optimum conditions for selective Sc extraction from BR were determined at lab scale in conjunction with the efficiency in Sc uptake by the resin. Under the SCALE research project, a

---

E. Balomenos (✉) · P. Davris
Research and Sustainable Development, Mytilineos SA, Metallurgy BU, Ag. Nikolaos Plant, 320 03 Viotia, Greece
e-mail: thymis@metal.ntua.gr; efthymios.balomenos-external@alhellas.gr

G. Nazari
Director of Technology, Lt 6 Blk 1 Phase 2 Cavite Economic Zone, Rosario, Cavite, Philippines
e-mail: Ghazaleh.nazari@ii-vi.com

G. Abrenica
Technical Manager, Lt 6 Blk 1 Phase 2 Cavite Economic Zone, Rosario, Cavite, Philippines
e-mail: gomer.abrenica@ii-vi.com

A. Pilihou · E. Mikeli · D. Panias
Laboratory of Metallurgy, National Technical University of Athens, Heroon Polytechniou 9, 157 80 Zografos Athens, Greece
e-mail: anastasiapilihou@gmail.com

E. Mikeli
e-mail: elenamikk@gmail.com

S. Patkar
Director of Strategic Technology Planning and Commercialization, 375 Saxonburg Blvd, Saxonburg, PA 16056, USA
e-mail: shailesh.patkar@ii-vi.com

W.-Q. Xu
General Manager of Platform Technology Development and Incubation, 375 Saxonburg Blvd, Saxonburg, PA 16056, USA
e-mail: Wen-Qing.Xu@II-VI.com

BR leaching pilot plant (Mytilineos) and Sc extraction pilot plant (II–VI) have been established and operated in Mytilineos's plant to demonstrate this process.

Keywords Red mud · Bauxite residue · Scandium · Recovery of scandium · Solvent impregnated resin · SIR · Composite-enhanced extractant polymer resin

Introduction

"Bauxite Residue" (BR) refers to the insoluble solid material, generated during the extraction of alumina (Al_2O_3) from Bauxite ore using the Bayer process. When bauxite ore is treated with caustic soda, the aluminum hydroxides/oxides contained within, are solubilized, with approximately 50% of the bauxite mass being transferred to the liquid phase, while the remaining solid fraction constitutes the bauxite residue. It is estimated that for each ton of alumina produced 0.9–1.5 tons of solid residue (on a dry basis) is generated depending on the initial bauxite ore grade and alumina extraction efficiency [16]. Bauxite residue contains various major (gr/kg) metal oxides of Fe, Al, Ti, Ca, Si, Na, as well as minor (mg/kg) metal oxides like V, Ga, REE/Sc and others (depending on the initial chemical composition of the bauxite ore) along with inclusions of unwashed sodium aluminate solution.

The worldwide typical concentration of REE in BR is 800–2500 mg/kg and is related to the initial bauxite ore and the operation conditions of the Bayer process [7]. Typically REE are hosted at alumina bearing minerals of the bauxite ore, that are dissolved during the Bayer process; the REE contained are incorporated into secondary precipitation phases known as "desilication products—DSP," a mineral matrix that contains a mixture of Fe, Ti, Si, Al Ca, and Na ions [18].

Scandium often differs from the other REE mineral behavior; especially in laterite bauxites and their corresponding BR, it is often correlated with iron and titanium and zircon minerals [11, 18, 23], which for the most part are unaffected through the Bayer process. This is also confirmed by the laterite deposits in Australia and the Greek BR [5] where the main mineral, with high concentration of Sc, is goethite [19]. However, there are cases of BR, where Scandium is found to be related to larger extent to the soluble Al-bearing minerals, as reported by researchers [17]. It has been estimated that 70% of the world's Sc resources might be found in bauxite minerals and bauxite residue [15].

Direct leaching of BR with mineral acids [1, 4, 10, 20], requires significant acid consumption, as BR is by nature highly alkaline, and results in leach solutions with low Sc concentrations (< 20 mg/l), while the major BR metals such as Fe, Al, Ti, Ca, Na, and Si dissolve extensively and are thus found in concentrations of several g/L. This complicates the Sc extraction and refining from such solutions. Applying selective leaching on BR is the first step to generate a REE solution with significantly lower concentration of unwanted co-dissolved metals. Mineral acid-leaching performance for the Greek BR in terms of increasing selectivity of Sc against Fe which usually is the main BR component follows the trend of $HNO_3 > H_2SO_4 > HCl$

[4]. In general recovery, rate of Sc during selective leaching is limited in the range of 20–50% with Fe dissolution <3% [2, 3, 14, 20]. In most cases, a relatively limited selective dissolution against Fe, Ti, and Si can be achieved but not against readily soluble Ca, Al, and Na. As this work focuses on developing an industrial operation for Sc extraction, H_2SO_4 was selected as the acid to be used based on its selectivity in Sc extraction, cost, and ease of handling.

II–VI, headquartered in Saxonburg, Pennsylvania, has research and development, manufacturing, sales, service, and distribution facilities worldwide. The Company produces a wide variety of application-specific photonic and electronic materials and components and deploys them in various forms. The company has developed the II–VI Selective-Ion Recovery (SIR) Technology includes the use of a composite extractant-enhanced ion-exchange resin to extract scandium (Sc) from acidic solution or slurries, and its subsequent recovery as a Sc concentrate [21, 22]. The factors that contribute to its performance have been studied to a great extent using Pregnant Leach Solution (PLS) from various Sc-containing sources. The leaching conditions of BR were evaluated and optimized to improve recovery of Sc using this technology.

The earlier exploratory leaching experiments [9, 12, 13] resulted in conclusions that (i) the dissolution of impurities increases with increasing acid concentration during leaching and (ii) the solubility of silica in solution increases as acidity decreases, leading to silica gel formation over time. For proper operation of ion exchange columns, it is important to ensure that there is no precipitation in the column to avoid pressure drop and fouling in the column. Therefore, to study the impact of each operating condition on the effectiveness and feasibility of the SIR process, it was decided to carry out two campaigns—low-acid leaching and high-acid leaching, in order to find the optimum conditions to maximize loading of Sc while maintaining steady operation of SIR. Leaching experiments were carried out at the National Technical University of Athens (NTUA) and II–VI and ion-exchange lab tests were carried out at II–VI. Based on the results, the optimum leaching conditions were established and upscaled to the industrial pilot at MYTILINEOS alumina refinery.

Experimental

Lab Experiments

Leaching experiments, using an aqueous solution of sulfuric acid as leaching agent, were conducted in glass reactors with mechanical stirring. The temperature (85–95 °C) was adjusted with thermocouple sensors immersed into the liquid and attached to the controller respectively.

After leaching, the slurry was filtered using glass fiber acid-resistant filters of 0.6 μm pore size and the PLS was further diluted for chemical analysis. The metals in the solution were measured by ICP-OES and AAS.

The PLS was delivered through a silicon tube to the column using a peristaltic pump at flowrate of 2 bed-volumes per hour (BV/hr) until the exhaustion point of the resin for Sc loading was attained.

All samples were analyzed for Sc and impurities content using ICP-OES. The determination of free acid was done by titration and pH by using TPS Aqua pH.

Pilot Plant

The MYTILINEOS acid leaching pilot plant consists of a series of PP (Polypropylene) reactor tanks, at 500 L capacity, with mechanical steering and heating/cooling through immerged coils. Filter pressed BR produced at Mytilineos alumina refinery is mixed with industrial water in the first reactor (100-TK-10) to produce a pulp of specific density measured through an inline Coriolis Mass Flow Meter. The pulp is pumped to the second reactor (100-TK-30) where it is heated and contacted with concentrated sulfuric acid. The leaching takes place at 85 °C with a retention time of 30 min. The pulp exiting the 100-TK-30 is driven to the cooling tank (200-TK-40) where it is cooled to 60 °C and is subsequently passed to the filter press circuit. The filter press separates the solids from the liquids, generating the final PLS to be used for the SIR process. Operating at final pulp containing 40% solids to liquids (wt/vol), in a single shift operation, the pilot plant can process up to 1000 kg of pulp containing 300 kg of BR. The PLS produced was analyzed using ICP-OES.

Materials

In all experiments, BR filter cake from the Mytilineos's Alumina Refinery was used. The chemical composition of the BR as determined by alkaline fusion and wet chemical analysis by ICP-OES and AAS at NTUA, is shown in Table 1.

Table 1 BR's chemical composition

wt%							mg/kg			
Fe_2O_3	Al_2O_3	SiO_2	CaO	Na_2O	TiO_2	LOI	Ce	La	Y	Sc
39.16	16.53	9.90	8.40	3.46	4.67	10.10	657	110	132	71

Results and Discussion

Preliminary Leaching Campaigns

The leaching tests conducted initially at NTUA focused on leaching at elevated temperatures of 85–95 °C at solution of 3 M and 1 M H$_2$SO$_4$, and pulp densities of 20%.

The results for Sc recovery at 3 M H$_2$SO$_4$ 85 °C are shown in Fig. 1. At the high sulfuric acid concentration leaching conditions, the Sc recovery is between 70 and 95% following the same leaching trend with Fe, while the Si recovery is extremely low as it precipitates at high acid concentration and elevated temperatures [6]. Titanium is dissolved massively initially but with time its concentration decreases from 23 to about 0.5 g/l in 4 days retention time, as it undergoes hydrolysis at the elevated temperature [8]. Iron and aluminum on the other hand remain in solution at high concentrations (49.7 and 9.7 g/l respectively), even after 4 days of retention time.

In order to reduce the concentration of impurities in the pregnant solution, leaching tests were then carried out at 1M H$_2$SO4 and 95 °C. As shown in Fig. 2, under these conditions, the Sc recovery is between 48 and 55%, resulting in Sc concentration ~8 mg/L in the PLS. Ti and Fe recovery are below 5%, and similarly Ti precipitates out of the solution over time. However, Si dissolution is substantial and generates solutions with more than 4 g/l dissolved Si. Extending the retention time to 24 h and maintaining the elevated temperatures allows the precipitation of Si.

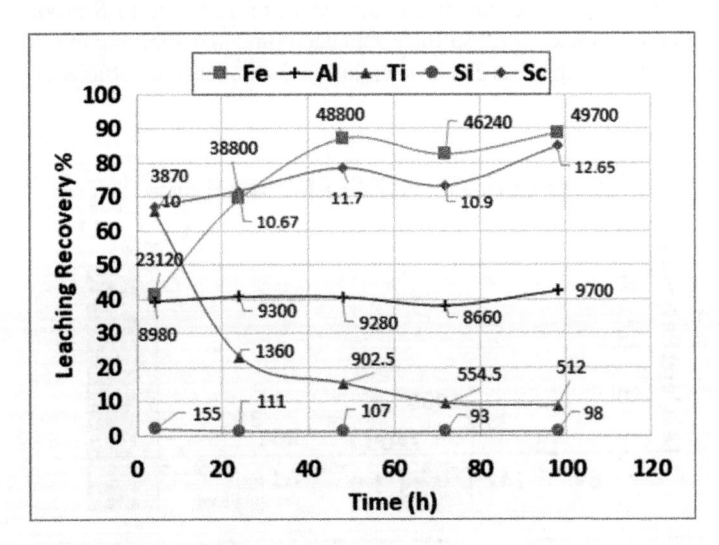

Fig. 1 Leaching yields per element at 3M H$_2$SO$_4$, 20 pulp density, 85 °C for different retention times. The respective data labels present the concentration of each element expressed in mg/L of PLS. (Color figure online)

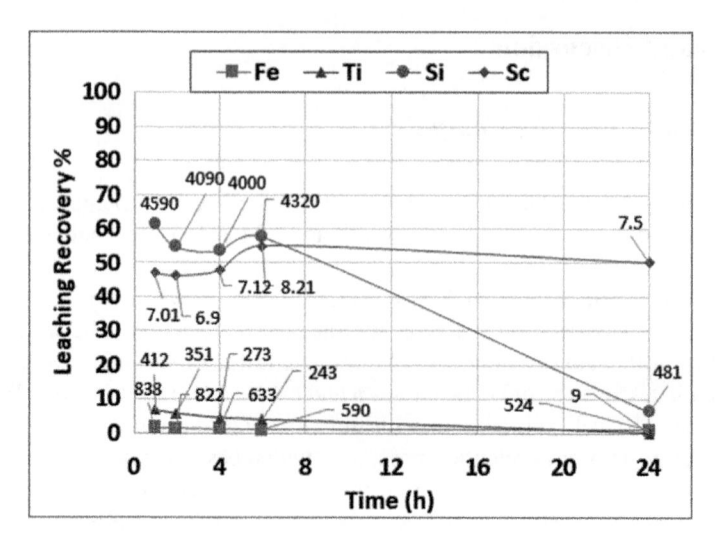

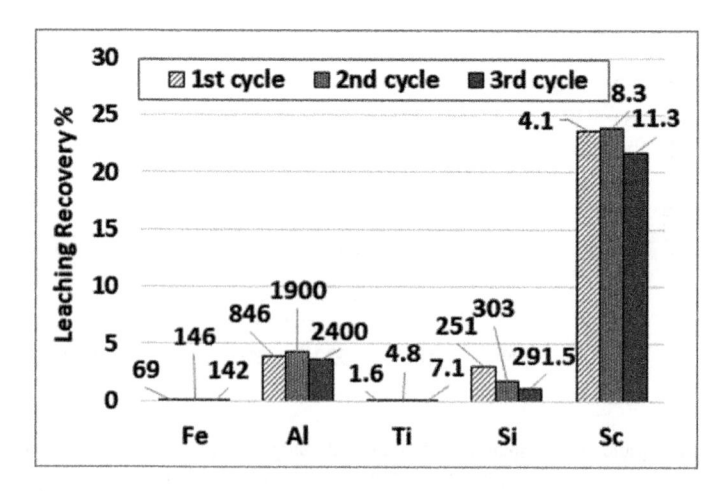

Fig. 2 Leaching yields per element at 1M H_2SO_4, 20 pulp density, 95 °C for different retention times. The respective data labels present the concentration of each element expressed in mg/L of PLS. (Color figure online)

Then in accordance with the results of the SIR resin presented below, a third option was also studied. In order to achieve higher Sc concentration with minimum impurities, consecutive leaching was examined at 0.4M H_2SO_4, 85 °C and 1 h retention time. Three leaching cycles were conducted, re-using the PLS produced with addition of concentrated 39 g/l sulfuric acid each time and applying this solution in an un-leached BR sample. As can be seen in Fig. 3, the leaching efficiencies in each

Fig. 3 Leaching yields per element at 0.4 M H_2SO_4, 20% pulp density, 85 °C, 1 h retention time. The respective data labels present the concentration of each element expressed in mg/L of PLS. (Color figure online)

cycle degrade slightly with Sc, Al and Ti exhibiting the highest relative concentration increase from first cycle to second and from second to third.

Effect of Impurities on the Performance of the II–VI SIR Resin

Due to low level of Sc in the PLS, it was important to reduce the impurity levels that would compete with Sc to load on the resin. The presence of Si, on the other hand, while it may precipitate on the resin and inhibit efficient adsorption to some degree, is more of a concern for the efficient functioning of the SIR columns.

High-Acid Concentration Leaching of BR—Campaign 1

The PLS prepared by leaching BR at initial H_2SO_4 concentration of 3M at 85 °C for 1 h and was contacted with the resin in the column to study the adsorption behavior of Sc. As shown in Fig. 4, the loading of Sc onto the resin from the PLS was low at about 60 mg/L resin due to significantly higher concentration of impurities relative to Sc.

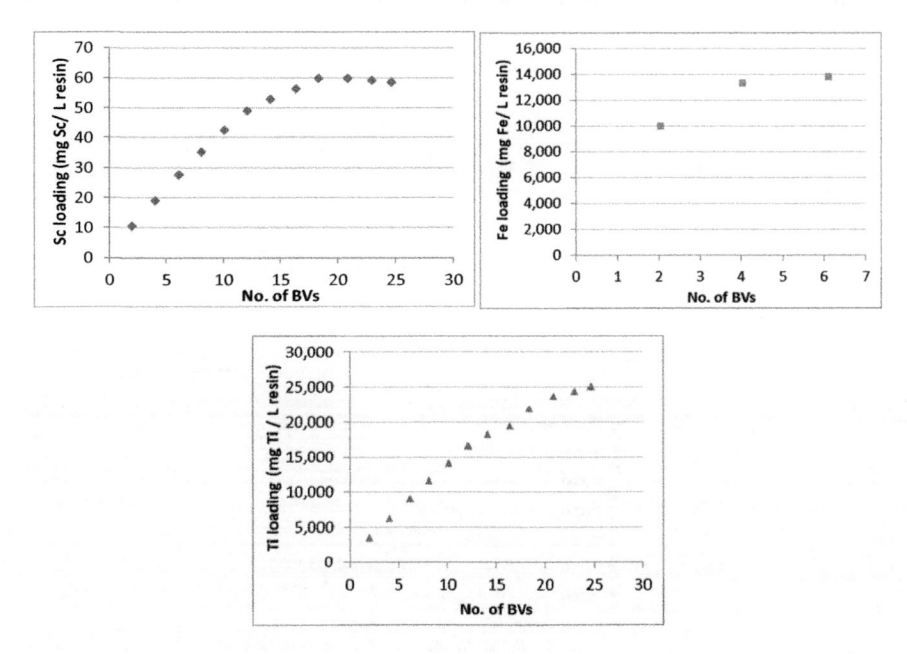

Fig. 4 Loading of (top left) Sc, (top right) Fe and (bottom) Ti onto the SIR resin from Campaign 1 PLS. (Color figure online)

Low-Acid Concentration Leaching of BR—Campaign 2

Low acid leaching tests (1M H_2SO_4) were also carried out at 85 °C and 4 h retention time with the aim of controlling the level of impurities in the PLS. The PLS was contacted with the SIR resin in a column to study the adsorption behavior of the critical metal ions. The high Si concentration (>4000 mg/L) led to silica gel precipitation in the column and created pressure drop and fouling. As discussed above, Si and Ti can be removed from the solution by extending the retention time of the leaching at elevated temperatures. Therefore, a second test carried out by leaching BR at at 85 °C with 16 h retention time, producing a PLS with less than 800 mg/L of Si and 43 mg/L of Ti. The PLS produced from the 16 h leaching process was fed to the SIR column. The loading profile of Sc is shown in Fig. 5. Although Si precipitation and gel formation were not observed in this test due to low level of silica, no significant improvement in the Sc loading was achieved relative to the PLS with high Si and Ti.

It was hypothesized that the low Sc loading could be due to extremely low level of Sc in the PLS (~8 mg/L in both cases), which reduces the driving force for loading the Sc in the presence of competing high levels of impurities. To evaluate this hypothesis, a known amount of Sc was added to the 16 h PLS to spike Sc concentration to 60 mg/L. The PLS was then fed through the SIR resin. In Fig. 5, at the exhaustion point, the loading capacity of the SIR for Sc was calculated at 2,640 mg/L, which indicated that increasing the Sc concentration in solution favored the Sc loading over the other cations even in the presence of high level of impurities.

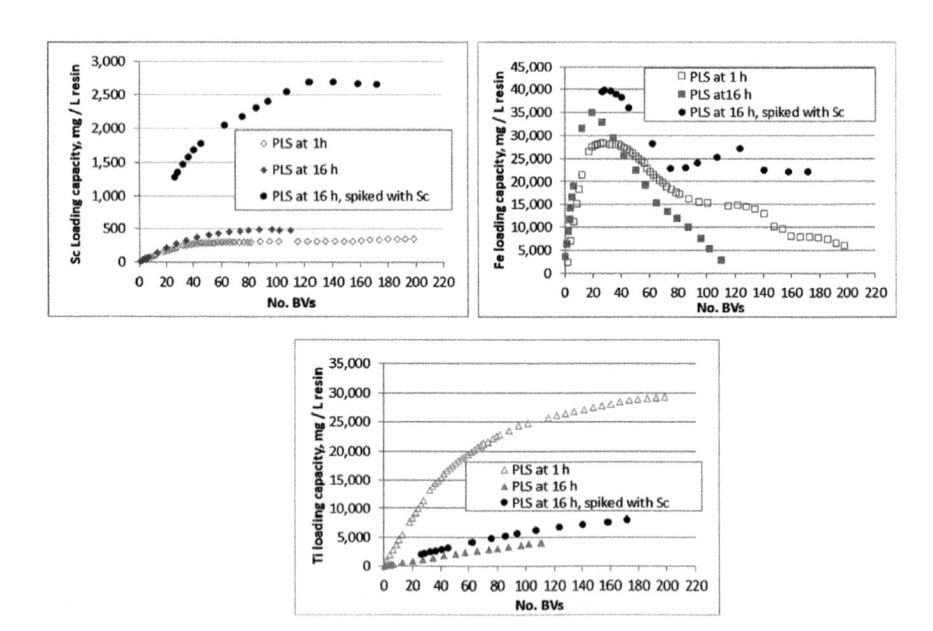

Fig. 5 Loading of (top left) Sc, (top right) Fe and (bottom) Ti into the SIR from Campaign 2 PLS. (Color figure online)

Scandium Extraction from Bauxite Residue ...

Table 2 Composition of the PLS from optimum conditions

Concentration in PLS, mg/L	Cycle 1	Cycle 2
Sc	8.8	8.2
Fe	98	95
Ti	5.9	5.5
Si	213	217

Optimum Leaching Conditions—Lower Than 1M H$_2$SO$_4$ Leaching of BR

Considering the results of Campaign 1 and 2, the optimum leaching conditions were identified to reduce impurity levels while maintaining a minimum Sc concentration of 8 mg/L. The BR was leached at 45 g/L H$_2$SO$_4$, which resulted in a PLS with low values of Sc, Fe, and Ti. In order to increase Sc concentration, the filtrate was recycled as starting solution while acid concentration was adjusted for the next batch of BR for leaching. To avoid problems with silica gel formation and possible Sc precipitation, the PLS was acidified prior to loading on the SIR resin.

The PLS produced under optimum conditions proceeded to the SIR process for Sc recovery (Table 2). The loading capacity increased to 5,000 mg/L resin as shown in Fig. 6, which was significantly higher than any previous tests that were conducted. After the first cycle, the resin was eluted, regenerated, and recycled. Figure 6 elucidates that the resin was efficiently recycled by the second cycle and the results were repeatable.

The resin from the optimum leaching conditions experiment was washed and eluted. The recovery was determined to be greater than 98%. The crude Sc concentrate produced contained about 25 wt% Sc, 5%wt Ti, and 0.1 wt% Fe as hydroxides.

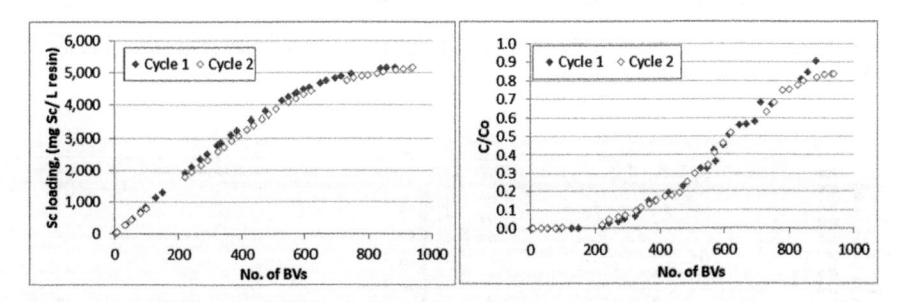

Fig. 6 Comparison between first and second loading cycles using PLS from the proposed leaching process, (Left) Sc loading curve, (Right) Sc adsorption isotherm. (Color figure online)

Pilot Plant

To achieve the production of the optimum PLS at industrial conditions, MYTILINEOS set up an acid leaching pilot shown in Fig. 7. Aiming at increasing productivity, the pilot plant was operated at a solids to liquids ratio of 0.4 (kg solids dissolved in 1 L of liquid) or at 0.3 kg solids per kg pulp (pulp density). A series of different acid additions were tested ranging from 0.25 to 0.40 kg of sulfuric acid addition per kg of BR (or equivalently leaching in solution between 1.0 and 1.6 M sulfuric acid). The results are presented in Fig. 8, where there is clear region between 0.25 and 0.28 kg of acid/kg of BR of high Sc selectivity over Fe, Ti, and Si, followed by a region of gradual increase of both Sc and major metals up to 0.34 kg acid/kg BR. Above 0.34 kg acid/kg BR, the recovery of Sc and Al has reached a plateau while Si and Ti start to be further dissolved. The leaching conditions of the BR that correspond

Fig. 7 Left: Leaching Pilot unit at MYTILINEOS; Right: II–VI SIR Pilot unit installed at MYTILINEOS. (Color figure online)

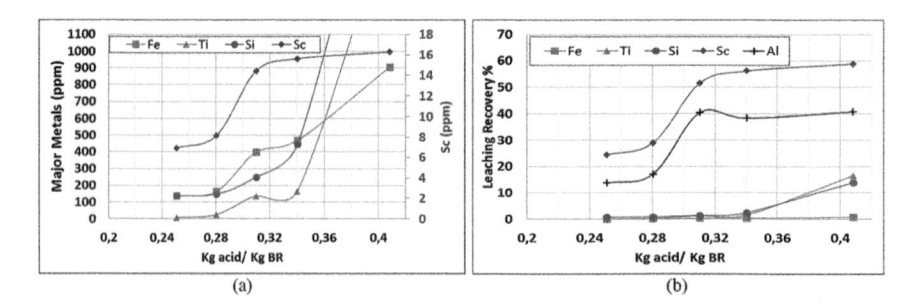

Fig. 8 Pilot Scale leaching results at MYTILINEOS **a** concentrations in final pregnant solution, **b** corresponding leaching yields. (Color figure online)

to the optimum conditions for the SIR process discussed above are found to be at 0.28–0.30 kg acid/kg BR.

The ongoing work at the MYTILINEOS pilot plant aims to generate 12 m^3 of PLS produced with leaching more than 10 t of BR with 0.28–0.30 kg of sulfuric acid/kg of BR. This PLS will be fed to the II–VI SIR pilot unit (Fig. 7), in order to reach the exhaustion point of the SIR resin contained within a 15L column therein. Based on the results presented above, it is expected that from the elution of this resin, a 25% wt Sc concentrate will be produced. Future work will focus in recycling the spent acid streams and optimizing the overall flowsheet, to further reduce the acid consumption of the process.

Conclusion

The performance of II–VI SIR in recovering Sc from BR was demonstrated. Due to low level of Sc concentration in the solution, it was important to lower the impurity levels in solution to achieve high Sc loading. With respect to operating efficiency, it was important that any particulates or solids in the PLS or those that may precipitate out of the PLS at any time during the process were eliminated upstream of the loading process. Consequently, the optimum leaching conditions for the BR were effectively identified. High Sc loading was achieved under these conditions and a resulting crude Sc concentrate was also produced. The significant increase in Sc content substantially reduces the sizing requirements for downstream purification steps and presents a pragmatic approach to the recovery of scandium from Bauxite Residue.

Acknowledgments The research leading to these results has received funding from the European Community's Horizon 2020 SCALE Programme (H2020/2014-2020/No. 730105). This publication reflects only the authors' views, exempting the SCALE Consortium from any liability for the information presented herein.

References

1. Alkan G et al (2018) Novel approach for enhanced scandium and titanium leaching efficiency from bauxite residue with suppressed silica gel formation. Sci Rep 8(1):5676
2. Borra CR et al (2015) Leaching of rare earths from bauxite residue (red mud). Miner Eng 76:20–27
3. Borra CR et al (2016) Smelting of bauxite residue (red mud) in view of iron and selective rare earths recovery. J Sustain Metall 2(1):28–37
4. Borra CR, Blanpain B, Pontikes Y, Binnemans K, Van Gerven T (2016) Recovery of rare earths and other valuable metals from bauxite residue (red mud): a review. J Sustain Metall 2(4):365–386
5. Chassé M, Griffin WL, O'Reilly SY, Calas G (2016) Scandium speciation in a world-class lateritic deposit. Geochem Perspect Lett 3(2):105–114

6. Davris P, Stopic S, Balomenos E, Panias D, Paspaliaris I, Friedrich B (2017) Leaching of rare earth elements from eudialyte concentrate by suppressing silica gel formation. Miner Eng 108:115–122. https://doi.org/10.1016/j.mineng.2016.12.011
7. Deady ÉA, Mouchos E, Goodenough K, Williamson BJ, Wall F (2018) A review of the potential for rare-earth element resources from European red muds: examples from Seydişehir, Turkey and Parnassus-Giona, Greece. Mineral Mag 80(01):43–61
8. Grzmil BU et al (2008) Hydrolysis of titanium sulphate compounds. Chem Papers 62(1):18–25. https://doi.org/10.2478/s11696-007-0074-8
9. Hatzilyberis K et al (2020) Design of an advanced hydrometallurgy process for the intensified and optimized industrial recovery of scandium from bauxite residue. Chem Eng Process 155:108015
10. Hatzilyberis K et al (2018) Process design aspects for scandium-selective leaching of bauxite residue with sulfuric acid. Minerals 8(3):79.3
11. Liu Z, Zong Y, Li H, Zhao Z (2018) Characterization of scandium and gallium in red mud with Time of Flight-Secondary Ion Mass Spectrometry (ToF-SIMS) and Electron Probe Micro-Analysis (EPMA). Miner Eng 119:263–273
12. Lymperopoulou Th, Georgiou P, Tsakanika L-A, Hatzilyberis K, Ochsenkühn-Petropoulou M (2019) Optimizing conditions for scandium extraction from bauxite residue using Taguchi methodology. Minerals 9:236
13. Ochsenkuehn-Petropoulou M, Tsakanika L-A, Lymperopoulou T, Ochsenkuehn K-M, Hatzilyberis K, Georgiou P, Stergiopoulos C, Serifi O, Tsopelas F (2018) Efficiency of sulfuric acid on selective scandium leachability from bauxite residue. Metals 8(11):915
14. Ochsenkühn-Petropoulou MT et al (2002) Pilot-plant investigation of the leaching process for the recovery of scandium from red mud. Ind Eng Chem Res 41(23):5794–5801
15. Petrakova OV et al (2016) Improved efficiency of red mud processing through scandium oxide recovery. In: Hyland M (ed) Light metals 2015. Springer International Publishing, Cham, pp 93–96
16. Power G, Gräfe M, Klauber C (2011) Bauxite residue issues: I. Current management, disposal and storage practices. Hydrometallurgy 108(1):33–45
17. Suss A, Panov A, Kozyrev A, Kuznetsova N, Gorbachev S (2018) Specific features of scandium behavior during sodium bicarbonate digestion of red mud. In: Light metals 2018. Springer International Publishing, Cham, pp 165–173
18. Vind J et al (2018) Rare earth element phases in bauxite residue. Minerals 8(2):77
19. Vind J et al (2018) Modes of occurrences of scandium in Greek bauxite and bauxite residue. Miner Eng 123:35–48
20. Wang W, Pranolo Y, Cheng CY (2013) Recovery of scandium from synthetic red mud leach solutions by solvent extraction with D2EHPA. Sep Purif Technol 108:96–102
21. Xu W-Q, Mattera V Jr, Abella MYR, A GM, Patkar S (2019) Selective recovery of rare earth metals from an acidic slurry or acidic solution. US20190078175A1 and WO2019099859A1
22. Xu W-Q, Mattera V Jr, Abella MYR, Abrenica GM, Patkar S (2017) Composite extractant-enhanced polymer resin, method of making the same, and its usage for extraction of valuable metal (s). WO 2017/074921A1
23. Zhang N, Li H-X, Cheng H-J, Liu X-M (2017) Electron probe microanalysis for revealing occurrence mode of scandium in Bayer red mud. Rare Met 36(4):295–303

Scandium – Leaching and Extraction Chemistry

Dag Øistein Eriksen

Abstract Scandium (Sc) is one of the key elements in the green economy due to its use in fuel cells and as alloying metal for aluminium, but the scandium market is not working in the sense that very little is offered at a high price making it impossible to gain use of the metal. Scandium is a rare earth element (REE) and as such, it is not very rare, but the concentration of it is always low making it a challenge to produce scandium at low cost. Compared with the other REE, Sc^{3+} is a much smaller ion giving it properties closer to Al^{3+}, Fe^{3+}, and Zr^{4+}. We therefore often do not find Sc together with the other REE, but instead in titanium-, aluminium-, and zirconium-containing minerals. Processes involving Sc separation are different from the usual REE processes. Exploitation of the mineral davidite is used as an example of small deposits, which may be utilized through efficient mining whereas other large operations like recovery from bauxite residues (red mud) are considered for comparison.

Keywords Scandium · Davidite · Bauxite · Vanadium · Titanium · Leaching · Solvent extraction · Red mud

An Important Metal for the Green Economy

In implementing environmentally friendly technology, the use of hydrogen as an energy carrier is imperative. The Solid Oxide Fuel Cell (SOFC) is a key technology using zirconia (ZrO_2) as a catalyst for transforming hydrogen gas to water and electricity. Scandium stabilized zirconia is so far the superior catalyst, which also is reducing the operating temperature [1]. In low temperature fuel cells (Proton Exchange Membrane-, PEM-cells), scandium–platinum alloys offer significantly improved performance [2]. The third and perhaps highest volume application is in scandium–aluminum–magnesium alloys that offer greatly increased welding and strength properties and thus significant opportunities in transport applications [3–5].

D. Ø. Eriksen (✉)
Primus.inter.pares AS, Kongsberggata 20, 0468 Oslo, Norway
e-mail: post@pipas.no; d.o.eriksen@kjemi.uio.no

© The Minerals, Metals & Minerals Society 2021
G. Azimi et al. (eds.), *Rare Metal Technology 2021*, The Minerals, Metals & Materials Series, https://doi.org/10.1007/978-3-030-65489-4_23

Riva et al. [6] have in their review shown that scandium is the superior alloying metal for Al and Al(Mg). Czerwinsky [7] reports assessment of the differences between use of cerium instead of scandium, and the author concludes that this has no benefit on the physical properties of the alloys, but CeO_2 is abundant and inexpensive.

The Chemistry of Scandium in Minerals

Scandium (Sc) is the lightest of REE. It has atomic number 21, thus being the neighbor of calcium (Ca). Unlike calcium, which is the fifth most abundant element in the Earth's crust, Scandium is scarce, but still more abundant than lead, tin, silver etc. Scandium has the electron configuration $[Ar]3s^23d^1$. This immediately indicates that the sole oxidation state is + 3 in aqueous media. Scandium is always found together with other elements, but not necessarily the other REE. The reason for this can be attributed to the small ionic size of Sc^{3+} compared with the sizes of the other REE. Figure 1 shows the ionic radii of all REE and, in comparison, also some other selected elements [8]. The trends are here important, not the exact numbers. The figure shows the lanthanide contraction of the 4f-elements and that yttrium (Y), although having atomic number 39, has an ionic radius comparable to dysprosium, atomic number 66. Scandium has an ionic radius closer to trivalent vanadium and iron, V(III) and Fe(III).

In the Hard-Soft Acid—Base (HSAB) theory, small ions with high valence will usually be harder than the opposite [9]. For example, the ferrous ion (Fe^{2+}) is larger than the ferric ion (Fe^{3+}) and will be a softer acid. It will therefore have affinity to a soft base like sulphide, S^{2-}. Ferric, on the other hand, will require a harder base like O^{2-}. This is what we find in the minerals pyrite, FeS (i.e. FeS_2) versus hematite, Fe_2O_3. From this, we can extrapolate that Sc^{3+} is a hard acid and will have affinity

Fig. 1 The ionic radii of the Rare Earth Elements and in addition some other selected elements for comparison. Data taken from Webelements.com and Wikipedia.org. (Color figure online)

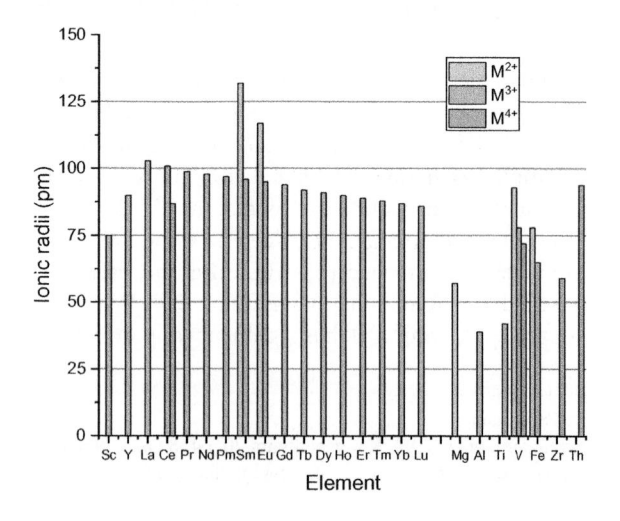

to oxygen, but not sulphide. This is correct, but empirical rules and semi-empirical theories are never absolute. For example, Eu^{2+} is much larger than Eu^{3+}, but the divalent ion can be strongly bonded to the hard base sulphate. Ti^{4+} is a hard acid, but it has such a high affinity for oxygen that the TiO^{2+} ion can be considered as a soft acid. Thus, joining forces with a ferrous ion they may form ilmenite, $FeTiO_3$. Still, the Ti ion is in the tetravalent state and as such it is a hard acid Sc^{3+} can substitute. In vanadium-containing minerals we can use similar reasoning. V(IV) appears as VO^{2+} and has a size close to Sc^{3+}. Nickel Ni^{2+} is a soft acid with strong affinity to sulphide, but there are also oxide-based minerals of nickel, laterites. In such we may also find Sc.

In addition, as a rule of thumb, the chemical properties of two elements are similar when going one period to the right and one row down because the increase in valence charge is counteracted by the increase in ionic radius. For example, Mg^{2+} has similarities with Sc^{3+} which again resembles Zr^{4+}. Again, this rule is not absolute, but can be used to indicate similarities between elements. For example, Zr^{4+} forms bonds to oxygen and as such is attractive to Sc^{3+}, but Zr^{4+} will also form negative chloride complexes while Sc^{3+} do not. The fact that scandium and the other REE do not form chloride complexes represent an option for separation of REE from elements that form such complexes.

Summing up, scandium is a member of the REE but is not always found together with the other REE. Instead, we find scandium in minerals containing aluminium, titanium, vanadium, zirconium, etc. Thus, when we are searching for Sc-containing ores we should not look for xenotime (YPO_4) and other REE-minerals, but instead look for ferric-, alumina-, zirconium- or titanium-containing minerals.

Sources for Scandium

Scandium (Sc), like other rare earth elements, is ubiquitous at low concentrations but very rarely found in high concentrations, making it a challenge to produce at low cost. Like the other REE, scandium is almost never found in nature in pure minerals. The best-known mineral for Sc is tortveitite, but some other Sc minerals are also known [10]. All these minerals are rare and found in minute amounts.

The most widely produced Al rock is bauxite. Bauxite is a mixture of oxidic minerals, mainly of Al and Fe(III) and thus can be expected to host Sc. In the production of alumina, Al_2O_3, by the Bayer process, the amphoteric property of Al is utilized. Ferric hydroxide, FeO(OH)—goethite, remains as solid while $Al(OH)_4^-$ is dissolved and is separated from the solid residue. Scandium follows the ferric hydroxide and is therefore found in the residual ponds. This residue is usually called red mud. As millions of tons of bauxite are processed annually, the ponds of residues are expanding and represent environmental challenges. Therefore, exploitation of red mud is environmentally beneficial and many ponds are considered potential sources for scandium. During the last years, there have been published a large series of reports and papers on scandium recovery from red mud, e.g. [11–18].

Ilmenite, $FeTiO_3$, and rutile, TiO_2, are the most widely exploited titanium minerals and it was recently announced (May 2020) by Rio Tinto (one of the world's largest mining companies) that it will utilize one of its large ilmenite process streams as a source of high volume – low cost scandium. The company will bring Sc to market utilizing their own technology [19]. An iron mineral that can host large resources of scandium is limonite, $FeO(OH) \cdot nH_2O$, and Scandium International Mining Corp., is targeting the Nyngan deposit in New South Wales, Australia, to produce the metal, expecting to start production in 2021 [20, 21].

Perhaps the most exotic mineral under consideration as a scandium resource is davidite, $(Ce,La)(Y,U)(Ti,Fe^{3+})_{20}O_{38}$ [22–24]. The content of the light rare earth elements, La and Ce, can vary while the Y, U, Ti, and Fe are more fixed. Scandium is often present in davidite, and so is vanadium [25]. The mineral is found in rather small deposits in Australia, Kazakhstan, and Norway. It has been noticed during explorations because of the content of uranium. Davidite is found in Norway in the Fennoscandian shield extending from the Kola Peninsula in Russia to Finnmark County in Norway. The two best-surveyed deposits are Biggjovagge and Biggejavri [23]. The mineral davidite is uncommon and no deposit has so far been considered a source of scandium that would significantly alter the scandium market.

Experimental Results and Discussion of Leaching of Davidite

Davidite concentrate from the gangue material from the chalcopyrite mine at Biggjovagge, Finnmark, Norway was received from the mine owner and used as such. This mine is now closed and the area brought back to its original condition. Leaching tests were conducted by Megon AS at University of Oslo, Department of Chemistry and elemental analyses were performed by ICP-AES at Institute for Energy Technology, Kjeller, Norway.

HCl or H_2SO_4 were tested as leaching agents. Both acids have potential as complexing agents for zirconium and uranyl; chloride also for ferric present, sulphate also for rare earths.

In addition to the two kinds of acids, pulp density, leaching time and temperature were varied. A selection of analyses of pregnant leach solutions (PLS) is listed in Table 1. For all these tests, the temperature was 100 °C and the leaching duration 1 h. The yields of the various elements are listed. As is shown, the rough conditions gave less than 60% yield of Sc. This was considered too low and the project was stopped. However, there are things to learn from Table 1:

- Recovery of scandium is strongly correlated with titanium and vanadium, correlation coefficients are higher than 0.95. Linear correlation coefficient is defined as [26]:

Table 1 Yields of selected elements as function of leaching parameters. Temperature was 100 °C and the leaching duration 1 h. The particle size was between 50 and 100 μm

Acids	Init.conc. (M)	Pulp dens. (g/L)	Dissolved mass (%)	Fe (%)	Ti (%)	Sc (%)	Y (%)	La (%)	Ce (%)	Nd (%)	V (%)	Zr (%)	U (%)
HCl	11	150	24	54.0	27.8	57.8	97.6	64.8	84.9	67.7	96.7	93.3	90
HCl	4	150	16.5	52.4	9.3	29.6	92.7	66.7	92.5	50.0	43.3	60.0	95
HCl	11	300	22	27.8	7.2	19.3	87.8	35.2	100.0	89.3	60.0	43.3	82.5
HCl	11	300	15	21.4	5.3	13.3	70.7	30.1	84.9	71.4	40.0	66.7	75
H_2SO_4	12	150	18	47.6	33.3	57.8	65.9	52.8	67.9	100.0	100.0	90.0	70
H_2SO_4	6	100	16.6	61.9	13.8	35.6	63.4	56.9	49.1	32.1	38.3	66.7	67.5
H_2SO_4	4.5	150	16.5	61.9	13.3	38.5	100	63.0	81.1	82.1	60.0	100.0	100
Correlation coefficient relative to Sc				0.3584	0.9590		0.2305	0.2394	0.1066	0.6412	0.9959	0.5765	0.0967

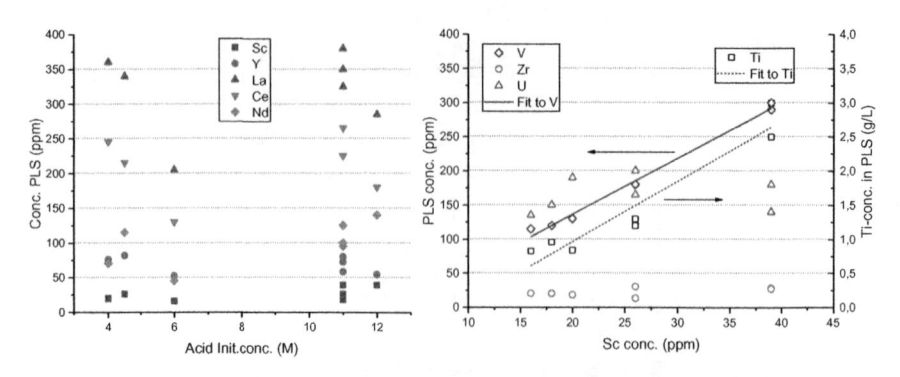

Fig. 2 Concentration of REE in PLS as function of initial concentration of acid (left panel) and the concentration of V, Zr, U and Ti in PLS as a function of scandium concentration (right panel). (Color figure online)

$$r \stackrel{\text{def}}{=} \frac{N \sum_i^N x_i y_i - \sum_i^N x_i \sum_i^N y_i}{\left[N \sum_i^N x_i^2 - \left(\sum_i^N x_i\right)^2\right]^{1/2} \left[N \sum_i^N y_i^2 - \left(\sum_i^N y_i\right)^2\right]^{1/2}} \quad (1)$$

- Recovery of scandium is only weakly correlated with the other REE and zirconium
- Recovery of scandium is almost uncorrelated with uranium
- The kind of acid is not important, but the acid concentration is

These results show that there is a strong correlation between the dissolution of titanium, vanadium and scandium, indicating that scandium and vanadium may be substituting Ti in the stoichiometric formula $(Ce,La)(Y,U)(Ti,Fe^{3+})_{20}O_{38}$. The other REE, as well as uranium, must be bonded at other sites in the crystal lattice. Figure 2 is a graphical presentation of the experimental results. The amount of dissolved material is rather constant and since the yields of the REE and U are constantly high, this may indicate that the REE and U are easily leached, while the titanium part is more difficult to leach. Leaching times up to 90 min did not improve the yield. The residue had properties from visual inspection that was quite similar to the input feed and it possessed very good settling and filtration properties.

Separation of Scandium from the Other Cations Present in the PLS

As shown in Fig. 1 Sc^{3+} is a smaller ion than the other trivalent REE. In chloride environments, REE are not complexed and remain as trivalent cations. This means that they may be extracted by cation exchanging solvent extraction agents like the phosphorous based ones. The harder acid the cation is, the more acidic the environment it is possible to extract it from.

The phosphorous-based extraction agents can be written as $(R_1X_1)(R_2X_2)POOH$, where R_i is an alkyl chain and X_i can be O (oxygen), S (sulphur) or nothing. Also, the oxygens in the POOH group can be substituted by S, one or both. The acidities are decreasing in the sequence $(RO)_2POOH$, $R(RO)POOH$, R_2POOH, R_2PSOH, R_2PSSH [27]. The thiol compounds are used to extract soft acids like Ni^{2+}, Cd^{2+} etc. For very hard acids like Sc^{3+}, it is possible to use the hardest bases, i.e., $(RO)_2POO^-$, $R(RO)POO^-$, the challenge is, however, to recover (strip) scandium from the extracting species. The best way is to use NaOH solution forming $Sc(OH)_3(s)$ in the organic phase and settle out. Then the hydroxide can be dissolved in acid and scandium reprecipitated by addition of oxalic acid. To form Sc_2O_3 calcination is employed [28]. The former company Megon AS employed this in their production of high purity Sc_2O_3 in the late 1980s using the phosphonic acid extractant PC88A produced by Daihachi Chemical Co., Japan. Sc^{3+} was extracted from ca 2 M HCl solution and stripped with a NaOH-solution. As can be deduced, such a process consumes acid and energy and will therefore be expensive.

Das et al. [29] have made a comparative study of many of the organo–phosphoric acid extractants. For Sc^{3+}, they recommend the phosphoric acid and the phosphinic acid extractants. The latter will extract more weakly, but be easier to strip, while, in fact, the former may extract Sc^{3+} so hard it cannot be stripped. The extraction properties of bis-(2-ethylhexyl) phosphoric acid (HDEHP or D2EHPA) are strongly dependent not only on the concentration but also on the diluent [27].

Ricketts [30] has published a review of industrial scandium solvent extraction. Among the solvent extraction processes reported, also the use of Primene JM-T is mentioned. This is a primary amine, and as such it is extracting anions. Ricketts does not reveal the aqueous phases but the main value elements were uranium and tungsten in two different processes, respectively. It is therefore concluded that sulphuric acid environments were feeds in both cases. The stripping solutions consisted of chloride solutions. Primene JM-T seems to be the preferred extractant in the process of Scandium International.

Conclusions

Scandium is a critical metal if the Green Economy is to be implemented with state of the art technologies [4]. There might be a huge amount of scandium on the market in near future at a reasonable price. However, there has been a tendency by mining companies to underestimate the cost of refining rare earth metals —and scandium is no exception. We will have to await the announced producers with their products before scandium can be withdrawn from the list of critical metals. Thus, there might still be demand for scandium and new sources.

Davidite is an interesting source material as it contains several critical elements and elements of high value, but the deposits are small. Thus, the need for effective and low cost mining operations is imperative. The co-extracted elements, e.g., Ti and V, may be cost carriers and add value to the operation. Compared with other sources, e.g., red mud and ilmenite gangue, davidite will also enable production of valuable elements like REE, Zr and U.

The interest for scandium chemistry is booming with many new papers published annually. The metal is, however, easy to separate from most other elements due to its hard acid nature. The challenge is to treat the rest of the feed and recover the scandium in an economic acceptable process. Compared with red mud and ilmenite gangue material, the volumes of scandium produced from davidite will presumably be much smaller, but the cost of the production may also be smaller due to the ease of leaching and selectivity of (scandium) separations.

References

1. Simoncic P, Navrotsky A (2007) Systematics of phase transition and mixing energetics in rare earth, yttrium, and scandium stabilized zirconia and hafnia.https://doi.org/10.1111/j.1551-2916.2007.01678.x
2. Garapati MS, Sundara R (2019) Highly efficient and ORR active platinum-scandium alloy—partially exfoliated carbon nanotubes electrocatalyst for proton exchange membrane fuel cell. Int J Hydrog Energy 44:10951–10963
3. Marscheider-Weidemann F, Langkau S, Hummen T, Erdmann L, Tercero Espinoza L, Angerer G, Marwede M, Benecke S (2016) Rohstoffe fur Zukunftstechnologien. DERARohstoffinformationen 28:353 S. Berlin (in German)
4. Ref. no.: 2016/227 Swedish Agency for Growth Policy Analysis, Studentplan 3, SE-831 40 Östersund, Sweden
5. The Aluminium-Scandium Alloy Advantage (2018) www.scandiummining.com
6. Riva S, Yusenko KV, Lavery NP, Jarvis DJ, Brown SGR (2016) The scandium effect in multicomponent alloys. Int Mater Rev 61(3):203–228. https://doi.org/10.1080/09506608.2015.1137692
7. Czerwinski F (2020) Critical assessment 36: assessing differences between the use of cerium and scandium in aluminium alloying. Mater Sci Technol 36(3):255–263. https://doi.org/10.1080/02670836.2019.1702775
8. Data taken from Webelements.com and Wikipedia.org
9. Choppin GR (2004) Complexation of metal ions. In: Rydberg J, Cox M, Musikas C, Choppin GR (eds) Solvent extraction principles and practice, 2nd edn. Marcel Dekker Inc., New York, Basel
10. Kristiansen R (2003) Scandium. mineraler i Norge (in Norwegian). www.nags.net
11. Archambo M, Kawatra SK (2020) Red mud: fundamentals and new avenues for utilization. Miner Process Extr Metall Rev. https://doi.org/10.1080/08827508.2020.1781109
12. Verna A, Suri NM, Kant S (2017) Applications of Bauxide Residue: A Mini-Review. Waste Manag Res 35(10):999–1012
13. Zhang N, Li H-X, Liu X-M (2016) Rare Met 35(12):887–900
14. Wang W, Pranolo Y, Cheng CY (2011) Metallurgical processes for scandium recovery from various resources: a review. Hydrometallurgy 108:100–108
15. Zhu X, Niu Z, Li W, Zhao H, Tan Q (2020) A novel process for recovery of aluminum, iron, vanadium, scandium, titanium and silicon from red mud. J Environ Chem Eng 8:103528

16. Liu Z, Zong Y, Li H, Jia D, Zhao Z (2017) Selectively recovering scandium from high alkali Bayer red mud without impurities of iron, titanium and gallium. J Rare Earths 35(9):896–905
17. Liu Z, Li H, Jing Q, Zhang M (2017) Recovery of scandium from leachate of sulfation-roasted Bayer red mud by liquid-liquid extraction. JOM 69(11). https://doi.org/10.1007/s11837-017-2518-0
18. Narayanan RP, Kazantzis NK, Emnett MH (2018) Selective process steps for the recovery of scandium from jamaican bauxite residue (red mud). ACS Sustain Chem Eng 6:1478–1488
19. https://www.pm-review.com/rio-tinto-reveals-breakthrough-scandium-extraction-from-titanium-dioxide-production-waste/
20. USGS MCS2020-scandium
21. Feasibility Study – Nyngan Scandium Project, NI 43–101 Technical Report
22. Pabst A (1961) X-ray crystallography of davidite. Am Mineral 46:700–718
23. Olerud S (1985) Mikrosondeundersøkelse av radioaktiv albittfels fra Biggejavri, KautokeinoNGU-rapport nr 85.159 (in Norwegian)
24. Eriksen DØ (1987) Rapport Nr. 1. Utlakingsforsøk av davidittkonsentrat, Bergvesenet rapport nr 5839 (in Norwegian)
25. Kompanchenko AA, Voloshin AV, BalaganskyVV (2018) Minerals 8:474. https://doi.org/10.3390/min8110474
26. Bevington PR (1969) Data reduction and error analysis for the physical sciences. McGraw-Hill, Inc.
27. Rydberg J, Cox M, Musikas C, Choppin GR (2004) Solvent extraction equilibria in solvent extraction principles and practice, 2nd edn. Marcel Dekker Inc., New York, Basel
28. Gao L-K, Rao B, Dai H-X, Hong Z, Xie H-Y (2019) Separation and extraction of scandium and titanium from a refractory anatase lixivium by solvent extraction with D2EHPA and primary amine N1923. J Chem Eng Jpn 52(11):822–828
29. Das S, Behera SS, Murmu BM, Mohapatra RK, Mandal D, Samantray R, ParhiPK, Senanayake G (2018) Extraction of scandium(III) from acidic solutions using organo-phosphoric acid reagents: a comparative study. Sep Purif Technol 202:248–258
30. Ricketts NJ (2019) Scandium solvent extraction. In: Chsonis C (ed) Light metals 2019. The minerals, metals & materials series. https://doi.org/10.1007/978-3-030-05864-7_174

Preparation of Industrial Sodium Chromate Compound from an Indigenous Chromite Ore by Oxidative Decomposition

Alafara A. Baba, Kuranga I. Ayinla, Bankim Ch. Tripathy, Abdullah S. Ibrahim, Girigisu Sadisu, Daud T. Olaoluwa, and Mustapha A. Raji

Abstract Pure chromate is an important raw material used extensively in chemical and metallurgical operations. Considering these uses, the oxidizing roasting process of an indigenous chromite ore to prepare industrial sodium chromate was investigated in this study. The effects of sodium carbonate to chromite mole ratio, reaction temperature, contact time and its thermodynamics-cum-kinetics behaviour were also discussed. The results showed that the reaction mechanism was greatly influenced as the temperature varies. A two-stage recovery process was found favorable for the conversion method and chromate conversion rate of 97.06% with minimal residual pollutants was achieved at optimal conditions. The residual products containing iron as characterized could be easily recovered to produce sponge iron, leading to complete detoxification and zero emission of chromium residue for defined industrial applications.

Keywords Chromate compound · Chromite · Indigenous · Oxidative decomposition

Introduction

Chromite, the most viable among the chromium minerals, is used continuously to produce high-quality chromium compounds. The economics of chromite recovery from the ore depends upon the occurrence and its chemical composition [1–3]. In Nigeria, for example, sufficient reserves of chromite ore are found at Awe Local Government of Nassarawa State, Benue through of Northeastern part of the country.

A. A. Baba (✉) · K. I. Ayinla · A. S. Ibrahim · G. Sadisu · D. T. Olaoluwa · M. A. Raji
Department of Industrial Chemistry, University of Ilorin, P.M.B. 1515, Ilorin 240003, Nigeria
e-mail: baalafara@yahoo.com

B. Ch. Tripathy
CSIR—Institute of Minerals and Materials Technology, Bhubaneswar 751013, India

In recent years, the global demand for chromium-related compounds has been increasing steadily, because of enormous use of the chromium (an important non-ferrous metal) in the metallurgical and chemical industries [4, 5]. Since chromium-containing compounds serve as an intermediate product with their diverse applications, its manufacturing process is thus an important branch of inorganic salt industry and has attracted rising public environmental concerns [6, 7]. Sodium chromate is one of the most important products used chiefly for manufacturing chromic acid and chromium pigment in leather tanning and corrosion control applications [8–10].

Conventionally, the industrial production of sodium chromate (Na_2CrO_4) through leaching of the chromite ore can be obtained by either pyrometallurgical or hydrometallurgical process [11, 12]. These conventional methods are the lime-based roasting process, which have many disadvantages including low resource utilization and serious environmental pollution among others [13, 14]. Such disadvantages hinder the widespread application of the defined process. Consequently, it is gradually being replaced by lime-free roasting that is cleaner and low resource utilization [15, 16]. Therefore, a low energy consumption and eco-friendly method is continuously needed to develop production of sodium chromate being a primary material needed to resuscitate the Nigerian leather industries, for example. At present, the sodium chromate is still being imported for large-scale use, which represents an economic burden for local industries in Nigeria. Consequently, the aim of this study is the development of a feasible approach for producing chromate locally from a Nigerian sourced chromite ore by oxidative decomposition technique. Experimental parameters such as sodium carbonate to chromite mole ratio, reaction temperature, contact time, and particle sizes to enhance optimal chromate yield were investigated.

Materials and Methods

Materials

The chromite ore used for this study was sourced from Tunga Kaduka, Awe Local Government, of Nigeria. The ore was crushed, grinded, and pulverized into three different particle sizes: $-106 + 75$ μm, $-253 + 106$ μm, $-315 + 253$ μm fractions. This concentrate as analysed by X-ray fluorescence contained 49.6 wt% Cr_2O_3, 7.8 wt% Fe_2O_3, 6.4 wt% FeO, 10.8 wt% Al_2O_3, 15.9 wt% MgO, 0.1 wt% CaO, 2.7 wt% SiO_2, and 6.7 wt% other oxides [17]. All reagents used were of analytical grade (BDH grades) and doubly distilled water was used to prepare all aqueous solution.

Chromite Ore Dissolution Process Investigation

Roasting Process

A 2 g of the pulverised ore together with the varied amounts of Na_2CO_3 (20, 50, 60, and 75% in mass fraction) was weighed mixed along with a small amount of water in a nickel crucible and placed inside a muffle furnace. The roasting temperature was varied from 400 to 900 °C, keeping the roasting time constant at 2 h. After the required roasting time, the crucible was cooled inside before removing the roasted product.

Treatment of Roasted Products

The roasted sample in Sect. 2.2.1 was leached for 30 min in doubly distilled water at 50 °C in a designed glass reactor. The mixture was agitated to a constant stirring speed of 350 r/min. At regular intervals, samples were taken and immediately filtered. The chromium content in the solution was determined using atomic absorption spectroscopy (AsecT 1335). The leach residues were washed thoroughly and oven-dried at about 80 °C for 24 h prior to the phase analysis by Perkin-Elmer Model 460X-ray diffraction (XRD) equipment.

Results and Discussion

Roasting-Leaching Studies

The chromium-bearing mineral can be simply represented as $FeCr_2O_4$, which rarely exists as pure iron protoxide (FeO) and chromic oxide (Cr_2O_3). The chemical reactions in the oxidation roasting of chromite ore with sodium carbonate can be represented by the following relations:

$$FeCr_2O_4 + 2Na_2CO_3 + 7/4O_2 \rightarrow 2Na_2CrO_4 + 1/2Fe_2O_3 + CO_2 \quad (1)$$

$$MgCr_2O_4 + 2Na_2CO_3 + 3/2O_2 \rightarrow 2Na_2CrO_4 + MgO + CO_2 \quad (2)$$

$$Cr_2O_3 + 2Na_2CO_3 + 3/2O_2 \rightarrow 2Na_2CrO_4 + CO_2 \quad (3)$$

Effect of Na$_2$CO$_3$ Concentration

The investigated optimal parameters were represented using Design expert demo graph as summarised in Fig. 1a–c.

Figure 1a shows increase in leaching efficiency as the concentration of Na$_2$CO$_3$ increases from 20 to 50 wt% but decreases beyond 50 wt% as it reaches 68% after 180 min. The decline in the chromium leaching from 68 to 63% could be attributed to oxidative leaching of chromite ore by dissolved O$_2$ instead of gaseous O$_2$ serves as the oxidant. It is obvious that the viscosity of the Na$_2$CO$_3$ in aqueous solution increases concentration monotonically. Increase in Na$_2$CO$_3$ concentration from 50 to 75 wt. %, the solubility of O$_2$ in Na$_2$CO$_3$ aqueous solution decreases and the mass transfer of O$_2$ worsens, resulting in a sharply decline of the chromium leaching efficiency [18, 19].

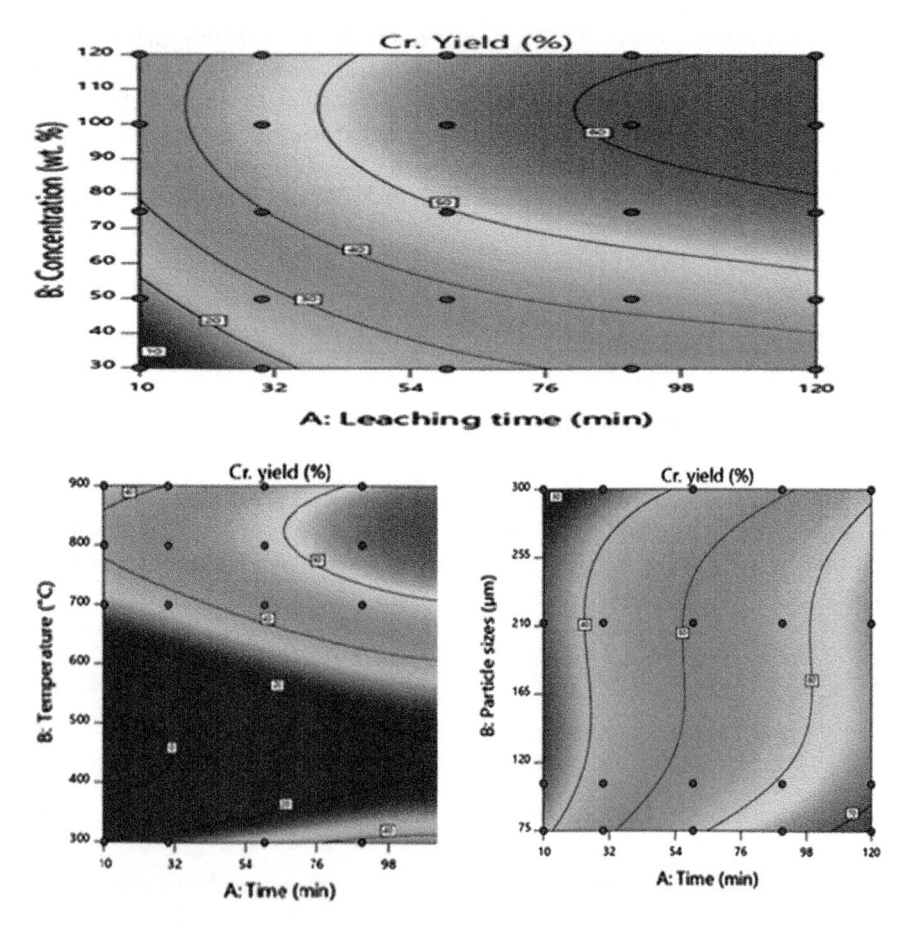

Fig. 1 Optimization profile of roasting-leaching investigation: **a** Effect of Na$_2$CO$_3$ concentration **b** Effect of temperature **c** Effect of particle sizes. (Color figure online)

Effect of Reaction Temperature

The chromium-leaching efficiency in the temperature range 300–900°C was carried out with particle size of $-106 + 75$ μm, stirring speed of 300 rpm, and 50 wt% Na_2CO_3 concentration as shown in Fig. 1b. The chromium leaching efficiency is greater at higher temperature in general. At optimum temperature of 900 °C, the chromium leaching efficiency was stable at about 97. 06% after 180 min. However, it is obvious that higher temperature changes the chemical equilibrium constant and boost the reaction rate constant, as higher temperature enhances the solubility of the product around the ore [20, 21].

Effect of Particle Sizes

The results for the effect of particle sizes on the extent of chromium recovery are presented in Fig. 1c. As expected, the smaller size of the particle, the faster the extent of ore dissolution. This is probably attributed to the active sites on the surface of chromite ore particles in the dissolution. In a situation when specific sizes get smaller, for example, more active sites on the solid surface are exposed to the solution. As the leaching proceeds, the reactants penetrate into micro-cracks and more soluble forms of chromite ore is expected to be quantitatively extracted [22]

Dissolution Mechanism and Residual Analysis

From the established leaching data in Fig. 1b, the chemical reaction rate-limiting step was applied. From the slope of the straight lines in Fig. 1b using fitted shrinking core model, Arrhenius equation $K = Ae^{-Ea/RT}$ was used to plot ln k versus (1/T) for each temperature and the activation energy was calculated from the slope—Ea/RT gave 23.2 kJ/mol. The calculated activation energy value further supports diffusion control mechanism model.

Furthermore, during water leaching, the separation of $NaCrO_2$ into Na_2CO_3 and Cr_2O_3 take place, following the precipitation of iron oxide which remained in the residue. The main phases identified in the residue after water leaching are MgO, Fe_2O_3, $MgFe_2O_4$, and sodium aluminate silicate ($NaAlSiO_3$). The SEM images of the leaching residues showing the macrostructure of the residue and the elemental analysis of the EDAX are shown in Fig. 2a, b and Table 1 respectively.

The micrograph of the solid particles before leaching presents a separate smooth mineral phases with porous surface. However, the micrograph of the leaching residue presents different surfaces. The chromite surface was almost dissolved and could not be observed in the leaching residues as obtained in Table 1.

As seen from Fig. 2a, b, the morphology of the residue derived from the oxidation of chromite is totally different from that of the ore before leaching and present a

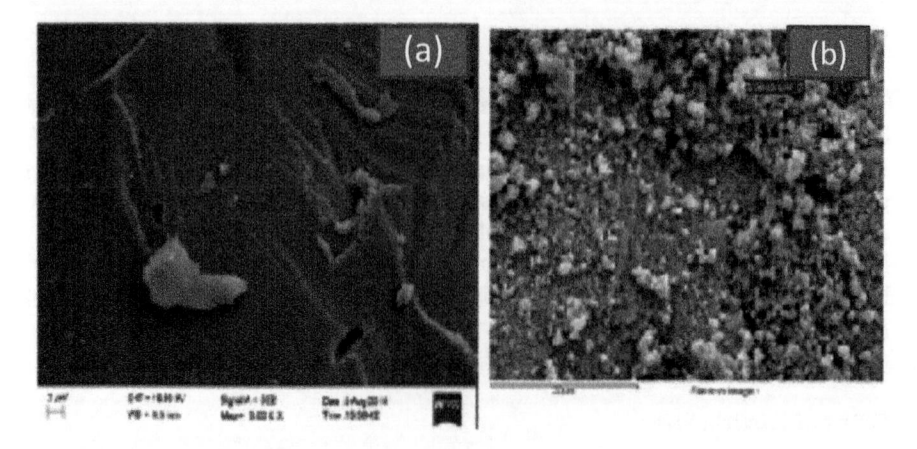

Fig. 2 SEM images of the leached residues: **a** Raw chromite ore **b** Leached chromite

Table 1 Summary of EDAX elemental composition of leach residue

Element	Na	Mg	Al	Si	Fe	O	C
Wt %	4.90	8.99	5.28	3.14	20.03	24.70	10.96

rough and porous surface. This apparently showed that the chromite can be oxidised and dissolved by water while roasting with Na_2CO_3.

Conclusion

In this study, the roasting of Tunga Kaduka (Nigeria) chromite ore concentrate with sodium carbonate in the absence of airflow, while preventing interference of nitrogen and other atmospheric gases in the air yielded maximum chromium recovery of 97.06% at 900 °C within 2 h. The results showed that the reaction mechanism was greatly influenced as the temperature varies. A two-stage recovery process was found favourable for the conversion method and chromate conversion rate of 97.06% with minimal residual pollutants were achieved at optimal conditions. The residual products containing iron as characterized could be easily recovered to produce sponge iron, leading to complete detoxification and zero emission of chromium residue for defined industrial applications.

References

1. Amer A (1992) Processing of Ras-Shait chromite deposits. Hydrometallurgy 28:29–43
2. Pascoe RD, Power MR, Simpson B (2007) QEMSCAN analysis as a tool for improved understanding of gravity separator performance. Miner Eng 20:487–49520
3. Rao RB, Reddy PSR, Prakash S, Ansari MI (1987) Recovery of chromite values from chromite ore beneficiation plant tailings. Trans IIM 40(3):203–206
4. Sonmez E, Turgut B (1998) Enrichment of low-grade Karaburhan chromite ores by gravitational methods. In: Atak S, Onal G, Celik T (eds) Innovation in mineral and coal processing. The Netherlands, Balkema, Rotterdam, pp 723–726
5. Soykan O, Eric RH, King RP (1991) The reduction mechanism of a natural chromite at 1416°C. Metall Trans 22B:53–63
6. Kashiwas K, Sato G, Narita E, Okabe T (1974) Studies on manufacturing process of sodium chromate. 1. Kinetics of oxidation reaction of chromite by NaOH– NaNO3 molten salts. Nippon Kagaku Kaishi 1:54–59
7. Kashiwas K, Sato GI, Atumi T, Okabe T (1974) Studies on manufacturing process of sodium chromate. 2. Appropriate oxidizing conditions of chromite with molten sodium salts. Nippon Kagaku Kaishi 3:469–473
8. Li X-B, Xu W-B, Zhou Q-S, Peng Z-H, Liu G-H (2011) Leaching kinetics of acidsoluble Cr(VI) from chromite ore processing residue with hydrofluoric acid. J Centenary South Univ Technol 18(2):399–405
9. Sun Z, Zhang Y, Zheng S-L, Zhang Y (2009) A new method of potassium chromate production from chromite and KOH–KNO3–H2O binary sub-molten salt system. AIChE J 55(10):2646–2656
10. Sun Z, Zheng S-L, Xu H-B, Zhang Y (2007) Oxidation decomposition of chromite ore in molten potassium hydroxide. Int J Miner Process 83(1–2):60–67
11. Sun Z, Zheng SL, Zhang Y (2007) Thermodynamics study on the decomposition of chromite with KOH. Acta Metallurgica Sinica (English Letters) 20(3):187–192
12. Tham MJ, Walker RD, Gubbins KE (1970) Diffusion of oxygen and hydrogen in aqueous potassium hydroxide solutions. The J Phys Chem 74(8):1747–1751
13. Vardar E, Eric RH, Letowski FK (1994) Acid leaching of chromite. Miner Eng 7(5–6):605–617
14. Wang CY, Qiu DF, Yin F, Wang HY, Chen YQ (2010) Slurry electrolysis of ocean polymetallic nodule. Trans Nonferrous Metals Soc China 20(Supplement 10):s60–s64
15. Xu H-B, Zhang Y, Li Z-H, Zheng S-L, Wang Z-K, Qi T, Li H-Q (2006) Development of a new cleaner production process for producing chromic oxide from chromite ore. J Clean Produ 14(2):211–219
16. Geveci A, Topkaya Y, Ayhan E (2002) Sulfuric acid leaching of Turkish chromite concentrate. Miner Eng 15:885–888
17. Ayinla IK, Baba AA Tripathy BC, Ghosh MK, Dwari RK, Padhy SK (2018) Enrichment of Nigeria chromite ore for metallurgical application by dense medium flotation and magnetic separation. Metall Res Technol 116(3240):1–8
18. Murthy ChVGK, Sripriya R, Rao PVT (1994) Enrichment of Cr/Fe ratio in chromite concentrate produced in chromite beneficiation plant at Tata Steel. Trans Indian Inst Met 47(6):413–416
19. Zheng SL, Zhang Y (1999) Thermodynamic analysis on new reaction system of liquid phase oxidation of chromite in molten salt. Chin J Nonferrous Metal 9:800–804
20. Xu HB (2003) Applied fundamental research on the separation operations in the chromate cleaner production process based on potassium sub-molten salt techniques, PhD Thesis, Beijing: Institute of Process Engineering, Chinese Academy of Sciences
21. Zheng SL (2000) Fundamental study and optimization of the liquid-phase oxidation in the cleanermanufacturing process for chromates, PhD Thesis, Beijing: Institute of Process Engineering, Chinese Academy of Sciences
22. Gupta CK, Mukherjee TK (1990) Hydrometallurgy in extraction processes, vol 1. CRC Press, USA

Part V
Recycling, Co, and REE

The Italian National Research Council Operations Within the EIT Raw Materials Framework

Paolo Dambruoso, Salvatore Siano, Armida Torreggiani, Ornella Russo, Stefania Marzocchi, and Vladimiro Dal Santo

Abstract The adopted strategy and the results achieved by the Italian National Research Council within the Knowledge and Innovation Community "Raw Materials" of the European Institute of Innovation and Technology (KIC EIT-RM) are presented in detail. We focus on activities dedicated to education as well as validation and acceleration actions of the EIT-RM. Regarding the former, activities tackling the awareness of citizens on the impact of RMs in our life, guiding pupils towards an informed engagement into RMs university carriers, and lifelong learning of professionals dedicated to methodologies to access, organize, and share scientific literature and data are presented. Regarding the validation and acceleration actions, two main activities are discussed: (1) development of Platinum–Group Metals free catalysts and the corresponding know-how transfer initiative towards East and South East

---

P. Dambruoso (✉) · A. Torreggiani
Institute of the Organic Synthesis and Photoreactivity, Italian National Research Council (ISOF-CNR), Research Area of Bologna, via P. Gobetti, 101, 40129 Bologna, Italy
e-mail: paolo.dambruoso@isof.cnr.it

A. Torreggiani
e-mail: armida.torreggiani@isof.cnr.it

S. Siano
"Nello Carrara" Institute of Applied Physics, Italian National Research Council (IFAC-CNR), Research Area of Florence, via Madonna Del Piano, 10, 50019 Sesto Fiorentino (FI), Italy
e-mail: S.Siano@ifac.cnr.it

O. Russo
CNR National Research Council, Institute of Microelectronics and Microsystems (IMM), Via P. Gobetti, 101, Bologna 40129, Italy
e-mail: ornella.russo@area.bo.cnr.it

O. Russo · S. Marzocchi
CNR National Research Council, Library of the Research Area of Bologna, Via P. Gobetti, 101, Bologna 40129, Italy
e-mail: stefania.marzocchi@area.bo.cnr.it

V. Dal Santo
Institute of Chemical Sciences and Technologies "Giulio Natta", Italian National Research Council (SCITEC-CNR), Via Camillo Golgi 19, 20133 Milan, Italy
e-mail: vladimiro.dalsanto@scitec.cnr.it

Europe (ESEE) countries; (2) development of novel analytical logging tools and portable devices for real-time compositional analyses based on laser technologies.

Keywords Raw materials · Education · Lifelong learning · Digital competences

Introduction

The Italian National Research Council

The Italian National Research Council (CNR) [1] is the largest public research institution in Italy. Founded as legal person on 18 November 1923, CNR's mission is to perform research in its own Institutes, to promote innovation and competitiveness of the national industrial system, to promote the internationalization of the national research system, to provide technologies and solutions to emerging public and private needs, to advice Government and other public bodies, and to contribute to the qualification of human resources.

The EIT RawMaterials

EIT RawMaterials (EIT RM) [2] is the Knowledge and Innovation Community (KIC) that has been funded by the EIT (European Institute of Innovation and Technology [3], a body of the European Union) for 7 years of operations (2016–2022). Since the awarding date, EIT RM is the largest consortium in the raw materials sector worldwide. The EIT RM vision is to develop raw materials into a major strength for Europe while the EIT RM mission is to enable sustainable competitiveness of the European minerals, metals and materials sector along the value chain by driving innovation, education and entrepreneurship. The EIT RM community is focused on innovation, and the introduction of innovative processes, products and services aiming to secure the supplies and improve the raw materials sector in Europe.

This ambitious objective is pursued investing in exceeding 350 projects in the field of education, validation, and acceleration of innovative technologies, products and services, internationalization and also supporting business creation.

EIT RawMaterials consortium consists of exceeding 300 partners (including project partners) established in Europe and representing the whole knowledge triangle (research organizations, universities, and industries). Partners of EIT RawMaterials operate over the entire raw materials value chain, spamming from exploration to recycling and circular economy.

CNR Operations Within the EIT RawMaterials Framework

After the start-up period of one year (2015), EIT Raw Materials operation started in January 2016 and, since then, CNR has been partner of the EIT RawMaterials as linked third party of ART-ER S. Cons. P.A. [4], (former ASTER S. Cons. P.A.) together with Alma Mater Studiorum—University of Bologna [5].

As a general strategy, CNR approached EIT RawMaterials challenges and opportunities, promoting within the consortium, a valuable set of assets focused on all EIT RawMaterials pillars and selected sectors of raw materials value chain.

Regarding the Education pillar, all operations were conceived and developed as a bridge joining schools to industries, also considering the impact on the society. Using this approach, CNR Education offer has been based on four main pillars: awareness, motivation, state of the art, knowledge and communication. From this perspective, guided by their teachers and raw materials experts, pupils should initially become aware of the importance of raw materials into their daily life, to become ambassadors within the society. The most motivated pupils will be funneled into raw materials university studies, after having offered them the opportunity to understand, through dedicated internships, what their professional life will be. On the other hand, professionals operating along the entire raw materials value chain should continuously refine and improve their knowledge, taking advantage of professional education.

To achieve these objectives and implement this strategy within the Education pillar, the knowledge gained by CNR researchers within two assets, namely "Il linguaggio della Ricerca" [6] (The research language) and "SperimEstate" [7] (Summer experiments) have been exploited within the wider society learning sets of activities of EIT RM, while the education initiatives dedicated to researchers and developed by the Digital Library of the Research Area of Bologna [8] were translated in lifelong learning activities dedicated to professionals operating along the entire raw materials value chain.

Along with those developed within the Education pillar, CNR proposed valuable activities also in the Validation & Acceleration EIT RawMaterials pillar. Accordingly, various thematic Network of Infrastructures (NoIs) were developed, taking advantage of the background gained within various assets: the ChemNanoCare [9] thematic area of the ISOF Institute, the LabCat [10] of the ISTM-Milan Institute (now SCITEC), the nM2Lab laboratory [11] of ISM-Montelibretti and the laser laboratory for cultural heritage of the IFAC Institute [12].

In a single concept, "Excellence exploitation" fully represents the CNR strategy within the EIT RM framework.

Details on Selected CNR Activities with the EIT RM Portfolio

Among the 34 activities coordinated or participated by CNR within the EIT RawMaterials projects portfolio, below is a summary of the most representative within all EIT RawMaterials pillars.

RAISE

RAISE, Raw mAterIals Students Internships, is an Education activity coordinated by CNR started on 01 July 2019 that will end on 31 December 2021. RAISE gives access to secondary school students internships within research laboratories/companies operating in the raw materials sector. RAISE tackle a carrier guidance demand originated within the schools, making youngsters aware of the career opportunities in the raw materials field, offering them tools allowing an informed university studies engagement into raw materials related disciplines. Motivated pupils are asked to access two/three weeks of internships focused on KIC thematic pillars.

Aim and Objectives

Secondary school pupils at the last 2 years of their school curricula are facing the most relevant choice in their life: decide their role within the society, identify how they can contribute to the wealth of the community to which they belong, matching their expectations with their talents and with the opportunities offered by specific sectors of the world economy.

This choice ultimately corresponds to the choice of their university studies and this process is often dream-driven, sometimes neglecting unpleasant feedbacks from the real world. On the contrary, due to a vague knowledge or the bad reputation of apparently tough disciplines, youngsters are frequently not at all aware how passionately and productively they may be involved in studies/professional activities dedicated to foggy sectors.

In this scenario, impulsive or uninformed choices may be taken, resulting in regrettable time-wasting decisions, future unproductive and frustrating studies, or lack of valuable talents due to unconsidered opportunities.

The objective of this project is to offer pupils a first but immersive direct contact with the raw materials sector, giving them a flavour of its real world in order to facilitate their informed and committed engagement into studies of raw material-related disciplines.

Implementation

RAISE consortium consists of nine partners coming from five different European countries, and representing the entire knowledge triangle. Research organization and university partners should recruit students and offer internships within their research laboratories, while industrial partners should offer plant visits within their facilities. Two internship cycles are foreseen during the RAISE project lifespan, one per year, and each cycle consists of four phases, namely Design, Implementation, Execution, and Exploitation phases. During the Design phase, tutors should be identified and topics harmonized within the consortium, in order to maximize pupils' involvement along the entire raw materials value chain. During the Implementation phase, schools should be contacted in order to formalize their involvement in RAISE activities and recruit students that should be selected for motivation. After the finalization of the internships calendar, students should be able to start their internships upon the signature of the Education project agreed by RAISE partners and schools. Internships should then be executed, and students should also prepare a video report of their experience, that should be used as a dissemination tool within peers and during the final conference of the internship cycle. Results of the students' satisfaction survey will be used by RAISE partners to improve the internships offer for the coming years.

Results

At least 5 pupils/year per each Research Organization and University (overall 30 pupils/year) should be recruited to perform internships. RAISE will then offer to recruited pupils both direct access into integrated research laboratories/companies internships focused on KIC raw materials thematic pillars, and a chance to become "raw materials teachers and reporters" for their classmates. An average of 20 pupils/class (600 pupils/year, 1200 pupils over the project lifespan) will be involved in RAISE initiatives at the end of the project.

RM@Schools

Raw MatTERS Ambassadors at Schools (RM@Schools) is a Wider Society Learning project focused on an innovative program to make science education and careers in raw materials attractive for youngster.

Aim and Objectives

The project offers an active learning to schools by RM Ambassadors (experts in some RM-related issues and trained teachers) by involving students in experiments with RM-related hands-on educational kits, in excursions in industries, and in science dissemination activities [13]. Various educational approaches are used to foster students' interest in science and technology, in particular in circular economy and RM-related topics, but the core element of the RM@Schools approach is to empower students to communicate with peers and wider society about critical concepts related to raw materials and their use. Students are asked to become Young RM Ambassadors themselves (science communicators) and to create dissemination products focused on issues related to RM (i.e. videos, cards, comics, etc.). Local Competitions for awarding the best communication products as well as an European Conference with delegates from European schools (students and teachers) are annually organized. The project's European Conference that takes place annually in Bologna plays a pivotal role in teaching science communication. It is not only a very important celebration of the work done by students but it also gives young people the opportunity to meet their peers from the other countries and explain by themselves on the stage what they have created in the framework of the project.

In addition, teachers are trained to become RM Ambassadors themselves in the future at school, and selected groups of students are trained on activities suitable to be proposed during Public Events in order to work together with RM Ambassadors.

Implementation

Several learning paths for pupils aged 10–18 years were developed in the framework of RM@Schools where different educational approaches (such as learning-by-doing, team working, peer-to-peer, gamification, etc.) are used to generate and foster students' interest in science and technology.

The learning pathway, covering the whole RMs value chain from geology to electronic waste management, has a modular structure:

- Lesson—introducing students to relevant content knowledge
- Activity—experiments with RM-related hand-on kits. The active learning approach is encouraged by involving students in experiments using raw material-related hands-on educational toolkits
- Visit—to industry or research centres, where students can experience its impact in the real world
- Create/Communicate—students are asked to communicate by creating a product designed to promote dialogue on a key message they have learnt
- Society—students are engaged in public events, such as science fairs, as well as in presenting their best dissemination products during an annual European Conference (Fig. 1).

Results

The project, born in 2016, becomes the flagship project in the Wider Society Learning segment of the EIT RawMaterials in 2018, and every year involves a large number of students and teachers from high schools. The best results of the last years have

The Italian National Research Council Operations ...   255

Fig. 1 Young RM Ambassadors in action at Ideen Expo 2019, a huge science fair in Hannover. (Color figure online)

been the strengthening of the international collaboration among the Consortium, the strong enlargement of the European Network (84 schools, 27 companies, 6 museums, 25 Universities and Research Institutes), and the major involvement of Young RM Ambassadors not only at local but also at the European level. This made possible, just to mention a few, the organisation of one international train—the trainer workshop in Berlin with the participation of teachers from four countries, the organisation of one stand completely devoted to RM@Schools at IdeenExpo, a national Science Fair in Hannover (DE), and tutored by Young RM Ambassadors, the participation to the IV European Conference in IT students and teachers from 14 countries (Fig. 2).

Strengthening the European and Local Networks made possible to organise about 70 open-access events only in 2019 by involving about 16.000 people (generic public, classes in our learning paths, teachers, etc.). In particular, one event was realised in

Fig. 2 International events realised in the framework of RM@Schools project during 2019. (Color figure online)

connection with the EU Researchers' Night directly in one Middle School with the involvement of about 100 Young RM Ambassadors allowing to host more than 400 people to the scientific stands. Moreover, the dialogue with other countries in particular in the ESEE regions was opened thanks to the RM@Schools inclusive action and from this dialogue a new project called RM@Schools-ESEE was born (2020). Finally, the European Conference, annually organised in Bologna (IT), connected with other EIT RawMaterials Projects (i.e., ESEE Education Initiatives, BetterGeoEdu, SmartPlace@Schools, etc.) involved delegates from 18 European countries and 500 participants, of which more than 70 young RM Ambassadors from 14 Countries.

Caronte

The CNR Library of the Bologna Research Area, in collaboration with a multi-faceted team composed of multidisciplinary professionals coming from all corners of the knowledge triangle, developed a set of education initiatives through a project called CARONTE Continuing educAtion and scientific infoRmatiON literacy on Raw MaTerials for profEssionals [14].

Aim and Objectives

The project team included seven full partners: two companies (ECODOM and Zanardi Fonderie) and two training providers from Italy (Centoform and CSCS), the General Council of the Catalan Chambers of Commerce from Spain, and DIMECC, a high-tech network from Finland. Besides this group, a number of collaborators (companies, universities, training providers) took part in the activities, thus facilitating systemic thinking and collaboration between research centres, industries and academia exploiting all synergies offered by the consortium.

The project, supported by EIT RM (July 2017–December 2019), addressed the training activities to researchers and young professionals operating along the entire raw materials value chain.

Several studies on information literacy at the workplace have pointed out that professionals waste time and money searching for information on the web and that information literacy is essential to have a competitive advantage. Moreover, the labour market requires people to be able to constantly update themselves and their knowledge.

The objective of this project is to design, pilot, and disseminate within companies, innovative approaches and strategies in retrieving, effectively organizing and properly sharing among colleagues scientific and technical information regarding the science and innovation frontiers.

Implementation

CARONTE training activities have been designed and developed using ADDIE (Analysis, Design, Develop, Implement and Evaluate) methodology. Starting from results emerged through a project designed survey on "Professionals' behaviour and perceptions about searching and managing information at work", the team identified a general "framework of Information digital competence for informed professionals" (Fig. 3), aimed at empowering the capability to critically use digital information, content and data and to effectively perform information-related job tasks.

The framework is a graphic model that synthesizes the most common information tasks that professionals face during their work (explore a topic, stay updated, write a document, summarize the state of the art, present results, make informed decisions) and the five areas of information competence (intended as the whole of knowledge, skills, and awareness) that they need to improve: find, organize, share, evaluate, apply.

On the basis of this framework, a blended learning program (based on case studies provided by companies' innovators) with a set of nine learning modules, all addressed to companies R&D managers and knowledge workers, was developed, finalized, and tested. Learning modules have been designed with the aim of creating the basic "bricks" for building customized courses. Some of them have been tested separately, in a single workshop specifically dedicated to that learning module, some others have been combined and tested within a more articulated training path.

CARONTE team piloted the modules with different target groups both in Italy and in Spain. Each pilot event has been an occasion to test one or more learning modules, with the aim of detecting criticalities and improving the overall training

Fig. 3 The framework of Information digital competence for informed professionals. (Color figure online)

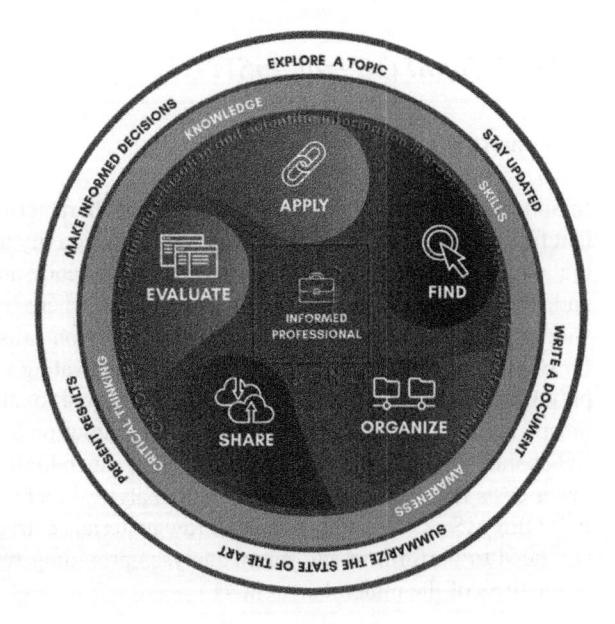

quality in terms of contents, expectations, usefulness, trainers' capability. Through the analysis of the emerged criticalities, the learning modules have been revised, in some cases redesigned, and issued in their final version, now available on the project website [15].

Results

At the end of the project, a total of 113 professionals (57,5% females) and 16 new trainers on Information management needs have been educated by CARONTE trainers.

Professionals discovered the unknown (for them) power of their day-by-day used tools to implement innovative approaches and strategies to search and access online scientific results, documents, and data.

In addition, some of the training modules and materials developed within the project are customized for companies of the Skillman network and made available also through the Skillman e-learning platform [16]. Skillman is a worldwide Transnational Platform of Centers of Vocational Excellence based in the EU and represented in several countries worldwide. It is designed to introduce skills, competences, and innovative curricula for the advanced manufacturing sector within the Vocational Education and Training pathways. It connects training providers with civil society and industry, giving support services that drive growth and effectiveness in the sector. Project results were also disseminated through International Conference (QQML, Infolog), the Skillmann International Forum at Didacta Fair and other public events: more than 800 people were reached.

PIMAS, InnoLOG, and InSITE

Aim and Objectives

In mineral exploration, the analytical material characterisation represents an important factor in order to guarantee sustainable productivity and advantageous profits of the associated mining activity by rising the efficiency and quality of mineral evaluation. Such a need gets crucial in many cases where high concentration deposits were mostly exploited and only low concentration tails are available, as well as for evaluating the possible convenience in re-activating mines abandoned along the past. However, traditional petrographic and compositional analyses, which have been mostly used to date, do not represent a suitable solution for the present specific material characterization problems. These require much higher analytical efficiency and lower costs than those of the traditional analytical techniques, such as microscopy, ICP-OES/MS, XRD, and other. In all raw materials extraction and processing, there is a need to perform in situ many analyses providing real-time results for prompt evaluations of the mineral content.

In this respect, CNR is focusing its activity within the KIC-RM in fostering a real breakthrough in mineral exploration practice. To date, it coordinated the Network of Infrastructure PIMAS *"Portable Instant Mineral Analysis Systems"* (2017–2019), participated to the upscaling project InnoLOG *"Innovative geophysical logging tools for mineral exploration"* (2017–2020, coordinated by CSIC, ES) and it is currently involved in the upscaling project InSITE "In situ *ore grading system using LIBS in harsh environments*" (2020–2023, coordinated by INESC TEC, PT). In all these projects, the technical contribution by CNR has been focused on its recognized excellence on laser-based techniques for material analyse and processing gained in various field of applications, such as those of cultural heritage, surgery, and industrial productions.

In PIMAS, a set of novel photonic sensors able to provide onsite and in situ real-time mineral evaluations during exploratory operations, mining, and allow large-scale characterization of raw material collections were investigated and tested in order to introduce innovative services within the RM sector. In particular, CNR focused on designing and testing LIBS *"Laser Induced Breakdown Spectroscopy,"* Raman spectroscopy, and optical setups, which were optimized in view of the practical analytical needs. The former, which is based on the detection of atomic optical emission upon excitation using a pulsed laser with high instantaneous power, can allow measuring the elemental composition. The Raman technique, which exploits a well-known non-linear scattering phenomenon of the monochromatic radiation, provides information on molecular vibrational levels and hence on the crystal structure. Thus, such a laser compositional approach can return the same type of information usually achieved using the well-established XRF-XRD analytical combination but the technological and analytical advantages it offers with respect to the latter are really significant. Raman and LIBS instruments can be miniaturized, their use does not require any sample preparation, allow high analytical speeds (up to some analyses/second and some hundreds of analyses/second, respectively), and compositional micro-depth profiling. Furthermore, LIBS calibration and/or suitable elaborations of the spectral data it provides allow quantitative evaluations of the elemental contents.

By exploiting this significant potential, InnoLOG aimed at rising the effectiveness of the exploration approach based on rapid borehole drilling and following characterization of its wall using suitable logging tools, against the alternative method based on slow core collections and following laboratory analyses, which is time consuming and closetful. Thus, it proposed to develop two novel tools based on hyperspectral imaging (by CSIC, ES) and the mentioned laser spectroscopies proposed by CNR, respectively.

Finally, the project InSITE has been fully dedicated to develop novel LIBS systems including robust technologies and advanced software based on artificial intelligence routines for material analysis in harsh environments. Among the latter a versatile portable device to be used in various operative scenarios and a LIBS equipment to be installed on a ROV for deep sea mineral exploration are currently under development.

Implementation

The mentioned NoI and upscaling projects have been implemented following a similar strategy based on the examination of the practical analytical needs of the RM sector, proposal of innovative laser-based and other photonic solutions, calibrations, and eventually validation of the whole in operative environments. Despite general purposes, handheld LIBS and Raman spectroscopy devices are available on the market, they cannot be considered as on-the-shelf tools for mineral exploration and processing. The lack of suitable calibrations, low sensitivity, low repeatability, as well unsuitable construction details provide insufficient reliability and limited versatility. Thus, innovative solutions have been aimed at overcoming such limitations and at developing dedicated tools, which are not available on the market. These have been defined developed by exploiting the most advanced photonic components such as diode lasers and diode pumped solid state (DPSS) lasers, miniaturized sensors, and optical fibers and then tested in operative conditions.

Results

A new set of photonic instruments and related services to characterize and mapping the chemical composition of minerals was developed and validated in the framework of PIMAS and InnoLOG, in collaboration with the partnerships of these projects. The main output of the former was a low-cost LIBS instrument (Fig. 4), to be used both as hand-held tool and in scanning mode to reconstruct compositional imaging.

Fig. 4 Testing the LIBS prototype developed in PIMAS. (Color figure online)

Fig. 5 Testing the novel hybrid Raman-LIBS Raman logging tool. (Color figure online)

Furthermore, other technologies combining LIBS with UV-NIR and fluorescence spectroscopies were developed within the foreseen service activities of the NoI.

A significant breakthrough was achieved within InnoLOG with the development and validation of an original hybrid Raman-LIBS logging system (Fig. 5) providing the compositional mapping of borehole walls. A related patent application was submitted and validation tests were carried out, in collaboration with the partnership of the project. The prototype is equipped with a relatively narrow probe (2″ ∅, 2.5 m length), umbilical and modules at the ground level. Presently, it can operate down to 100 m depth but there are no relevant optical limitations to extend this limit to 200 m and even more.

FREECATS

FREECATS (critical raw materials **free cat**alysts) have been the "seminal" project of CNR-SCITEC (former CNR-ISTM) research group, selected at the very first call of EIT RawMaterials in 2016. The project was active for 36 months, from 1.1.2016 to 31.12.2018, with seven partners covering all corners of the Knowledge Triangle, from five different EU countries.

Aim and Objectives

FREECATS main aim was the setup of a Network of Infrastructure (NoI) capable of providing a state-of-the-art service (trough a Single Point of Contact, a SPoC) with the scope of supporting LEs and SMEs in the development of critical raw materials-free* catalysts to be employed in various fields of industry, such as chemical industry, energy, mobility, and cleantech. Reduction, recycling or elimination of CRMs as active phase of the catalyst is critical for the economic feasibility and long-term sustainability of many applications in the fields of energy and mobility. Moreover, some recycled CRMs are not part of product/value chains and are not marketable. The development of catalysts based on them will improve economics of recycling processes.

Results

Among the results obtained during FREECATS activities deployment, it is worth to highlight the collaboration with Bracco Imaging SpA, a leading Italian pharma company. Bracco Imaging SpA was seeking new visible light-active photocatalysts for the degradation of some by-products in its production chain. Under the leadership of CNR-ISTM, FREECATS NoI joined forces and developed a series of iron- and titanium-based photocatalysts prepared through simple synthetic procedures (Fig. 6) that demonstrated to degrade pollutants in water under sunlight, thus meeting the requests of the company. Noteworthy, the collaboration was selected by EIT RM and included in "Success Stories" [17].

Indirect results of FREECATS had even more impact: a number of follow-up projects stem from the very collaborative core network, including not only EIT RM

Fig. 6 Picture of the three photo-catalysts supplied to Bracco Imaging SpA. (Color figure online)

projects (BloW UP [18] and RAISESEE [19]) but also H2020 MSCA-ITN project BIKE [20], all coordinated by CNR-SCITEC.

Blow UP successfully transferred the NoI model to Balkan area (ESEE Countries included in the Regional Innovation Scheme of EIT) for the desiloing of new waste-derived raw materials and developing new applications. By focused capacity building activities, by the inclusion of leading universities and research centers, BloW UP resulted in the development of new materials derived from local mining waste: a BiH patent has been submitted.

References

1. https://www.cnr.it/en
2. https://eitrawmaterials.eu
3. https://eit.europa.eu/
4. https://en.art-er.it
5. https://www.unibo.it/en/homepage
6. https://ldr-network.bo.cnr.it/Bologna/
7. http://sperimestate.bo.imm.cnr.it/index.html
8. http://biblioteca.bo.cnr.it/index.php/en/
9. https://www.isof.cnr.it/content/chemnanocare
10. https://labcat.istm.cnr.it
11. https://nm2lab-dissemination.ism.cnr.it/?lang=en
12. https://www.ifac.cnr.it/index.php?lang=en
13. http://rmschools.isof.cnr.it/
14. https://www.caronteproject.eu
15. https://www.caronteproject.eu/learning_modules
16. http://learn.skillman.eu
17. https://eitrawmaterials.eu/freecats-project-offers-state-of-the-art-service-supporting-the-development-of-critical-raw-materials-free-catalysts/
18. https://blowupris.istm.cnr.it/
19. https://raiseseeris.istm.cnr.it/
20. https://www.bike-msca.eu/

Experimental Determination of Liquidus Temperature and Phase Equilibria of the CaO–Al₂O₃–SiO₂–Na₂O Slag System Relevant to E-Waste Smelting

Md Khairul Islam, Michael Somerville, Mark I. Pownceby, James Tardio, Nawshad Haque, and Suresh Bhargava

Abstract The recovery of valuable and critical metals from electronic wastes (e-waste) via the pyrometallurgical route has some challenges including high processing temperatures. Designing appropriate slag systems based on the major elemental components in e-waste could bring operational advantages by lowering the liquidus temperature. In this study, the quaternary slag system CaO–Al₂O₃–SiO₂–Na₂O was investigated to determine the liquidus temperature and phase equilibria of slags relevant to e-waste smelting. The slags were thermally equilibrated at different temperatures inside a vertical tube furnace followed by rapid quenching. The quenched slags were examined by SEM to observe the phase formed and the equilibrium compositions were determined using energy dispersive (ED) spectrometry. The liquidus temperature of the slags in the anorthite (CaO·Al₂O₃·2SiO₂) phase field was significantly decreased with increasing levels of Na₂O. The slag composition moved towards the pseudo wollastonite (CaO·SiO₂) region upon the addition of Na₂O.

Keywords Liquidus · Phase equilibria · E-waste · Smelting · Slag · Na₂O · Flux

Introduction

Electronic wastes (e-waste) represent one of the fastest-growing and biggest waste streams in the world with around 50 million metric tonnes generated annually across the world [1, 2]. These wastes are a growing concern for every nation as they require proper management to avoid the detrimental effects of contained hazardous

---

M. K. Islam (✉) · J. Tardio · S. Bhargava
Centre for Advanced Materials & Industrial Chemistry (CAMIC), School of Science, RMIT University, GPO Box 2476, Melbourne, VIC 3001, Australia
e-mail: s3737858@student.rmit.edu.au

M. K. Islam · M. Somerville · M. I. Pownceby · N. Haque
CSIRO Mineral Resources, Private Bag 10, Clayton South, VIC 3169, Australia

M. K. Islam
Bangladesh Council of Scientific and Industrial Research (BCSIR), IMMM, Joypurhat 5900, Bangladesh

substances. On the other hand, e-waste has the potential to be a great source of secondary production of valuable and critical metals through recycling and recovery of the valuable materials. This type of secondary metal production is also termed "urban mining." Although several proposed methods are available to process e-wastes, so far, the pyrometallurgical technique seems to be the most-effective route when implemented at an industrial scale [3]. The pyrometallurgical route has kinetic advantages and can handle a large amount of feed compared to the alternative chemical and/or biological routes of metal recovery from e-wastes.

Printed circuit boards (PCBs) form the major value component of most e-waste. They contain valuable metals such as copper (Cu), gold (Au), silver (Ag), tin (Sn) etc. Based on the contained compositions, the $CaO-Al_2O_3-SiO_2$ (CAS) slag system is the most relevant to the recovery of metals from waste PCBs [4]. The CAS system is a highly refractory system requiring elevated temperatures to create a liquid slag. However, alkali oxides such as Na_2O added to the system presents a possible way to reduce the liquidus temperature. There is little systematic information of the phase equilibria in the $CaO-Al_2O_3-SiO_2-Na_2O$ slag system although a few studies are available which essentially focus on the physical, mechanical, and electrical properties of the glass (quenched slag) phase [5–7]. An earlier study focusing on the glass-forming region of the $CaO-Al_2O_3-SiO_2-Na_2O$ system reported phase equilibria with isoplethal sections at 5%, 10% and 15% Al_2O_3 and more than 50% SiO_2. The findings showed that with increasing Al_2O_3 content the primary phase field of plagioclase ($NaAlSi_3O_8$, $CaAl_2Si_2O_8$) expanded greatly, while that of sodium disilicate ($Na_2O \cdot 2SiO_2$) decreased [8]. Zhang and co-workers [9] studied the phase equilibria at liquidus temperatures of the pseudo ternary $CaO \cdot SiO_2-Na_2O \cdot SiO_2-Na_2O \cdot Al_2O_3 \cdot 6SiO_2$ slag system which is relevant to the applications of ash generated from the process of waste-to-energy combustion or incineration of municipal waste. In addition, the effect of alkali oxides on the liquidus temperatures of slag systems relevant to zinc and lead smelting has been studied [10]. However, because of the very low contents of alkalis in these slags, the effect of only 1 wt% Na_2O and 1 wt% K_2O in the $ZnO-FeO-Al_2O_3-CaO-SiO_2$ slags on the liquidus temperatures and phase equilibria was determined. Hence, this did not provide a broader picture in terms of the effect of alkali addition in aluminosilicate slags.

In this study, we have accurately determined the primary phases present and measured the liquidus temperature of $CaO-Al_2O_3-SiO_2-Na_2O$ slags at two different CaO/SiO_2 (C/S) ratios. The slag composition was chosen based on the contained components of mobile phone PCBs. The final slag generated after smelting PCBs contains around 45–55 wt% SiO_2, 20 wt% Al_2O_3, and 25–35 wt% CaO [11]. However, the slag composition can vary significantly depending on the previous processing steps (i.e., shredding, physical separation etc.), source and type of PCBs. In the present work, Na_2O was added to this slag system as a flux to lower the liquidus temperature. The experimental technique involved equilibration of small slag samples at the required temperature, subsequent quenching in a cold water bath followed by optical and electron microscopy to identify the primary crystals and to determine phase compositions.

Experimental

Materials

Two master slags in the CaO–Al$_2$O$_3$–SiO$_2$ system were prepared by melting reagent grade Al$_2$O$_3$, SiO$_2$, CaCO$_3$ powders at C/S ratios of 0.3 (S100) and 0.6 (S200). The required amount of CaO in the slag was obtained from the decomposition of pure CaCO$_3$. Na$_2$CO$_3$ was used to obtain the desired amount of Na$_2$O at levels of 5, 10, 15, and 20% Na$_2$O in subslags made from the two master slags (i.e., eight subslag compositions in total). Platinum foil for containing the slags in the drop quench experiments was supplied by Cookson Dental, UK.

Methods

The slags were equilibrated at high temperature in a vertical tube furnace and rapidly quenched using a water bath situated at the open end of the furnace. In this way, the high temperature structure and phase assemblage of the system were retained through rapid solidification. The equilibrated and solidified slag samples were analysed with a Scanning Electron Microscope (SEM) equipped with an Energy Dispersive (ED) X-ray spectrometer for preliminary measurement of the composition of crystal and/or liquid phases. The liquidus temperature of the slags was estimated within an uncertainty range of 10–15 °C. An iterative approach was used which started the quenching from a temperature at which the slag was completely liquid. The experimental temperature was lowered in a systematic way before quenching. After each temperature, the quenched samples were examined by SEM to identify any solid phase precipitation. Based on the presence or absence of any solid phase, the liquidus temperature was determined. The uncertainty range represents the difference between the equilibration temperatures of the single liquid phase and the two-phase (primary solid phase + liquid).

The procedure for studying the slag system is essentially divided into two parts: these include (a) slag making and (b) drop quench equilibration testing.

Drop Quench Testing—Phase Equilibria Determination

Capsules made from thin platinum foil (0.025 mm thickness) were filled with 200–300 mg of the slag mixtures and hung from the top of a vertical tube furnace by platinum wires (0.5 mm diameter). The top end of the capsules was open and exposed to the furnace (air) atmosphere. The sample was lowered to the hot zone in the centre of the furnace tube. The length of the constant temperature hot zone was 3–4 cm. The temperature of the samples was continuously measured by a B-type thermocouple inserted in an alumina sheath. The Type-B thermocouple used in the experiments was calibrated against the melting temperature of pure copper and was shown to have an accuracy of ±5 °C. The tip of the thermocouple was placed in the same zone where the slag samples were located. The thermocouple and sample holder were inserted through the top endcap and sealed, while the bottom endcap was closed with a thin polypropylene sheet just touching with the water surface of the quenching bath. The furnace arrangement for the drop quench test is schematically shown in Fig. 1.

Based on a procedure developed by Rait [12], the furnace was programmed to heat to an initial temperature 100 °C above the experimental temperature at a heating rate of 200 °C/hour. The samples were held for 30 minutes at that temperature to completely melt and homogenize the molten slag. The temperature was then reduced to 200 °C below the experimental temperature, held for 30 minutes before being heated to the final experimental temperature. The samples were allowed to equilibrate for 4–24 hours at the experimental temperature to complete the reaction. Once the equilibration time was reached, the samples were released from the top holder

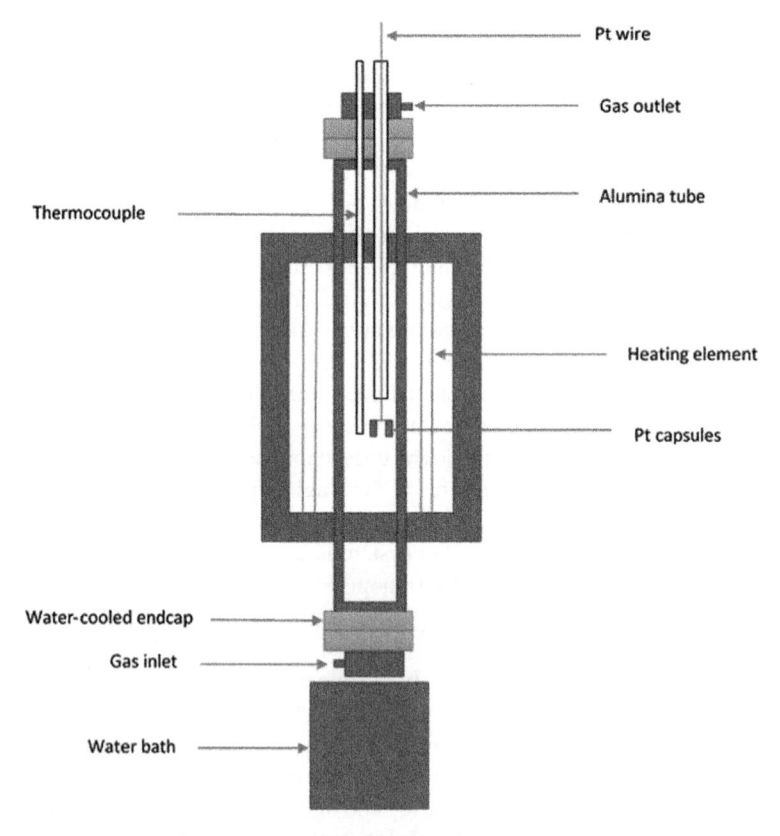

Fig. 1 Schematic diagram of the drop quench test furnace and experimental setup. (Color figure online)

Table 1 Chemical composition of the slags investigated determined via XRF

Sample ID	Sum (%)	Al$_2$O$_3$ (%)	CaO (%)	SiO$_2$ (%)	Na$_2$O (%)	C/S ratio	Equilibration time (hours)
S100	99.8	18.7	20.0	61.1	0	0.3	24
S105	99.8	17.8	19.1	58.0	4.9	0.3	24
S110	98.7	16.5	17.6	53.2	11.4	0.3	4
S115	99.5	16.1	17.3	52.8	13.3	0.3	4
S120	99.8	15.5	16.7	51.0	16.6	0.3	4
S200	99.7	19.1	29.6	51.0	0	0.6	24
S205	99.5	18.0	28.3	48.5	4.7	0.6	4
S210	99.4	17.1	26.9	46.1	9.3	0.6	4
S215	99.6	16.4	25.8	44.2	13.2	0.6	4
S220	99.3	15.7	24.8	42.3	16.5	0.6	4

to drop directly into the quenching bath to ensure rapid solidification. Achievement of equilibrium was confirmed by experimenting at different equilibration times and analyzing the compositions of the phases present (solids and liquid). Preliminary experiments suggested that most slag compositions reached equilibrium within 4 hours, although some Na-free slags took longer to reach equilibrium and therefore longer equilibration times were required (up to 24 hours).

The samples obtained from the drop quench tests were dried, separated from the platinum capsules and mounted in epoxy resin. After standard grinding and polishing of the resin blocks, the specimens were examined using a SEM to determine the phase assemblage and phases were quantitative analysed using ED X-ray spectrometry.

Confirmation of the Slag Composition

Compositions of the quenched samples of two master slags (S100 and S200) and the eight Na-doped sub-slags were analysed by X-ray Fluorescence Spectroscopy (XRF). The compositions of the investigated slags are shown in Table 1. The resulting slag compositions were close to the target composition and showed negligible Na loss during melting. The composition of the two master slags and their primary phase fields are shown in the ternary phase diagram in Fig. 2.

SEM–ED Analysis

The quenched slag samples were mounted in polymer resin blocks. These were initially ground using SiC abrasive papers, finely polished with diamond suspensions

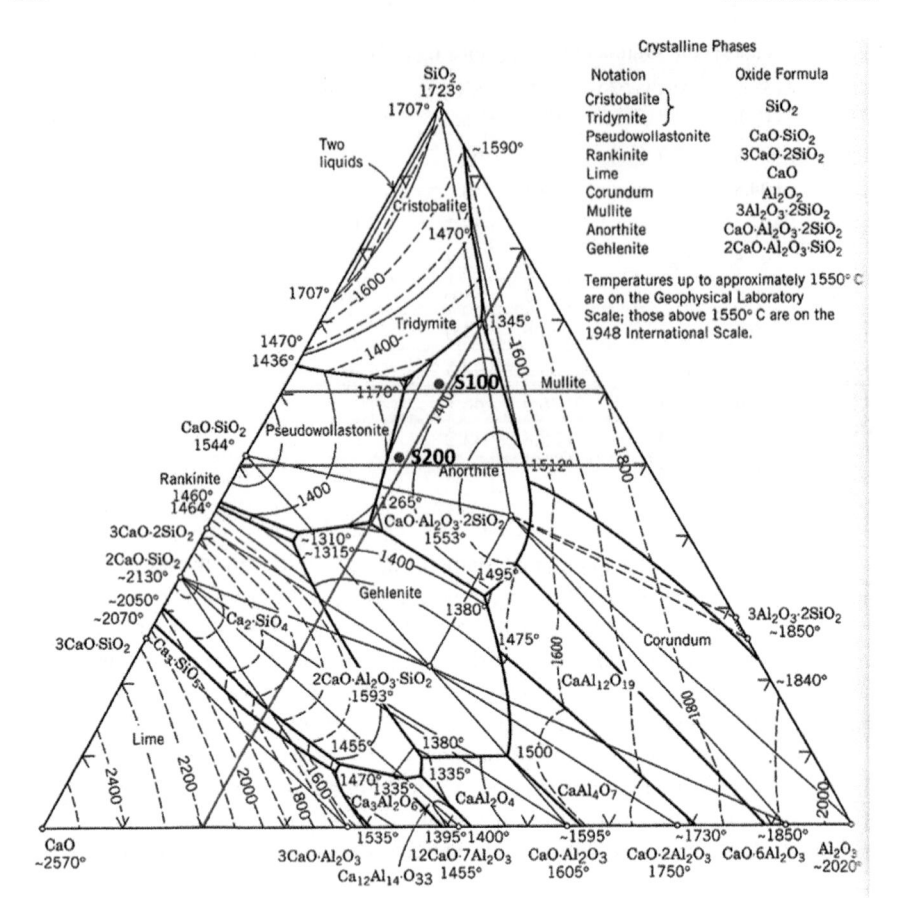

Fig. 2 Ternary CaO–Al$_2$O$_3$–SiO$_2$ phase diagram showing the composition of the two master slags investigated [13]. (Color figure online)

of different particle sizes (6, 3, and 1 μm) and then finally with colloidal silica. The polished samples were carbon coated before examination in the SEM to avoid charge build up on the surface. A Bruker SEM equipped with an ED X-ray spectrometer was used to investigate the microstructure and identify the phases present. The SEM analysis was done using an accelerating potential of 15 kV with a working distance of 10.0 mm. The preliminary compositions of the solid and/or liquid phases present were measured by ED X-ray analysis.

Confirmation of Equilibrium

Reaction at the experimental temperature was carried out for different times ranging from 4 to 24 hours. It was observed that most of the slag samples reached equilibrium in less than 4 hours. Some of the slags however, especially near the ternary eutectic compositions, showed difficulties in crystallization and took a longer time to nucleate and precipitate crystals. To ensure equilibrium was reached, two different measurements were done. Slag of a particular composition was equilibrated for different times and the composition of phases at several locations were measured. For example, the S200 master slag was equilibrated at 1300 °C for 4 hours and 24 hours. The quenched slag showed anorthite ($CaO \cdot Al_2O_3 \cdot 2SiO_2$) crystals and a liquid phase. The average liquid phase and solid phase compositions were compared with their respective phase compositions from both the 4 and 24 hours equilibrated samples. The homogeneity of the crystals and the liquid slag in different locations confirmed that equilibrium was reached after 4 hours of equilibration, since after 24 hours equilibration time, there was no significant variation in the composition of the individual phases. It was therefore decided that 4 hours equilibration was sufficient for reaction to be completed.

Results and Discussion

Primary Phase Identification

The slags were equilibrated at a range of temperatures to identify the liquidus temperature. In all the samples, the composition of liquid and crystal phases in equilibrium were determined. Anorthite ($CaO \cdot Al_2O_3 \cdot 2SiO_2$) and pseudo wollastonite ($CaO \cdot SiO_2$) primary phases in equilibrium with the liquid slag were identified. Figure 3a–f shows the representative microstructures of the primary phases in equilibrium with the liquid. Figure 3a shows the microstructure of a slag quenched from a pure liquid meaning the liquidus temperature of this slag was below the experimental temperature from which it was quenched. The anorthite primary phase in equilibrium with the liquid phase (glass) is shown in Fig. 3b, f, while Fig. 3c–e shows the precipitation of pseudo wollastonite coexisting with a liquid.

Microscopic observation of the quenched slag samples revealed the high-temperature phases present. Sometimes the liquid phase and the precipitated solid crystals had very similar compositions and were difficult to distinguish in BSE imaging. A good balance of brightness and contrast and searching for crystals (if any) throughout the whole sample surface was required to determine the presence of crystals precipitated from the liquid. Such an example is anorthite in equilibrium with liquid in the S100 sample at 1300 °C is shown in Fig. 3f. It was necessary to use a very high contrast to distinguish anorthite from the matrix during examination

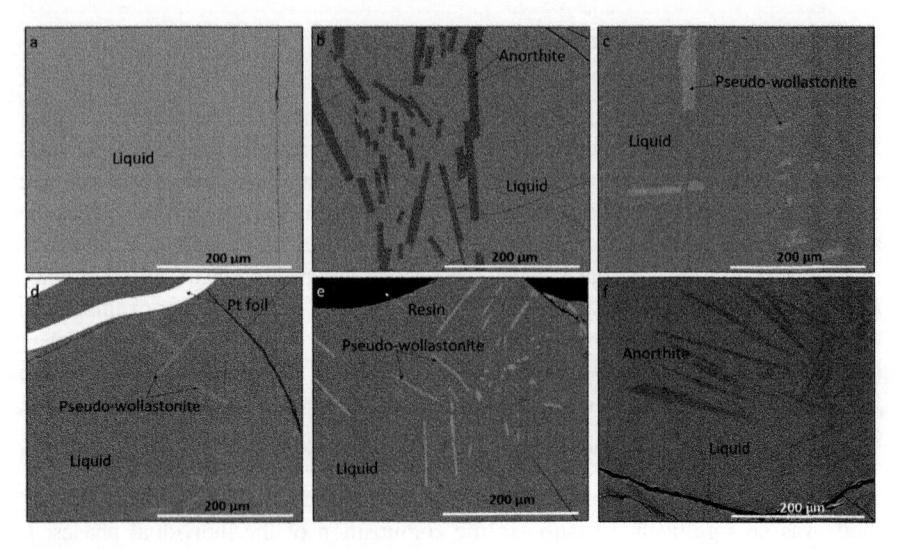

Fig. 3 Representative Back-Scattered Electron (BSE) images of the quenched slags (**a**) only liquid phase of sample S205 at 1300 °C (**b**) anorthite + liquid phase of sample S200 at 1300 °C (**c**) pseudo wollastonite + liquid phase of sample S205 at 1280 °C (**d**) pseudo wollastonite + liquid phase of sample S215 at 1170 °C (**e**) pseudo wollastonite + liquid phase of sample S110 at 1244 °C (**f**) anorthite + liquid phase of S100 at 1300 °C taken at high contrast

with SEM. However, at a very high contrast, the image can become noisy and a very slow scan rate is required to obtain good quality images.

For the Na_2O containing samples, at conditions close to the liquidus, the phase fraction of solid crystals (pseudo wollastonite) was very low. Hence, it was necessary to examine the whole specimen surface to find the crystals. In many samples, the crystals were found near the platinum foil edges. These edges may create nucleation sites for crystallization. Consequently, in many cases at or near the liquidus temperature, the crystals were often found in gaps within the Pt foil while the rest of the specimen was a homogenous glassy structure.

Estimation of the Liquidus Temperature

The drop quench test work required an iterative approach to liquidus determination. For a particular slag composition, many tests were required to narrow the temperature range between fully liquid samples and samples containing crystals. Fresh powdered slag samples were used for every test run. In this study, the liquidus temperature for the 8 different quaternary slag compositions within the $CaO–Al_2O_3–SiO_2–Na_2O$ system was experimentally determined. The liquidus temperature was estimated within an uncertainly limit of 10–15 °C. For the two master slags S100 and S200, the liquidus temperatures were determined to be 1320–1335 °C and 1335–1340 °C, respectively. The liquidus temperature of slag compositions midway between the measured compositions can be estimated from the trends of the two series of slags.

Effect of Na$_2$O on the Liquidus Temperature of Quaternary Slag System

Generally, the liquidus temperature decreased with increasing Na$_2$O content in the slags of all compositions examined. However, the extent to which the liquidus temperature was affected depended on the C/S ratio of the bulk slag, the Al$_2$O$_3$ content, and the primary phase field of the bulk slags. The main reason behind the lowering of liquidus is the destabilization of the silica network due to the depolymerization caused by Na$_2$O incorporation [14].

The experimentally obtained liquidus temperatures of the master slags and the eight Na-doped subslags were plotted against the Na$_2$O content. The results are shown in Fig. 4. The decline in liquidus temperature with increasing Na$_2$O content followed different trends depending on the C/S ratio of the slags and the primary phase field of the starting composition. It was noted that the trend of decreasing liquidus was greater for the S100 and S200 slag systems (C/S ratio 0.3 and 0.6 at Al$_2$O$_3$ content ranging from 15.5 to 19.1 wt%). These ternary compositions lie in the anorthite primary phase field of the ternary CaO–Al$_2$O$_3$–SiO$_2$ (CAS) slag system.

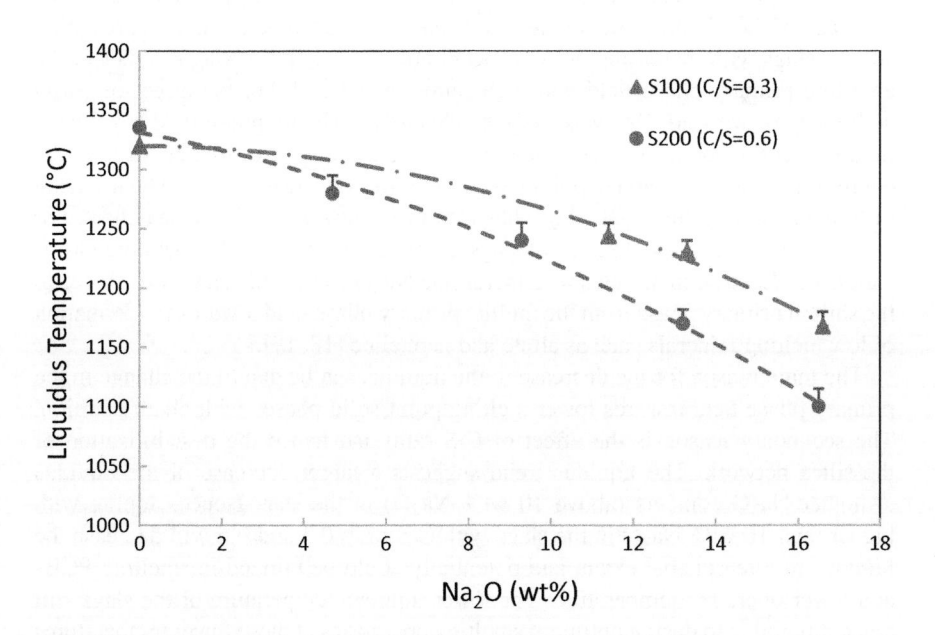

Fig. 4 The liquidus temperature of slags with varying Na$_2$O content. (Color figure online)

From Fig. 4, it can be seen that the liquidus temperature decreased to 1170 °C when doped with 16.5 wt% Na_2O for the slag series S100 (C/S ratio 0.3), while the liquidus temperature of the S100 master slag without Na_2O was around 1350 °C [15]. This indicated a 180 °C reduction in the slag liquidus temperature due to the incorporation of Na_2O. The trend line of the liquidus for this series of slags with varying Na_2O showed a gradual and consistent decrease of the liquidus with increasing Na_2O content.

Similarly, with the S200 slag series (C/S ratio 0.6), where the bulk composition before adding Na_2O was in anorthite primary phase field, the liquidus temperature followed a steady fall with increasing Na_2O content. Here the lowest liquidus temperature dropped to around 1100 °C at a C/S ratio = 0.6 and 15.7 wt% Al_2O_3 and 16.5 wt% Na_2O. This was the lowest melting minerals such as albiteliquidus temperature of the slag composition range investigated in this research.

The microstructure of the quenched slag revealed that the primary phase field shifted from anorthite to pseudo wollastonite in the case of slags with C/S ratios of 0.3 and 0.6. Consequently, this shifted the primary phase field from a higher melting temperature solid phase towards a lower melting solid phase.

Due to the gradual increase of Na_2O in the bulk slag, the proportion of Al_2O_3 decreased for a fixed C/S ratio. This change in bulk composition shifted the equilibrium primary phase from anorthite to pseudo wollastonite for the S100 and S200 slag series (Fig. 3b–e). The alkali oxide (Na_2O) acts as a network modifier in the silicate structure by breaking the covalent bonds between Si and O (bridging oxygen bonds) and forming new ionic bonds between Na and O (non-bridging oxygen bonds). This breakage of the silicate structure decreases the liquidus temperature of the resulting silicate slags with increasing Na_2O concentration [16, 17]. For slag samples in the anorthite primary phase field with C/S ratios of 0.3 and 0.6, the effect of Na_2O addition was two-fold. One was from the change of primary phase field from anorthite to pseudo wollastonite and the other from the presence of higher Na_2O in the equilibrium liquid slag and depolymerization of the silicate networks. The lowering of liquidus temperature with Na_2O has also been shown by other researchers. For example, in work on the liquidus of slags generated during high-temperature gasification, the decrease in liquidus with increasing Na_2O content of ash was attributed to the shift in primary phase from the mullite primary phase field towards the formation of low melting minerals such as albite and nepheline [18, 19].

The main reason for the decrease in the liquidus can be due to the change in the primary phase field towards lower melting point solid phases with alkali addition. The secondary reason is the effect of C/S ratio and hence the destabilization of the silica network. The liquidus trend suggests a larger decrease of the liquidus at higher Na_2O contents (above 10 wt% Na_2O) in the slag. Hence, doping with higher than 10 wt% Na_2O in the slags with C/S ratio 0.3 and 0.6 will decrease the liquidus to a reasonable extent and potentially could be utilized in smelting PCBs at a lower operating temperature. The lower liquidus temperature of the slags will make it possible to design optimum smelting operations at these lower temperatures bringing flexibility to the design and operation of smelters.

Conclusion

The primary phase field was identified and the liquidus temperature was determined to within 10–15 °C for a range of compositions of the quaternary CaO–SiO$_2$–Al$_2$O$_3$–Na$_2$O slag system by drop quench experiments followed by scanning electron microscopy.

The primary phase field shifted from anorthite to pseudo wollastonite due to the addition of Na$_2$O into the ternary CAS slags with C/S ratios of 0.3 and 0.6. The liquidus temperature significantly decreased with increasing Na$_2$O content in the CaO–SiO$_2$–Al$_2$O$_3$–Na$_2$O quaternary slag system. In the case of compositions in the anorthite primary phase field with ternary CAS slag C/S ratios of 0.3 and 0.6, the addition of Na$_2$O up to 16.5 wt% lowered the liquidus temperature by around 150 °C and 240 °C, respectively.

Accurate determination of the liquidus data reported in this study will enrich the thermodynamic database of the CaO–SiO$_2$–Al$_2$O$_3$–Na$_2$O quaternary slag system and help design efficient smelting operations involving this slag system. The significant drop of the liquidus temperature with Na$_2$O doping will reduce the operating temperature in e-waste smelting using these slags. Lower smelting temperatures reduce the overall energy consumption leading towards a potentially low-cost processing technique to recover value metals from e-waste.

References

1. Chen M, Avarmaa K, Klemettinen L, O'Brien H, Sukhomlinov D, Shi J, Taskinen P, Jokilaakso A (2020) Recovery of precious metals (Au, Ag, Pt, and Pd) from urban mining through copper smelting. Metall Mater Trans B 51(4):1495–1508
2. Zeng X, Mathews JA, Li J (2018) Urban mining of e-waste is becoming more cost-effective than virgin mining. Environ Sci Technol 52(8):4835–4841
3. Zhang L, Xu Z (2016) A review of current progress of recycling technologies for metals from waste electrical and electronic equipment. J Cleaner Prod 127:19–36
4. Park HS, Kim YJ (2019) A novel process of extracting precious metals from waste printed circuit boards: Utilization of gold concentrate as a fluxing material. J Hazard Mater 365:659–664
5. Ozawa S, Susa M (2005) Effect of Na$_2$O additions on thermal conductivities of CaO–SiO$_2$ slags. Ironmaking Steelmaking 32(6):487–493
6. Zhang GH, Zheng WW, Jiao S, Chou KC (2017) Influences of Na$_2$O and K$_2$O additions on electrical conductivity of CaO–SiO$_2$–(Al$_2$O$_3$) melts. ISIJ Int 57(12):2091–2096
7. Esteban-Tejeda L, Zheng K, Prado C, Cabal B, Torrecillas R, Boccaccini AR, Moya JS (2016) Bone tissue scaffolds based on antimicrobial SiO$_2$–Na$_2$O–Al$_2$O$_3$–CaO–B$_2$O$_3$ glass. J Non-Cryst Solids 432:73–80
8. Moir GK, Glasser FP (1976) Phase equilibria in the glass-forming region of the system Na$_2$O–CaO–Al$_2$O$_3$–SiO$_2$. Phys Chem Glasses 17(3):45–53
9. Zhang Z, Xiao Y, Voncken JHL, Yang Y, Boom R, Wang N, Zou Z, Li G (2014) Phase equilibria in the CaO·SiO$_2$–Na$_2$O·SiO$_2$–Na$_2$O·Al$_2$O$_3$·6SiO$_2$ system. J Eur Ceram Soc 34(2):533–539
10. Zhao B, Hayes PC, Jak E (2013) Effects of Na$_2$O and K$_2$O on liquidus temperatures in ZnO–'FeO'–Al$_2$O$_3$–CaO–SiO$_2$ system in equilibrium with metallic iron. Miner Process Extr Metall 121(1):32–39

11. Park HS, Han YS, Park JH (2019) Massive recycling of waste mobile phones: Pyrolysis, physical treatment, and pyrometallurgical processing of insoluble residue. ACS Sustain Chem Eng 7(16):14119–14125
12. Rait R (1997) Phase equilibria in iron bath smelting type slags. Ph.D. Thesis, University of Melbourne
13. Levin EM, McMurdie HF (1975) Phase diagrams for ceramists. 1975 supplement, Fig. 630
14. Mills K, Yuan L, Li Z, Zhang G, Chou K (2012) A review of the factors affecting the thermophysical properties of silicate slags. High Temp Mater Process 31(4–5):301–321
15. Rankin GA (1915) The ternary system $CaO-Al_2O_3-SiO_2$. Am J Sci 39(229):1–79
16. Mills KC (1993) The influence of structure on the physico-chemical properties of slags. ISIJ Int 33(1):148–55
17. Mysen BO (1990) Relationships between silicate melt structure and petrologic processes. Earth-Sci Rev 27(4):281–365
18. Chen X-D, Kong L-X, Bai J, Bai Z-Q, Li W (2016) Effect of Na_2O on mineral transformation of coal ash under high temperature gasification condition. J Fuel Chem Technol 44(3):263–272
19. Ilyushechkin AY, Hla SS, Chen X, Roberts DG (2018) Effect of sodium in brown coal ash transformations and slagging behaviour under gasification conditions. Fuel Process Technol 179:86–98

How to Prepare Future Generations for the Challenges in the Raw Materials Sector

Armida Torreggiani, Alberto Zanelli, Alessandra Degli Esposti, Eleonora Polo, Paolo Dambruoso, Renata Lapinska-Viola, Kerstin Forsberg, and Emilia Benvenuti

Abstract Today some raw materials (RMs) have become essential in the manufacturing of common goods and technologies (i.e., mobile phones, computers, automobiles). Readily accessible raw materials, such as rare earth elements (REEs), indium, neodymium, and others are important to industries and allow the transition towards a low-carbon economy. With the future global resource use projected to double by 2030, addressing raw materials through the entire value chain becomes a priority as well as transferring these ideas to youngsters. Some learning paths for pupils from 10 to 18 years old were developed in the framework of a European project, Raw Matters Ambassadors @Schools (RM@Schools), funded by the Knowledge and Innovation Community "Raw Materials" of the European Institute of Innovation

---

A. Torreggiani (✉) · A. Zanelli · A. Degli Esposti · E. Polo · P. Dambruoso · R. Lapinska-Viola
Institute of the Organic Synthesis and Photoreactivity – Italian National Research Council (ISOF-CNR), Via P. Gobetti, 101, 40129 Bologna, Italy
e-mail: armida.torreggiani@isof.cnr.it

A. Zanelli
e-mail: alberto.zanelli@isof.cnr.it

A. Degli Esposti
e-mail: alessandra.degliesposti@isof.cnr.it

E. Polo
e-mail: tr3@unife.it

P. Dambruoso
e-mail: paolo.dambruoso@isof.cnr.it

R. Lapinska-Viola
e-mail: renata.lapinska-viola@isof.cnr.it

K. Forsberg
KTH - Royal Institute of Technology, Teknikringen 42, 11428 Stockholm, Sweden
e-mail: kerstino@kth.se

E. Benvenuti
Institute of Nanostructured Materials - Italian National Research Council (ISMN-CNR), Via P. Gobetti, 101, 40129 Bologna, Italy
e-mail: ebenvenuti@bo.ismn.cnr.it

and Technology (KIC EIT-RM). It aims to increase among youngsters the understanding of how RMs are needed in modern society and to make careers in RM attractive. Thanks to a strategic European Partnership among the three sides of the knowledge triangle (research, education, and business), RM@Schools has developed learning pathways where different educational approaches are used to foster students' interest in science and technology, in particular in circular economy, and RM-related topics. The pathways are oriented toward a common goal: students are guided to become Young RM Ambassadors (science communicators) and create a "product" to be communicated outside of the class. By doing this, students develop twenty-first century learning skills such as creativity, critical thinking, awareness of responsibility, and teamwork.

Keywords Raw materials · Sustainability · Education · Schools · Science dissemination

Introduction

The use of natural earth materials has been crucial to the development and advancement of societies through time. Materials derived from the Earth's crust are integral to our daily life and have served many purposes in our development. In particular, many of our current technologies (mobile phones, computers, automobiles), and sources of renewable green energy (solar and wind) have become more and more reliant on metals (i.e., REEs, indium, etc.).

Supply problems or limited availability of these RMs can prevent the development and dissemination of renewable energy projects to address issues like climate change and transition to a low-carbon economy [1, 2]. Raw materials are ranked according to the economic importance of the material compared with the risk in accessing the material (the supply risk) [3]. If the supply risk is high and the material has high economic importance, then it will be considered a critical raw material for the functioning of the European economy. This list can shift as supply risk or importance change, for example when sources for material are found or it is substituted by another material.

Critical raw materials (CRMs) are essential for high-tech products and breakthrough technologies [4]. For example, a smartphone might contain up to 50 different metals, all of which provide different properties such as lightweight and user-friendly small size. In the photovoltaic sector, cadmium (Cd), germanium (Ge), gallium (Ga), tellurium (Te), selenium (Se), and indium (In) are necessary for photovoltaic panels in the cells that convert solar radiation into electrical current. Permanent magnets, used in wind turbines, require neodymium–iron–boron high-strength magnets containing three REEs: Nd—neodymium, Dy—dysprosium, and Tb—terbium, that are categorized as critical metals (CRMs). Electric and hybrid vehicles [5] also require a large number of REEs in component sensors, electric motors, and generators, Liquid Crystal Display (LCD) Screens, catalytic converters, glass, and mirrors.

The prospect of doubling the global resources use by 2030, the priorities are to address raw materials through the entire value chain (sourcing, use and recycling) and to foster a change of mind-set in young people.

To trigger the students' interest in raw materials and a sustainable society, several learning paths for pupils aged 10–18 years were developed in the framework of Raw Matters Ambassadors at Schools (RM@Schools), an European project funded by the European Institute for Innovation and Technology (EIT), the largest consortium in the raw materials sector worldwide [6, 7].

RM@Schools project is based on a successful national project, "Research Language" (http://www.bo.cnr.it/linguaggiodellaricerca/), launched in Italy in 2003 by scientists of the National research Council of Italy, based on a close collaboration among schools, scientists, and experts in science communication and dissemination. EIT RawMaterials provided the necessary support to scale up this project so that it can be successfully conducted in many countries.

RM@Schools was established in 2016 and became the flagship project in the Wider Society Learning segment of the EIT RawMaterials in 2018 [8] and it will run until 2022, due to its relevance as frontrunner project. Moreover, the project has been selected in 2020 as an example of Responsible Research and Innovation (RRI) project for its contribution to the ethics pillar, since it teaches on the utility and indispensability of minerals and mining, focusing on the consequences of their uses and production systems, the importance of the consumer's behaviour, the effects of consumption, as well as the recycling opportunities and the importance of research for a transition towards circular economy.

The RM@Schools consortium includes scientific organisations from European countries with expertise in the fields of raw materials and education (Fig. 1). It is led by National Research Council of Italy (CNR) and works to develop cooperation between the three sides of the knowledge triangle: research, education, and business in order to introduce students to issues around the value of raw materials while promoting new professional careers in this sector.

A network of research institutes, museums, and companies works together to create attractive activities for secondary schools.

RM@Schools—Aim and Objectives

Raw MatTERS Ambassadors at Schools (RM@Schools) is a Wider Society Learning project focused on an innovative program to make science education and careers in raw materials attractive for youngster [8]. In the project name, "Raw MatTERS," the key message resonates: raw materials are the key focus, whatever their origin and "Tackling European Resources Sustainably (TERS)" underlines that it is necessary to actively tackle the challenge ahead of us. Sustainability and the efficient use of all natural resources are of central importance in everything we do. Achieving this vision really matters to us.

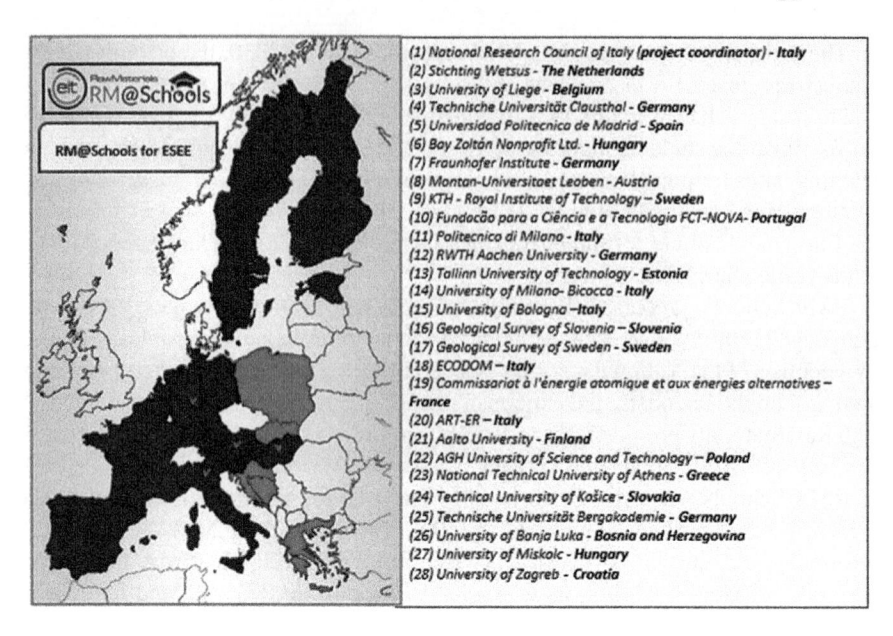

Fig. 1 RM@Schools Consortium (2019) [9]. (Color figure online)

The project offers an active learning to schools by RM Ambassadors (experts in some RM-related issues and trained teachers) by involving students in experiments with RM-related hands-on educational kits, in excursions in industries, and in science dissemination activities [9].

Different educational approaches are used to foster students' interest in science and technology, in particular in circular economy and RM-related topics, but the core element of the RM@Schools approach is to empower students to communicate with peers and wider society about critical concepts related to raw materials and their use, so the students are asked to become Young RM Ambassadors themselves (science communicators).

Local Contests for awarding the best communication products as well as an annual European Conference with delegates from European schools (students and teachers) are annually organized.

In addition, teachers are trained to become RM Ambassadors themselves in the future at school, and selected groups of students are trained on activities suitable to be proposed during Public Events in order to work together with RM Ambassadors.

Implementation

RM@Schools has developed a specific Learning Path Methodology and the learning pathways cover the whole RMs value chain from geology to electronic waste management, and a combination of educational approaches are used, such as frontal lessons, open discussion, learning by doing, peer-to-peer education, creating communication material and gamification [10–12].

The main steps of the RM@Schools learning pathway (Fig. 2), addressed to pupils aged 10–18 years, are the following:

1. **Lesson**—It begins by introducing the students to the relevant knowledge to understand the topic. This can be done starting with a frontal lesson provided by the teacher or a guest speaker or by working with resources (active-learning), followed-up by a lesson. It should also involve a sharing of the personal experiences of the students on the lesson's topic and a final discussion. This can be accomplished by using triggers, such as the smartphone example shown earlier.
2. **Activity**—Students are then engaged in a practical activity—laboratory experiments, serious games, etc.—to support or extend their learning.
3. **Visit**—After the activity, students have the opportunity to visit a place that is relevant to the lesson, so they can experience its impact in the real world.
4. **Create/Communicate**—Afterwards, the students are asked to communicate what they have learnt through the creation of a product designed to promote dialogue around a key message. This element has two major parts: (a) Create and (b) Communicate. The teacher during this phase can use a guided- or a student-centred approach. Different educational approaches such as teamwork, peer-to-peer learning, cooperative learning, and gamification can be used in order to foster creative thinking. Once students have developed a product (i.e., video, comic, poster, lab activity, etc.), they are expected to share it within their school and then in a larger community (Fig. 3).
5. **Society**—Students engagement with society can be done through participation in public events such as science fairs and festivals. Students can also choose to further engage with society through raising public awareness or taking social action.

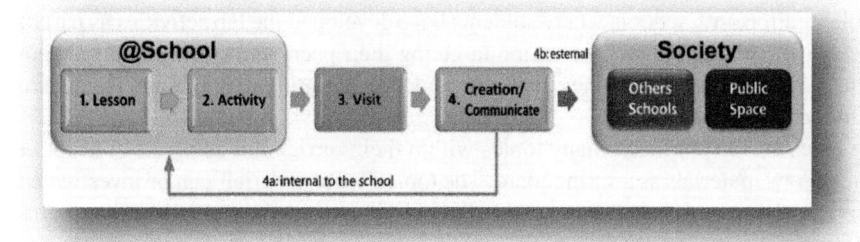

Fig. 2 The main steps of the RM@Schools learning pathway. (Color figure online)

Fig. 3 (left) Development of a tutorial video by Liceo N. Copernico—Bologna, Italy; (right) stand led by students from Bosnia and Herzegovina during a Public event. (Color figure online)

The learning paths are oriented toward the common goal of guiding students to create a "product" to be communicated outside of the class. By doing this, students develop skills such as creativity, initiative, critical thinking, awareness of responsibility, and teamwork, in line with the *Entrepreneurial Skills Pass document* [13]. With the end goal of creating a communication tool targeting an external audience, students are actively involved in the learning process.

Students benefit from such an approach more widely as the skills involved are increasingly viewed as important skills in the workplace. Many jobs now require creative thinking and communication skills, including positions in the world of business and science. People usually think that creativity is an inbuilt talent, but it is not entirely so. As with any skills, creativity can be nurtured in the classroom with simple strategies and practices that trigger creative thinking and provide opportunities for communicative action. Ideas generate new ideas.

RM@School Hands-on Toolkits

An active learning approach is encouraged by involving students in experiments using raw material-related hands-on educational toolkits and in communication actions. These toolkits have been developed by the consortium's experts, in some cases in collaboration with schools where students have developed the lab activities as part of their practical communication action targeting their peers and society around them [9]. In Table 1, some activities available under Exploration and Mining Theme are summarised as example.

Teachers can approach many topics within their curriculum using the concept of critical raw materials as a springboard. The topic of raw materials can be investigated from a science and technology perspective, but it can also be studied from a sociopolitical or economic viewpoint.

Table 1 Toolkits for schools available under the theme Exploration and Mining

Toolkit's title	Target age (years)	Learning objectives	Content	Subject links
MineralCheck	10–19	Students understand what minerals are and how to identify them, how they move between biotic and abiotic systems	Lab activity to test types of minerals and identify them	Geology; Geography; Chemistry; Magnetism
RockCheck	10–19	Students understand relationship between minerals and rocks, learn the rock cycle, identify different rock types	Computer app and detailed information on rocks and rock cycle	Geology; Geography: Chemistry; Magnetism
RockGame—rock the rock cycle	10–19	Students connect minerals and rocks with raw mineral resources	Board game: board template and cards provided: need to print out and prepare	Geology; Geography; Chemistry; Magnetism
RawMatCards	14–19	Students appreciate the role of minerals in modern technologies	Detailed information for each critical element provided for the teacher in pdf form, printable cards provided	Geology; Geography; Technology; Economics
RAWsiko—materials around us	12–19	Students understand relationship between CRMs and their distribution in the world, and the CRMs' strategic importance for many applications	Board game: board template (based on real RM distribution world map) and cards provided Provides information on why CRMs are economically and strategically important for the European economy and have a high risk associated with their supply	Geography; Technology; Economics

(continued)

Table 1 (continued)

Toolkit's title	Target age (years)	Learning objectives	Content	Subject links
RM@Art—colouring with minerals	12–19	Students learn about the usage of ores and minerals in everyday life and as pigments	Lab activity to learn about different proprieties of minerals and to make their own colours for painting	Geology; Art

Fig. 4 Example of hands-on toolkit and tutorials available on the RM@Schools website (Keratin extraction from wool waste for use in metal mining) by Liceo L. Galvani—Bologna, Italy. (Color figure online)

Science Communication by Students

Encouraging students to communicate within their class and to other audiences is a key part of their learning. In order to improve soft skills such as critical and creative thinking, teamwork capability, etc., and promote inclusion of all individuals—all talents necessary in science learning—students are asked to become societal Ambassadors by creating dissemination, by giving discursive argumentation with "peers" ("talking science") and participating to public events as active players.

The final communication product developed can take several forms, i.e., (a) making presentations, videos, surveys, posters, etc. (i.e. *A treasure in my pocket*—click here); (b) Participating in public science dissemination events; (c) Developing lab experiments for peers and creating tutorial videos with explanation, supporting text, etc. (Fig. 4). Good examples of communication products and experiments developed by students can be found on the RM@Schools website under "gallery."

Results

The project becomes the flagship project in the Wider Society Learning segment of the EIT RawMaterials in 2018, and every year involves a huge number of students and teachers from high schools. The best results of the last years have been the strengthening of the international collaboration among the Consortium, the strong

Fig. 5 European Conference RM@Schools project during 2019, CNR Research Area, Bologna, IT. (Color figure online)

enlargement of the European Network, and the major involvement of Young RM Ambassadors not only at local but also at the European level.

Strengthening the European and Local Networks made possible to organise about 70 open-access events only in 2019 by involving about 16.000 people (generic public, classes in our learning paths, teachers, etc.). Moreover, the dialogue with other countries, in particular, in the ESEE regions was opened thanks to the RM@Schools inclusive action and from this dialogue a new project called RM@Schools-ESEE was born (2020).

Finally, the European Conference, annually organised in Bologna (IT), is connected with other EIT RawMaterials Projects (i.e., ESEE Education Initiatives, BetterGeoEdu, SmartPlace@Schools, etc.) [14, 15] and involves delegates from 18 European countries and 500 participants. More than 70 young RM Ambassadors from 14 Countries took part to the 2019 EU Conference (Fig. 5). This project's European Conference plays pivotal role in teaching science communication. It is not only a very important celebration of work done by students but it also gives young people the opportunity to meet their peers from the other countries and explain by themselves on the stage what they have created in the framework of the project.

Feedback Survey

In order to evaluate the impact of the project among youngsters, a simple survey, limited to the students' prospective, was set up and the filled questionnaires were then processed. The first question focussed on a general evaluation of the project. At the item "*How do you evaluate the RM@Schools project?*" on a scale from 1 ("poor") to

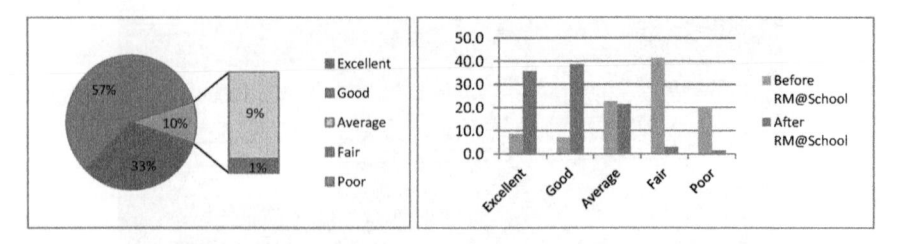

Fig. 6 (left) Percentage distribution of approval scores to the project; (right) Distribution of the "awareness about the raw materials" scores, before and after the project (absolute values). (Color figure online)

5 ("excellent"), the mean score was 4.2 (d.s. 0.7). 90% of the respondents assigned a high score (Fig. 6-left).

It is interesting to note the answers to the Item numbers 2 and 3: *"How do you evaluate your awareness about the existence of critical raw materials before participating in RM@Schools project"* and *"How do you evaluate your awareness about the existence of critical raw materials after participating in RM@Schools project"* respectively. Indeed, the evaluation of the awareness before–after the project significantly increased from "fair," to "good." The percentage of "excellent" and "good" increases from 8.6 and 7.1 to 35.7 and 38.6, respectively, as shown in Fig. 6-right.

Conclusions

Raw material is a great topic for students to investigate policy, consumerism, and the interaction between economics, politics, and product use and development.

Education and awareness of the uses of raw materials can lead to changes in governance as the values and the voice of citizens are listened to. By gaining a better understanding of these complex issues, we hope that students will become responsible and active citizens. Thus, the RM@Schools methodology could be helpful in changing societal perceptions of RMs from "indifference" to "involvement and responsibility" as well as for ensuring a next generation of well educated, innovative, and multidisciplinary experts so much needed in 21st century industry.

References

1. Report on critical raw materials and the circular economy, https://ec.europa.eu/transparency/regdoc/rep/10102/2018/EN/SWD-2018-36-F1-EN-MAIN-PART-2.PDF
2. Vidal O, Goffé B, Arndt N (2013) Nat Geosci 6
3. Frenzel M, Kullik J, Reuter MA, Gutzmer J (2017) Raw material 'criticality'—sense or nonsense? J. Phys D: Appl Phys 50, 123002 https://iopscience.iop.org/article/10.1088/1361-6463/aa5b64/pdf)

4. https://ec.europa.eu/growth/sectors/raw-materials/specific-interest/critical_en
5. https://www.universiteitleiden.nl/en/research/research-projects/science/cmlrare-earth-supply-chain-and-industrial-ecosystem-a-material-flow-assessment-of-european-union
6. https://eit.europa.eu/
7. https://eitrawmaterials.eu
8. Torreggiani A, Zanelli A, Canino M, Sotgiu G, Benvenuti E, Forini L, Aluigi A, Polo E, Lapinska-Viola R, Degli Esposti A (2020) RM@Schools: Fostering Students' Interest in Raw Materials and a Sustainable Society. In: 10th International conference the future of education virtual edition, Firenze (Italy) virtual conference, 18-19/6/2020
9. http://rmschools.isof.cnr.it/
10. Tobin K, Fraser BJ (1990) What does it mean to be an exemplary teacher? J Res Sci Teach 27(1):13–25
11. Alsop S, Bencze L, Pedretti E (eds) (2005) Analysing Exemplary Science Teaching. Bell and Bain Ltd, London
12. Colburn A (2003) The Lingo of Learning: 88 Education Terms Every Science Teacher Should Know. NSTA Press, VA, Arlington
13. Entrepreneurial Skills Pass, https://www.wko.at/site/bildungspolitik/entrepreneurial-skills-pass.pdf
14. https://www.bettergeoedu.com/
15. https://ecoceo.eu

Part VI
V, Ce, Mo, Cr, and Fe

Transformation and Distribution of Vanadium Phases in Stone Coal and Combustion Fly Ash

Deng Zhi-gan, Tang Fu-li, Wei Chang, Fan Gang, Li Min-ting, Li Xing-bin, and Li Cun-xiong

Abstract Stone coal is a kind of carbonaceous shale that contains vanadium. The sequential chemical extractions were adopted to analyze the vanadium phases of stone coal and its combustion fly ash. The results showed that the vanadium in stone coal mainly consisted in aluminosilicate. The vanadium in fly ash mainly existed in organic matter, aluminosilicate and Fe–Mn oxides, and the other vanadium existed in exchangeable fraction. Through the burning process, the vanadium was released and enriched in fly ash. Because of particular vanadium migration behaviors, the distribution of vanadium in phases became more scattered, and the leaching of vanadium in fly ash was easier than that in stone coal.

Keywords Stone coal · Combustion fly ash · Vanadium phases · Sequential chemical extraction

Introduction

Stone coal is a kind of carbonaceous shale that contains vanadium. In China, the reserves of vanadium in stone coal are 1.18×10^5 t in terms of V_2O_5, accounting for more than 87% of the Chinese reserve of vanadium [1–5]. The mineral compositions of stone coal are very complex, and the occurrence state and valency of vanadium are diversified [6]. Most of the vanadium in stone coal exist as V(III), and the others exist as V(IV) and V(V) [7]. Most of the vanadium in stone coal is exist in form of isomorphism in the lattice of mica. Trivalent vanadium (V(III)) substitutes for trivalent aluminum (Al(III)) in the octahedron crystal lattice of mica minerals [8, 9]. In order to utilize its contained calorific value, stone coal was widely used for power generation in south China. Vanadium is enriched in the fly ash, which could be exploited as

---

D. Zhi-gan (✉) · T. Fu-li · W. Chang · F. Gang (✉) · L. Min-ting · L. Xing-bin · L. Cun-xiong
Faculty of Metallurgical and Energy Engineering, Kunming University of Science and Technology, 253 Xuefu Road, Kunming, Yunnan 650093, China
e-mail: dengzhigan83@163.com

F. Gang
e-mail: kgfangang2008@163.com

vanadium resource [10–12]. During the fluidized-bed combustion process, phases of vanadium will occur significant transformations, which determined the existing state of vanadium in fly ash [13]. Vanadium in stone coal has organic affinity and mineral affinity, and the migration of vanadium during combustion process depends on their affinities and on the physical changes and chemical reactions [14–17]. Most of the vanadium with low valency(V(III) and V(IV)) is oxidized to V(V) in high temperature and oxidizing atmosphere [18, 19]. After many mineral phases, transformations during combustion, the inorganic minerals translate into fine fly ash (<250 um) [20]. Vanadium and other trace elements are released in form of volatiles and concentrated in fly ash in the form of crystalline compounds [21].

The transformation of vanadium phases leads to a significant effect on leaching vanadium from fly ash. Therefore, it is necessary to study the transformation behaviors of vanadium phases in combustion process of stone coal. The aim of this work is to research the transformation of vanadium phases during combustion. In order to determine the vanadium phases in stone coal and fly ash, an analytical method put forward by A. Tessier et al. was adopted in this analysis[22]. Combined with the characteristics of stone coal combustion and vanadium migration in fluidized bed, the transformation behaviors of vanadium phases were determined.

Experimental

Materials

The stone coal and the fly ash in this study were collected from a power plant in Sichuan Province of China. The main chemical compositions of the materials are shown in Table 1. The content of vanadium in terms of V_2O_5 existing in stone coal is 0.712%, and 1.137% in fly ash, and included a certain amount of iron, aluminum, silicon in the stone coal and fly ash.

The analytical grade reagents in this study included magnesium chloride, hydroxylamine hydrochloride, hydrochloric acid, hydrogen peroxide, nitric acid, ammonium acetate, acetic acid, sulfuric acid, and hydrofluoric acid. All aqueous solutions were prepared using deionized water.

Table 1 Main chemical compositions of materials (mass fraction, %)

Materials	Si	Ca	Al	Fe	Mg	K	Na	V
Stone coal	19.74	0.043	3.81	2.32	0.041	0.77	0.29	0.39
Fly ash	31.77	11.88	4.98	5.21	0.059	0.76	0.002	0.64

Methods

The appropriate chemical reagents could react with the specific minerals, and the specific minerals would be disintegrated, which is the base of extraction of some elements from minerals. An analytical procedure involving sequential chemical extractions put forward by A. Tessier et al., which was adopted to analyze the vanadium phases in stone coal and fly ash. In the analytical procedure, the trace elements were divided into five fractions: exchangeable, carbonates, Fe–Mn oxides, organic matter, and others [22]. After the fluidized-bed burning process, carbonates were almost disintegrated and existed barely in fly ash. It is not necessary to investigate carbonates in vanadium phases analysis of fly ash. The flowsheet of vanadium dissolution from stone coal and fly ash is shown in Fig. 1.

Fig. 1 Flowsheet of vanadium dissolution from stone coal and fly ash

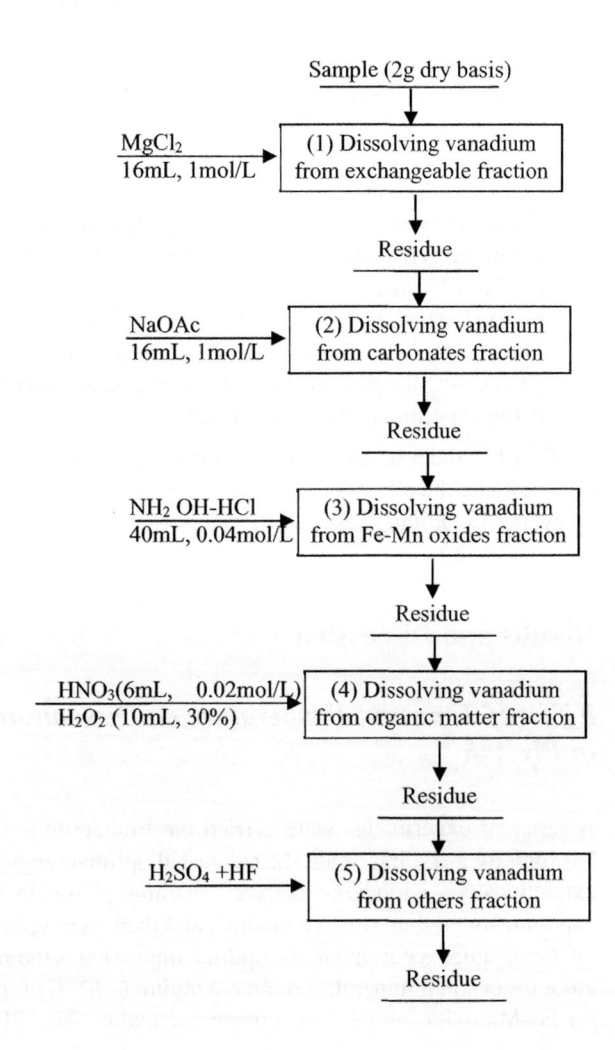

(1) Exchangeable. 2 g of sample was weighed for the experiments. The material was extracted at room temperature for 1 h with 16 mL of magnesium chloride solution (1 M MgCl$_2$, pH 7.0) with continuous agitation.
(2) Bound to carbonates. The residue from (1) was extracted at room temperature for 2 h with 16 mL of 1 M NaOAc (adjusted to pH 5.0 with acetic acid). Continuous agitation was maintained.
(3) Bound to Fe–Mn oxides. The residue from (2) was extracted with 40 mL of 0.04 MNH 2 OH–HCl (in 25% (v/v) HOAc). The experiment was performed at 96 ± 3 °C in water bath for 1.5 h with occasional agitation.
(4) Bound to organic matter. The residue from (3) was added 6 mL of 0.02 M HNO$_3$ and 10 mL of 30% H$_2$O$_2$ (pH 2.0 with HNO$_3$), and the mixture was heated to 85 ± 2 °C in water bath for 2 h with occasional agitation. A second 6 mL of 30% H$_2$O$_2$ (pH 2.0 with HNO$_3$) was then added and the sample was heated again to 85 ± 2 °C in water bath for 3 h with intermittent agitation. After that, 10 mL of 3.2 M NH$_4$OAc (in 20% (v/v) HNO$_3$) was added and the mixture was diluted to 40 mL and agitated continuously for 30 min.
(5) Others. The trace elements included in this fraction are stored in the mineral crystal structures. Because it is difficult to break the mineral crystal structures, the trace elements are difficult to be released. The residue from (4) was first digested in a high-purity graphite crucible with a solution of concentrated HClO$_4$ (4 mL) and HF (20 mL) to near dryness; then a second addition of HClO$_4$ (2 mL) and HF (20 mL) was evaporated to near dryness. Finally, HClO$_4$ (2 mL) alone was added and the mixture was evaporated until the appearance of white fumes. The residue was dissolved in 12 M HCl and diluted to 50 mL. Considering HClO$_4$ with the property of volatility and strong corrosivity, H$_2$SO$_4$ was adopted in the experiment instead of HClO$_4$.

Fly ash is the product of combustion, but is not original mineral, so it is necessary to investigate the effects of time and temperature on dissolution of vanadium before analyzing vanadium phases.

Results and Discussion

Effect of Time and Temperature on Dissolution of Vanadium in Fly Ash

A series of experiments were carried out to determine the optimal conditions for leaching fly ash. First of all, the suggested optimal temperature on original mineral extractions was adopted to analyze vanadium phases in fly ash. Optimal condition experiments on time were conducted, and then were optimal condition experiments on temperature according to the optimal time. That is the suggested leaching temperature for original mineral: room temperature (~20 °C) for exchangeable, 96 ± 3 °C for Fe–Mn oxides, 85 ± 2 °C for organic matter. According to the investigation on

Fig. 2 Effect of time on vanadium concentration in extracting solution of exchangeable fraction

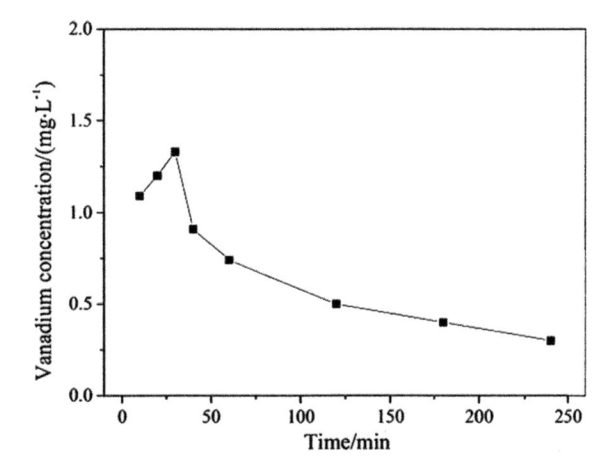

time above, the optimal leaching time was chosen for carrying out the experiments on temperature.

Effect of Time and Temperature on Leaching Exchangeable Fraction

The effect of leaching time on leaching exchangeable fraction was studied from 10 min to 240 min. The results are shown in Fig. 5, the leaching percent of vanadium was increased with the leaching time from 10 min to 30 min, and the leaching percent reaches the highest level after 30 min. After that, the leaching percent decreases rapidly, and then to be slowly. The possible reason is that some precipitation reactions take place in $MgCl_2$ solution and generate some insoluble vanadium precipitation, which leads to the decrease of vanadium concentration in extracting solution. 30 min was chosen as the optimal leaching time in the subsequent experiments (Fig. 2).

Figure 3 shows the effect of temperature on leaching exchangeable fraction. As Fig. 3, the leaching temperature exerted the influence obviously on the leaching percent and the leaching percent of vanadium was increased with the rise of temperature. The higher temperature could increase the speed of molecular motion and the collisions between the vanadium atoms and leaching agent atoms [23]. It could be inferred that the leaching reaction is a endothermic process and raising temperature could improve the reactivity of vanadium in exchangeable fraction. When the temperature reached 80 °C under the condition of natural pressure, and the leaching percent almost remains stable with the increase of temperature.

Effect of Time and Temperature on Leaching Fe–Mn Oxides Fraction

In order to study the effect of time on leaching Fe–Mn oxides fraction, the time was varied from 0.5 to 12 h. The results are shown in Fig. 4. It could be concluded that

Fig. 3 Effect of temperature on vanadium concentration in extracting solution of exchangeable fraction

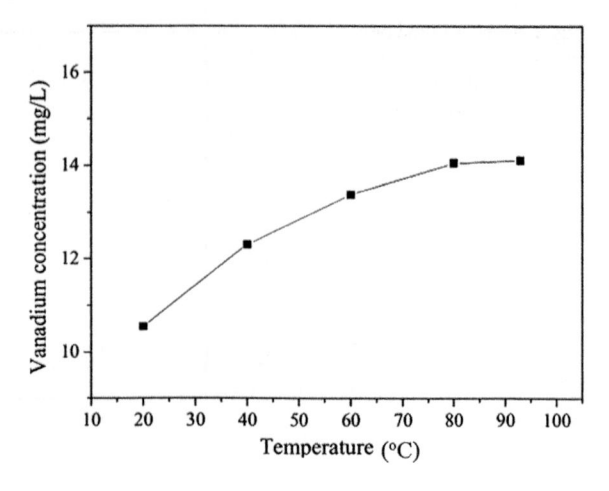

Fig. 4 Effect of time on vanadium concentration in extracting solution of Fe–Mn oxides fraction

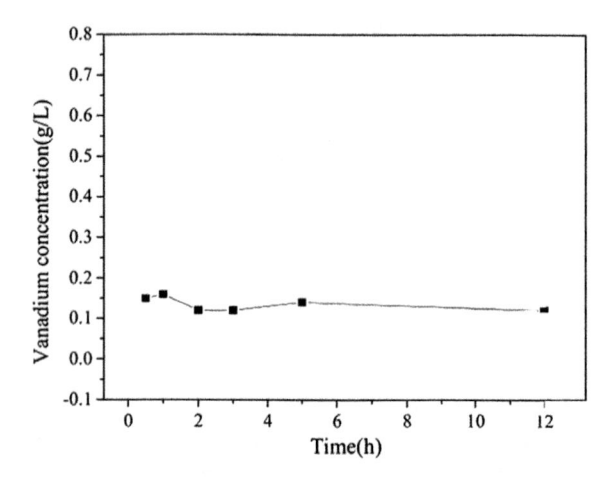

the leaching time was almost no influence on the leaching percent in Fe–Mn oxides fraction. With the leaching time varied from 0.5 to 12 h, the vanadium concentration in extracted solution almost remained stable. It could be thought that it is easy to leach the vanadium from Fe–Mn oxides fraction by $NH_2OH–HCl$, and 0.5 h should be the suitable leaching time.

The effect of temperature on leaching Fe–Mn oxides fraction was studied from 40 to 93 °C (93 °C is the boiling point of water in Kunming). The results are shown in Fig. 5. With the increased temperature, the vanadium concentration in extracting solution almost remained unaltered. It could be considered that temperature almost exerted no influence on the leaching percent. 60 °C was regarded as the optimal leaching temperature.

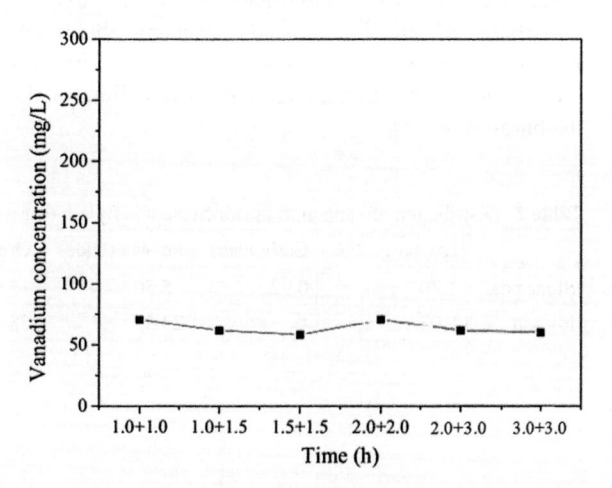

Fig. 5 Effect of temperature on vanadium concentration in extracting solution of Fe–Mn oxides fraction

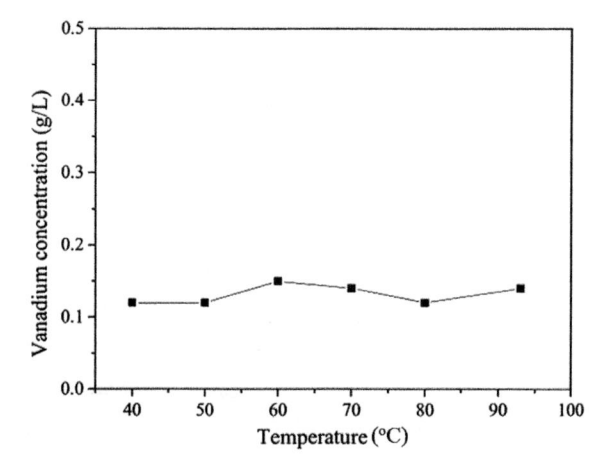

Effect of Time and Temperature on Leaching Organic Matter Fraction

Figure 6 shows the effect of time on leaching organic matter fraction. It could be seen that it is easy to leach the vanadium from organic matter fraction, and the vanadium concentration almost remained stable with the leaching time varied from 1.0 + 1.0 h to 3.0 + 3.0 h. It could be thought that 1.0 + 1.0 h was the optimal leaching time.

Figure 7 shows the effect of temperature on leaching organic matter fraction. It could be seen that temperature has a profound effect on increasing vanadium extraction. From 40 to 70 °C, the leaching percent increases significantly with the increase in temperature. When the temperature reaches 70 °C, the leaching percent almost remains stable. In order to leach this part of vanadium completely, 90 °C was chosen as the leaching temperature temporarily.

Fig. 6 Effect of time on vanadium concentration in extracting solution of organic matter fraction

Fig. 7 Effect of temperature on vanadium concentration in extracting solution of organic matter fraction

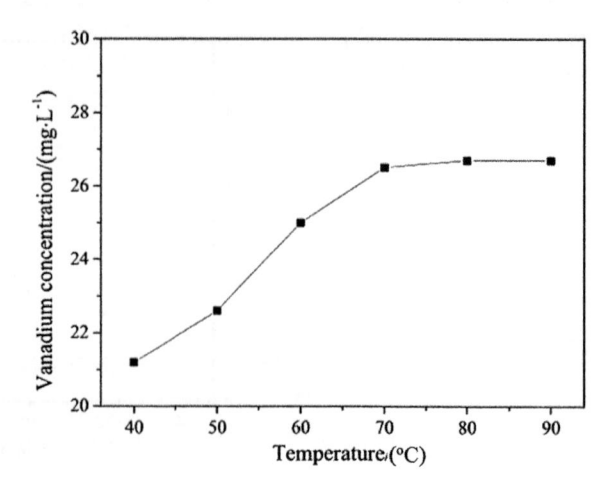

Distribution of Vanadium in Stone Coal and Fly Ash

In order to analyze the phases of vanadium, the suggested sequential leaching conditions were adopted to leach stone coal, and the optimal conditions which got from 3.1 were adopted to leach fly ash. The result is shown in Table 2.

It could be seen that the distribution of vanadium in stone coal was concentrated relatively, and the vanadium mainly consisted in other fraction. According to some researches [8, 9], this part of vanadium was mainly stored in crystal structures of aluminosilicate. This part of vanadium was difficult to release, and the proportion of vanadium reaches 77.28%. About 13.60% of vanadium consisted of organic matter; the others vanadium consisted in exchangeable fraction, carbonates, and Fe–Mn oxides. However, the distribution of vanadium in fly ash became more scattered. The vanadium mainly existed in organic matter, aluminosilicate, and Fe–Mn oxides, and a small part of vanadium existed in exchangeable fraction in the form of adsorption. Because the proportion of vanadium in aluminosilicate varied from 77.28 to 29.48%, it could be inferred that a large part of vanadium in aluminosilicate was released in the burning process.

Table 2 Distribution of vanadium in stone coal and fly ash (mass fraction, %)

	Exchangeable	Carbonates	Fe–Mn oxides	Organic matter	Others	Total
Stone coal	2.70	0.92	5.50	13.60	77.28	100.0
Fly ash	8.02	–	23.59	38.91	29.48	100.0

Transformation of Vanadium Phases in Combustion Process

In stone coal, the vanadium mainly consisted of aluminosilicate minerals, and the other vanadium consisted of organic matter, exchangeable fraction, carbonates and Fe–Mn oxides. During the fluidized-bed burning process, the decomposition behaviors of vanadium minerals occurred as follows: The mineral crystal structures of aluminosilicate, such as vanadium mica, were destroyed; the organic matter like vanadium porphyrin was burned and decomposed; the carbonates and Fe–Mn oxides were decomposed and converted. During the decomposed process above, the vanadium was released and oxidized to high valency [18, 19]. The vanadium was volatilized from stone coal and concentrated into fly ash. The formation of fly ash contained matter migration and multicomponent interactive reactions during the burning process. The main composition of fly ash was anhydrite, Fe–Mn oxides, and other organic matter.

In conditions of high temperature and strong oxidizing atmosphere, the vast majority of V(III) and V(IV) were oxidized to V(V) [24].

A small part of vanadium could react with Na and Cl and generate $NaVO_3$, $VOCl_3$, and Na_3VO_4 etc. [25]. These vanadium compounds were mainly consisting in exchangeable fraction, which could be dissolved easily by water.

Vanadium with high valency could also react with Fe, Na, and Ca and generate $FeVO_4$, $NaCaVO_4$, and $Ca_3(VO_4)_2$ etc. [26]. These vanadium compounds could not be dissolved easily by water.

Most of the vanadium in fly ash consisted of these unsolvable vanadium compounds. During the fluidized-bed burning process, these unsolvable vanadium compounds were concentrated in organic matter, aluminosilicate, and Fe–Mn oxides. As it is difficult to break the crystal structures of aluminosilicate, this part of vanadium is difficult to be leached. Through the burning process, a large part of vanadium in aluminosilicate was released, which resulted in that most of the vanadium in fly ash did not exist in the lattice structure. The distribution of vanadium in fly ash became more scattered, and the leaching of vanadium in fly ash was easier than that in stone coal. In order to extract vanadium from fly ash, some leaching agents such as sulfuric acid, hydrochloric acid, alkali and ammonia could be used to dissolve the unsolvable vanadium compounds.

Conclusions

(1) The vanadium in stone coal mainly consisted of aluminosilicate, and the other vanadium consisted of organic matter, exchangeable fraction, carbonates, and Fe–Mn oxides. The vanadium in fly ash mainly exists in organic matter, aluminosilicate, and Fe–Mn oxides, and a small part of vanadium existed in exchangeable fraction. Through the burning process, the distribution of vanadium in phases became more scattered.

(2) During the fluidized-bed burning process, the mineral crystal structures of aluminosilicate were destroyed, the organic matter was burned and decomposed, the carbonates and Fe–Mn oxides were decomposed and converted. The vanadium was released and oxidized to high valency. Most of the vanadium reacted with Fe, Na, and Ca and generates unsolvable vanadium compounds. These unsolvable vanadium compounds were concentrated in organic matter, aluminosilicate, and Fe–Mn oxides. The leachability of vanadium in fly ash was better than that in stone coal.

Acknowledgements This work was supported financially by the National Natural Science Foundation of China (Grant Number 51474115) and the Analysis and Testing Center of Kunming University of Science and Technology, China.

References

1. ZHU Yang-ge, ZHANG Guo-fan, FENG Qi-ming, LU Yi-ping, OU Le-ming, HUANG Si-jie (2010) Acid leaching of vanadium from roasted residue of stone coal [J]. Trans Nonferrous Met Soc China 20:107–111
2. Zhi-gan DENG, Chang WEI, Gang FAN, Min-ting LI, Cun-xiong LI, Xing-bin LI (2010) Extracting vanadium from stone-coal by oxygen pressure acid leaching and solvent extraction [J]. Trans Nonferrous Met Soc China 20:118–122
3. Yao-zhong LAN, Jin LIU (2005) Review of vanadium processing in China [J]. Eng Sci 3(3):58–62
4. Moskalyk RR, Alfantazi AM (2003) Processing of vanadium: A review [J]. Miner Eng 16(9):793–805
5. Cun-xiong LI, Chang WEI, Min-ting LI, Gang FAN, Zhi-gan DENG (2008) Process optimization of Vanadiumextraction from black shale by acidic oxidizing pressure leaching [J]. Chin J Nonferrous Metals 18:84–87 (in Chinese)
6. Li Min-ting, WU Hui-ling, DENG Zhi-gan, LIANG Yan-hui, LI Cun-xiong, WEI Chang (2008) Relationship betweendistribution of vanadium in stone coal and vanadium leaching percent. J Chin Rare Earth Soc 26:556–559. (In Chinese)
7. Chang-lin LI, Yun-feng ZHOU, Hai-xia FEI, Xiang-yang ZHOU, Jie LI (2012) Recent research and applications of vanadium extraction processes from carbonaceous shale. Rare Met Cemented Carbides 40(6):9–14 (In Chinese)
8. Yang Jian, Yi Fa-cheng, Li Hu-jie, Hou Lan-jie. Genesis and petrogeochemistry characteristics of lowercambrian black shale series in northern Guizhou [J]. Acta Mineral Sinica, 2004, 24(3): 285–289. (InChinese)
9. Hu Yang-jia, Zhang Yi-min, Bao Shen-xu, Liu Tao (2012) Effects of the mineral phase and valence of vanadiumon vanadium extraction from stone coal [J]. Int J Miner, Metall Mater 19(10):893–898
10. Alfantazi AM (1994) Hydrometallurgical processing of Egyptian black shale of the Quseir-Safaga region [J]. Hydrometallurgy 36(1):95–107
11. Qiang REN, Jian-zhong LIU, Jun-hu ZHOU, Lin YE, Ke-fa CEN (2006) Dynamic characteristic of sulfur releaseduring stone coal combustion [J]. J China Coal Soc 31(1):99–103 (in Chinese)
12. Habashi F (2006) Two hundred years of vanadium [J]. Metall 60(12):804–808
13. LIU Jian-zhong, ZHANG Bao-sheng, ZHOU Jun-hu, FENG Zhan-guan, CEN Ke-fa (2007) Combustion characteristics and classify attribute of stone coal [J]. Proc CSEE 27(29):17–21. (In Chinese)

14. LIN Hai-ling, FAN Bi-wei. Study on mechanism of phase transformation during roasting and extractionvanadium from Fangshankou bone coal [J]. Chinese Journal of Rare Metals, 2001, 25(4): 273–277. (In Chinese)
15. JIANG Jia-ji, XU Guo-zhen, HUANG Wenying (1991) Mossbauer spectroscopic study in V-bearing stone coal [J]. Geoscience 5(2):92–200. (In Chinese)
16. LI She-feng, FANG Meng-xiang, WANG Qin-hui, SHI Zheng-lun, LUO Zhong-yang, NI Ming-jiang (2010) Extraction of vanadium pentoxide from stone coal pellets by roasting in a fluidized bed combustor [J]. J Combust Sci Technol 16(4):317–322. (In Chinese)
17. Xavier Q, Roberto J, Angel LS (1996) Mobility of trace elements from coal and combustion wastes [J]. Fuel 75:821–838
18. WANG Ping, FENG Ya-li, LI Hao-ran, LIU Xin-wei, ZHANG Ping (2012) Leaching rate improvement ofvanadium by high-carbon stone coal fluidized through oxidizing roasting [J]. Chin J Nonferrous Metals 22(2):566–571. (In Chinese)
19. Yao-bing XU, Zhong-yang LUO, Qin-hui WANG, Jin-song ZHOU, Zheng-lun SHI (2010) Leaching kinetics ofvanadium pentoxide from ash of stone coal by sulfuric acid [J]. Chin J Process Eng 10(1):60–64 (In Chinese)
20. Ilham D, Randall EH, Philip JD (2001) Formation and use of coal combustion residues from three types ofpower plants burning Illinois coals [J]. Fuel 80:1659–1673
21. Steenari BM, Schelander S, Lindqvist O (1999) Chemical and leaching characteristics of ash fromcombustion of coal, peat and wood in a 12 MW CFB-a comparative study [J]. Fuel 78:249–258
22. Tessier A, Campbell PGC, Bisson M (1979) Sequential extraction procedure for the speciation ofparticulate trace metals [J]. Anal Chem 51:844–851
23. CHEN Xiang-yang, LAN Xin-zhe, ZHANG Qiu-li, MA Hong-zhou, ZHOU Jun (2010) Leaching vanadium by highconcentration sulfuric acid from stone coal [J]. Trans Nonferrous Metals Soc China 20:123–126
24. Bi-wei FAN, Hai-ling LIN (2001) Effects of roasting process on extraction of vanadium from stone coal atFangshankou [J]. Hydrometallurgy of China 20(2):79–83 (In Chinese)
25. NING Hua, ZHOU Xiao-yuan, SHANG De-long, DU Hai-ying (2010) Analysis on roasting process of vanadium extraction from stone coal [J]. Nonferrous Metals 62(1):80–83. (In Chinese)
26. LI Jing, LI Zhao-jian, WU Xue-wen, ZHONG Xiao-ling, WANG Hai -hua, LIU Su-qin, HUANG Ke-long (2007) Technology of roasting process on extraction of vanadium from stone coal and mechanism discussion [J]. Hunan Nonferrous Metals 23(6):7–10, 68. (In Chinese)

Solvo-Chemical Recovery of Cerium from Sulfate Solution Using Cyanex 923 and Oxalate Precipitation

Sadia Ilyas, Hyunjung Kim, and Rajiv Ranjan Srivastava

Abstract The solvo-chemical recovery of Ce(IV) from sulfate leach liquor was investigated using Cyanex 923 in kerosene. The quantitative solvation of Ce(IV) could be achieved by performing three stages of counter-current extraction at an organic-to-aqueous phase ratio of 2:3, while using 0.15 mol/L Cyanex 923 into the organic phase. The spectroscopy analysis of the organic phase revealed the formation of solvated species $\overline{[Ce(SO_4)_2].2[Cyanex923].[HSO_4^-]}$ into the organic phase. Stripping of Ce(IV) in its reduced form as Ce(III) was conducted in H_2SO_4 + H_2O_2 solution that yielding 1.3 g/L Ce(III) back into the aqueous phase. Finally, the recovery of rare metal was conducted via Ce(III) precipitation with oxalic acid, which exhibited different characteristics with changing temperatures. Precipitation kinetics showed good fits to the Avrami equation, while the determined activation energy (E_a, 8.6 kJ/mol) indicated to follow a diffusion-controlled mechanism.

Keywords Rare earth metals · Solvating extraction · Ce(IV)-Ce(III) conversion · Phosphine oxide mixture · Oxalate precipitation

Introduction

The rare-earth elements (REEs) are a group of 15 lanthanides plus scandium and yttrium that chemically exhibiting the similar properties [1]. They are becoming strategically important in the transition to a low-carbon green economy due to their inevitable role in lamp phosphors, permanent magnets, catalysts, rechargeable batteries, and other applications [2, 3]. Given that the demand for REEs is continuously soaring on contrary to its tightened export quota by China that supplies ~ 90% of the global REEs production, confronting the worldwide supply crunch of REEs [4]. Henceforth, the mining companies are actively seeking new exploitable reserves of REEs while a few old mines are being reopened [5]. Traditionally, South Korea is not known for REEs production but in the present geopolitical scenarios the country is looking for the exploitable reserves and the development of indigenous technology for the processing of vein deposit monazites (co-exists with iron ore mines) is high on the agenda [6, 7].

Cerium is the highest occurring REEs in vein deposits [8] that can uniquely form the higher oxide as Ce(IV) than other REEs exist in the trivalent form [9]. This property presents the possibility of its prior separation easier than others [6]. The most of cerium is currently extracted through the alkali roasting of monazite, which consumes high energy and requires additional acid for neutralization [8]. Therefore, the study on alternative processing route via H_2SO_4 digestion is timely especially on its extraction and recovery behavior from the leach liquor. The application of solvating extractant is one of the possible routes that can potentially be applied to the extraction of cerium as Ce(IV) [6, 8]. Among others, Cyanex 923 is advantageous to use in terms of its availability in liquid form and low aqueous solubility.

In the present study, we examined the extraction behavior of Ce(IV) from the sulfate leach liquor of vein deposit Korean monazite using Cyanex 923 in kerosene. Moreover, the recovery of its oxalate precipitates in its reduced form Ce(III) was also investigated with emphasis on kinetics and mechanism and the physical properties of the precipitates. The extraction behavior with varying parameters (like concentrations of Cyanex 923 and H_2SO_4, temperature, and organic-to-aqueous (O/A) phase ratio) followed by stripping were studied. Subsequently, the precipitation of Ce(III)-oxalate was examined from the stripped liquor under the varying parameters of oxalic acid, temperature, and time. The study will be useful not only to process the vein deposit Korean monazite but also to achieve the desired precursor properties for making the final product.

Experimental

A sulfuric acid processed stock leach liquor containing 890 mg/L Ce(IV), 32 mg/L Y(III), 300 mg/L La(III), and 1.9 mol/L free acid was used in this study for investigating the extraction behaviour with Cyanex 923 (sp. gr. 0.88 g/mL and purity, 93%

supplied by Cytec (Canada). Distilled kerosene (Junsei Chemical Co), hydrogen peroxide (purity 30%, Merck), sulfuric acid (purity 96%, Daejung), and oxalic acid (purity 90%, Junsei Chemical Co.) were used without further purification. Unless specified, the extraction studies were conducted at an organic-to-aqueous (O:A) phase ratio of 1:1 by taking a 30-mL volume of each phase in a 100 mL separating funnel, while maintaining the contact time (5 min), temperature (20 ± 2 °C), and settling time (10 min) constant. Metals' concentrations were analyzed into the aqueous phase for knowing the extraction efficiency into the organic phase through the mass balance of metals. Ce(IV)-loaded organic underwent for stripping after a water washing at an O:A of 2:1. Thereafter, stripping was examined at different H_2O_2 concentrations into sulfuric acid strip solution. The precipitation studies were conducted with 100 mL (preheated at the temperatures range of 25–90 (±2)°C) stripped solution in a 250-mL glass beaker. Solution pH was adjusted at 1.5(±0.2) by adding 20 (v/v) ammonia solution and maintained at the agitation speed of 150 rpm. Then, the oxalic acid of a predetermined concentration was introduced into the solution and contacted for the desirable duration (up to 15 min). The slurry settled for 1 h was filtered and both solid (after a water washing) and liquid stream (after proper dilution) were collected for further analysis using an inductively coupled plasma spectroscopy (ICP-OES, iCAP 7400 Duo, Thermo Scientific).

Results and Discussion

Ce(IV) Extraction with Cyanex 923

Effect of Extractant Concentration

Ce(IV) extraction with Cyanex 923 was investigated as a function of extractant concentration into the organic phase varied in the range of 0.05–0.25 mol/L at an O:A 1:1. Results in Fig. 1a depicted the progress in Ce(IV) extraction from 37

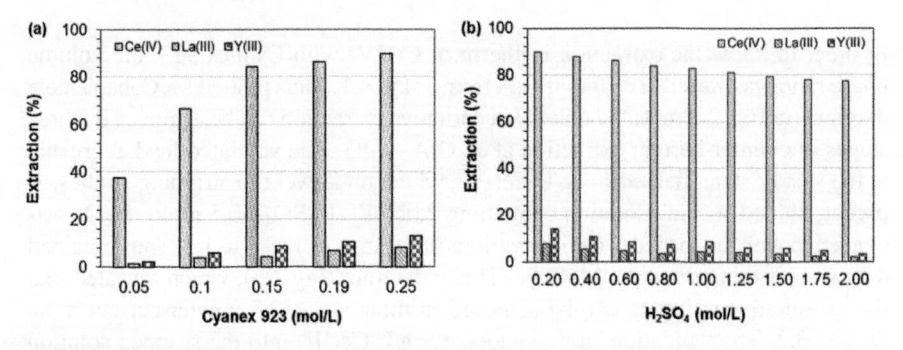

Fig. 1 Effect of extractant (**a**), and acid (**b**) concentrations on extraction of REEs. (Color figure online)

to 89%, which indicates for a shifting of distribution curve that solvating Ce(SO$_4$)$_2$ from aqueous solution to the immiscible phase [10, 11]. Under the studied conditions, the co-extraction of Y(III) and La(III) showed an increasing trend with increasing concentration of the extractant molecules, albeit very low as ~12% Y(III) and 8.2% La(III). The extraction stoichiometric was determined by plotting the logarithmic distribution of Ce(IV) [$D_{Ce(IV)}$] versus concentration of Cyanex 923. The slope value with a R^2 value >0.99 indicated that about two moles of extractant are required to extract one mole of Ce(IV) species.

Effect of Acid Concentration

Furthermore, the extraction of Ce(IV) as a function of acid concentration varied in the range of 0.2–2.0 mol/L H$_2$SO$_4$ was examined using 0.15 mol/L Cyanex 923 into the organic phase. Results in Fig. 1b depicted a decline in Ce(IV) extraction from 89 to ~77%, while the co-extraction of Y(III) and La(III) could also decrease from 7.8% to 2.1% and 14.2% to 4.4%, respectively. It indicates the competition between REEs and bisulfate ions in higher acidic condition [12, 13]. The plot of log [$D_{Ce(IV)}$] versus log [H$_2$SO$_4$] showed the participation of one mole acid with one mole of Ce(IV) extracted into the organic phase. Furthermore, the spectral analysis showed a new characteristic peak appeared for HSO$_4^-$ at 559 cm^{-1} (using FTIR technique) and a new band appeared at 6.54 ppm assigned to the proton of HSO$_4^-$ (using ^{1}H NMR technique) [14]. This also supports the involvement of one mole of proton (as HSO4$^-$ ion) with solvated species into the organic phase. Hence, the extraction reaction at equilibrium can be written as:

$$Ce(SO_4)_2 + 2\overline{[Cyanex\ 923]} + HSO_4^- \leftrightarrow \overline{[Ce(SO_4)_2].2[Cyanex\ 923].[HSO_4^-]} \tag{1}$$

Extraction and Stripping Isotherms

In order to know the extraction isotherm of Ce(IV) with Cyanex 923, the volume phase ratio was varied at different O:A from 1:4 to 4:1. Thus plotted McCabe–Thiele diagram (in Fig. 2a) revealed that the quantitative extraction can be achieved by three stages of counter-current extraction at an O:A = 2:3. The validated loaded organic at the steady-state yielded 1.3 g/L Ce(IV), which underwent to stripping studies. A predetermined striping solution containing 2.0mol/L H$_2$SO$_4$+ 0.5 mol/L H$_2$O$_2$ was contacted with the loaded organic at various O:A ratios from 3:1 to 1:3. Thus obtained data were used to draw the McCabe–Thiele diagram (Fig. 2b), which revealed that the quantitative stripping can be achieved in three stages of counter-current at an O:A = 3:2. The validation study yielded 1.9 g/L Ce(III) into the stripped solution with an efficacy of >99%.

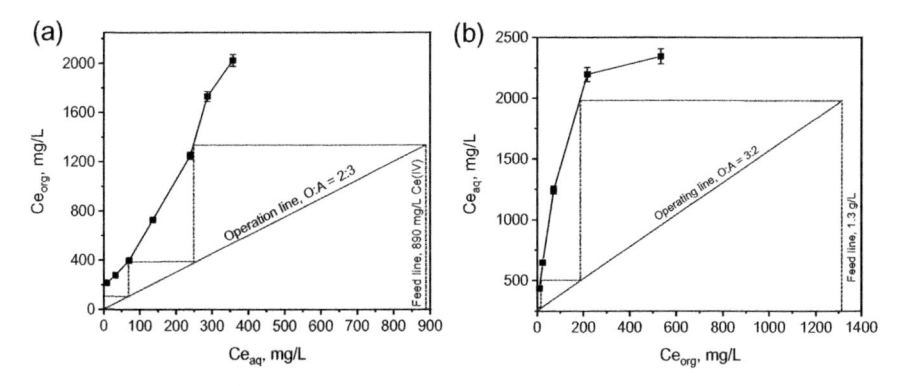

Fig. 2 McCabe–Thiele diagrams for Ce(IV) extraction with Cyanex 923 (**a**), and Ce(III) stripping with 2.0mol/L H_2SO_4 + 0.5 mol/L H_2O_2 (**b**)

Cerium(III) Precipitation with Oxalic Acid

Effect of Oxalic Acid Addition

The precipitation of Ce(III) sulfate with oxalic acid can be written as the following reaction:

$$Ce_2(SO_4)_3 + 3(H_2C_2O_4) + xH_2O \rightarrow Ce_2(C_2O_4)_3 \cdot xH_2O + 3H_2SO_4 \qquad (2)$$

Eh–pH diagram of Ce–$C_2H_2O_4$–H_2O system revealed that $Ce_2(C_2O_4)_3$ forms at a wider pH range above zero, while oxalic acid exhibits two pKa values at 1.0 and 4.2 [8]. Hence, the concentration of acid becomes important factor, which was determined under the varied concentration of oxalic acid in the range of Ce:$H_2C_2O_4$ = 1:0.5 to 1:5. Figure 3a depicted that a low precipitation (~23%) at the stoichiometric

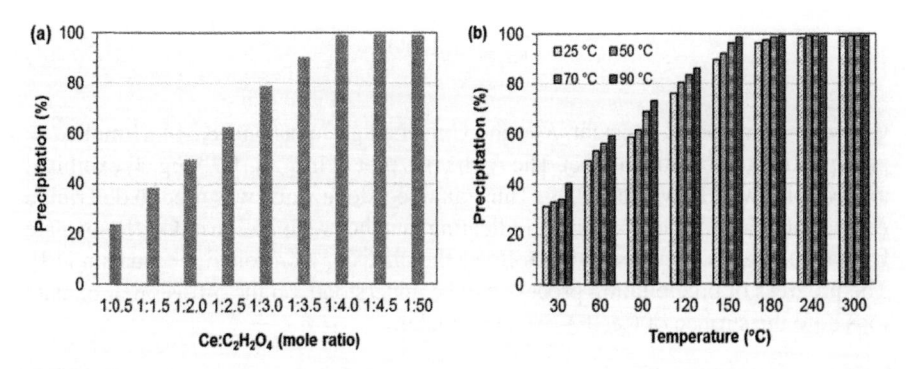

Fig. 3 Effect of extractant (**a**) and acid (**b**) concentrations on extraction of REEs. (Color figure online)

dosage of oxalic acid. Then, the increasing dosage Ce:H$_2$C$_2$O$_4$ could improve the precipitation yields and ≥99% efficiency was obtained at Ce:H$_2$C$_2$O$_4$ = 1: ≥ 10.

Effect of Precipitation Temperature

Furthermore, the precipitation was examined with different temperatures (25–90 °C) as a function of time (in between 30 and 300 s). Figure 3b depicted that the progress of precipitation improved with increasing temperature and prolong duration. The precipitation showed a rapid complexation as ≥50% of Ce(III) could be precipitated within 60 s. As an influential effect of temperature, >99% efficiency was obtained in 150 s at 90 °C, as compared with 300 s required at a lower temperature of 25 °C.

Precipitation Kinetics and Mechanism

Furthermore, the data were used to get the information on the kinetics and mechanism through the Avrami equation by plotting $\ln\{-\ln(1-x)\}$ vs. $\ln(t)$ as the equation below [15–17].

$$[\ln\{-\ln(1-x)\}] = \ln k_a + n\ln(t) \tag{3}$$

where x= precipitation fraction. t = time in sec, k_a = rate constant/sec, and n = constant that reflecting the nature of phase transformation that reflected by the intercept and slope values, respectively. The plots showed linear fits to the Avrami equation with R^2 values >0.99. The constant values (k_a and n) obtained at different temperatures are given in Table 2. The constant values of n were obtained to be < 1, indicating for an instantaneous nucleation and flakes-shape precipitation growth. Whereas, the rate constant values (k_a) were used to evaluate the apparent activation energy (E_a) by the Arrhenius equation [18].

$$k_a = Ae^{\left(\frac{-E_a}{RT}\right)} \tag{4}$$

where A = Arrhenius constant, R = the Universal gas constant (8.314 kJ/mol), T= precipitation temperature (in K). The Arrhenius plot of $\ln k_a$ vs. $1/T$ (Fig. 4) exhibited a linear plot with R^2 value > 0.99. Thus obtained slope value was used to determine E_a and calculated to be 8.6 kJ/mol, indicating that the precipitation of Ce(III)-oxalate from the sulfate stripped solution follows the diffusion-controlled mechanism [18]. The progress of precipitation process can be understood via the diffusion of oxalate ions onto the surface of Ce(III)-oxalated nuclei.

Fig. 4 The Arrhenius plot of $\ln k_a$ versus $1/T$. (Color figure online)

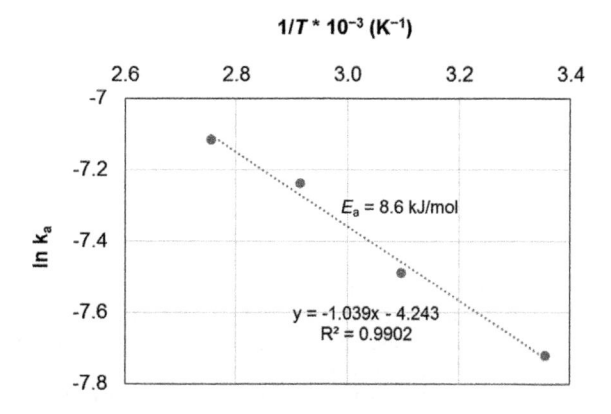

Conclusions

The solvo-chemical equilibria of Ce(IV) extraction with Cyanex 923 followed by the precipitation kinetics of Ce(III)-oxalate has been demonstrated. The parametric optimization studies for Ce(IV) extraction showed an inhibited solvation with increasing concentration acid while the degree of solvation gets improved with increasing molecules of Cyanex 923 into the organic phase, forming $\overline{[Ce(SO_4)_2].2[Cyanex923].[HSO_4^-]}$ as the solvated species. The extraction and stripping isotherms predicted by McCabe–Thiele plots were quantitatively validated in counter-current contacts at an O:A = 2:3 and 3:2, respectively. Stripping solution containing 2.0 mol/L H_2SO_4+ 0.5 mol/L H_2O_2 could quantitatively enrich 1.9 g/L Ce(III) solution. The subsequent precipitation of Ce(III) was optimized at a Ce:H2C2O4 ratio = 1: ≥ 10 with ~99% recovery. The precipitation kinetics followed the Avrami equation with constant values that indicating instantaneous nucleation growth followed by the flakes-shape morphology. The calculated activation energy (E_a, 8.6 kJ/mol) could underpin the precipitation mechanism involving the diffusion-controlled phenomenon through the surface site saturation.

Acknowledgements This work was supported by the Brain Pool Program through the National Research Foundation of Korea (NRF) funded by the Ministry of Science and ICT (Grant number 2019H1D3A2A02101993) and Basic Science Research Program through the National Research Foundation of Korea (NRF) funded by the Ministry of Education (Project number 2020R1I1A1A01074249).

Conflict of Interest The authors declare that they have no conflict of interest.

References

1. Kato T, Granata G, Tsunazawa Y, Takagi T, Tokoro C (2019) Mechanism and kinetics of enhancement of cerium dissolution from weathered residual rare earth ore by planetary ball

milling. Miner Eng 134:365–371
2. Habashi F (2013) Extractive metallurgy of rare earths. Can Metall Quart 52:224–233
3. Krishnamurthy N, Gupta CK (2002) Rare earth metals and alloys by electrolytic methods. Min Proc Ext Met Rev 22:477–507
4. Hearty G (2019) Rare earths: next element in the trade war?. In: Center for Strategic and International Studies. https://www.csis.org/analysis/rare-earths-next-element-trade-war#:~:text=With%20the%20trade%20war%20having,next%20salvo%20in%20the%20conflict.&text=A1%3A%20Rare%2Dearth%20elements%20are,which%20possess%20similar%20chemical%20properties. Accessed 14 August 2020
5. Binnemans K, Jones PT, Blanpain B, Van Gerven T, Yang Y, Walton A, Buchert M (2013) Recycling of rare earths: a critical review. J Clean Prod 51:1–22
6. Ilyas S, Kim H, Srivastava RR, Choi S (2021) Cleaner production of rare earth elements from phosphorus-bearing sulfuric acid solution of vein deposit monazite. J Clean Prod 278:123435
7. Park J-M (2010) South Korea discovers rare earths deposit. In: Science News. Reuters. https://www.reuters.com/article/us-korea-rareearths/south-korea-discovers-rare-earths-deposit-idUSTRE6A706B20101108. Accessed 10 August 2020
8. Ilyas S, Kim H, Srivastava RR (2021) Extraction equilibria of cerium(IV) with Cyanex 923 followed by precipitation kinetics of cerium(III) oxalate from sulfate solution. Sep Purif Technol 254:117634
9. Patnaik P (2003) Handbook of inorganic chemicals, vol 529. McGraw-Hill, New York
10. Ritcey GM, Ashbrook AW (1984) Solvent extraction part I. Elsevier, Amsterdam
11. Srivastava RR, Ilyas S, Kim H, Tri NLM, Hassan N, Mudassir M, Talib N (2020) Liquid–liquid extraction and reductive stripping of chromium to valorize industrial effluent.JOM 72(2):839–846
12. Jun L, Zhenggui W, Deqian L, Gengxiang M, Zucheng J (1998) Recovery of Ce (IV) and Th (IV) from rare earths (III) with Cyanex 923. Hydrometallurgy 50:77–87
13. Liao W, Yu G, Li D (2001) Solvent extraction of cerium (IV) and fluorine (I) from sulphuric acid leaching of bastnasite by Cyanex 923. Solvent Extr Ion Exc 19:243–259
14. Pavia DL, Lampman GM, Kriz GS, Vyvyan JR (2001) Introduction of spectroscopy. Brooks/Cole Cengage Learning, Belmont, USA
15. Avrami M (1939) Kinetics of phase change, I General theory. J Chem Phys 7:1103–1112
16. Avrami M (1940) Kinetics of phase change. II transformation-time relations for random distribution of nuclei. J Chem Phys 8:212–224
17. Avrami M (1941) Kinetics of phase change. III: Granulation, phase change and microstructure. J Chem Phys 9:177–184
18. Habashi F (1969) Extractive metallurgy. Gordon and Breach Science Publishers Inc, New York

Recovery of Molybdenum from Low Concentration Molybdenum-Containing Solution with Addition of Fe(III)

Bei Zhang, Bingbing Liu, Yuanfang Huang, Guihong Han, and Shengpeng Su

Abstract Molybdenum (Mo) is a strategic metal element, and recovery of Mo from low concentration Mo-containing solution is significant to alleviate the Mo resources shortage. In this study, the treatment of low-concentration Mo-containing solution by Fe^{3+} addition was investigated by batch experiment, and the effects of main parameters including molar ratio of n(Mo):n(Fe), pH value, initial concentrations, and reaction time on the recovery of Mo were studied. The results showed that Mo is precipitated rapidly, with a Mo recovery of 100% in 5 min when the pH range is 4–6 and molar ratio of n(Mo):n(Fe) is greater than 1:3, indicating the reaction between Mo and Fe readily approaches chemical equilibrium. Moreover, the initial concentration of Mo in the solution has little influence on the recovery of molybdenum. This technology has potential to be applied for the enrichment and recycling of molybdenum, while at the same time meeting the industrial discharge standard for the effluent.

Keywords Molybdenum · Recovery · Precipitation · Fe(iii)

Introduction

Molybdenum, due to its high boiling point and high melting point, is an important non-renewable strategic resource for developing new and high-tech technologies [1–3]. Molybdenite (MoS_2) concentrate is the material to produce molybdenum in industry, and it is usually processed by pyrometallurgy [4–6]. In the process of pyrometallurgy, molybdenite is oxidized and roasted to form calcined molybdenite at 550 °C –560 °C [7], and then solvent extraction, ion exchange, and other methods

---

B. Zhang · B. Liu (✉) · Y. Huang · G. Han · S. Su
School of Chemical Engineering, Zhengzhou University, Zhengzhou 450001, Henan, China
e-mail: liubingbing@zzu.edu.cn

G. Han
e-mail: hanguihong@zzu.edu.cn

are used for purification. During the process of purification, a small amount of molybdenum still exists in the solution, causing environmental pollution and resource waste [8, 9].

Molybdenum can be hydrolyzed in solution at pH < 6, which can form a series of hydroxides, such as $HMoO_4^-$, H_2MoO_4, and etc. What's more, Fe^{3+} can react with MoO_4^{2-} to form $Fe_2(MoO_4)_3$, which was poorly dissolved in water. Furthermore, Fe^{3+} can be hydrolyzed with the increase of pH value to form $Fe(OH)^{2+}$, $Fe(OH)_2^+$, $Fe(OH)_3$, and etc. which can interact with anions to form various complexes by electrostatic adsorption [10–12]. Pengfei Lin [13] et al. studied the treatment effect of Fe(III) on molybdenum-containing wastewater, and the results showed that pH had a great influence, and the optimal pH range was 4–4.5. Jiaxi Zhang [14] et al. compared the results of recovering molybdenum by coprecipitation and adsorption and concluded that the effect of coprecipitation was better than that of adsorption. Yue Ma [15] et al. investigated the effects of ferric chloride and poly-aluminum chloride on the recovery of molybdenum under different pH values, and the results showed that ferric chloride has the better effect than poly aluminum chloride in removing molybdenum. However, the initial concentration of molybdenum used in the study was low, which limited the application scope, and the obtained sediment has not been further explored [13–16].

$$MoO_4^{2-} + H^+ = HMoO_4^- \tag{1}$$

$$HMoO_4^- + H^+ = H_2MoO_4 \tag{2}$$

$$Fe^{3+} + MoO_4^{2-} = Fe_2(MoO_4)_3 \tag{3}$$

In this paper, the removal of molybdenum from the solution by adding Fe(III) was studied, as Formula (3). The influence of time, pH value, addition of Fe(III), and initial molybdenum concentration were evaluated with respect to the concentration of molybdenum. The Mo concentration in the solution was analyzed by the ICP–MS; the properties of the precipitate were evaluated by SEM and particle size analysis.

Experimental

Materials

The molybdenum-containing solution used in this study was prepared by dissolving sodium molybdenum dihydrate ($Na_2MoO_4.2H_2O$, Tianjin Yongda Chemical Reagent Co., Ltd. China) in pure water. The Fe(III) as the precipitant was prepared by iron chloride hexahydrate ($FeCl_3.6H_2O$, www.rhawn.cn). The concentration of Mo (VI) in the experiment was determined by Avio500 inductively coupled plasma mass spectrometer (ICP-MS). The polarizing microscope produced by Karl Zeiss Company in Germany was used to investigate the morphology and particle size statistics of the sediments formed by the interaction between MoO_4^{2-} and Fe^{3+}. SEM was used to detect the morphology, composition, and structure of the sediment.

Precipitation and Separation Process

The process flow chart is shown in Fig. 1. The HCl was added to the molybdenum-containing solution with the initial concentration of 100 mg/L, to adjust the pH at 6 or below. After stirring for 30 min, the molybdenum in the solution was present as polymer (MoO_4^{2-}, $HMoO_4^-$, H_2MoO_4). Fe(III) was added to the solution to form

Fig. 1 The process flow chart of recovery molybdenum with Fe(III). (Color figure online)

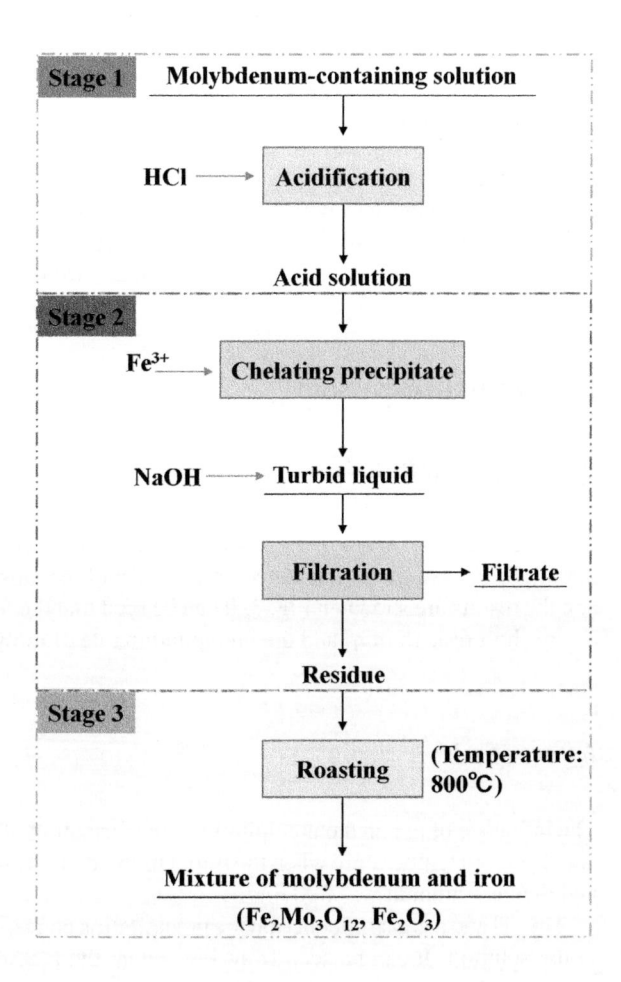

Fig. 2 The influence of reaction time on precipitation and transformation of molybdenum radical. (n(Mo):n(Fe) = 1:5; stirring speed = 150 r/min; pH = 5; 25 °C)

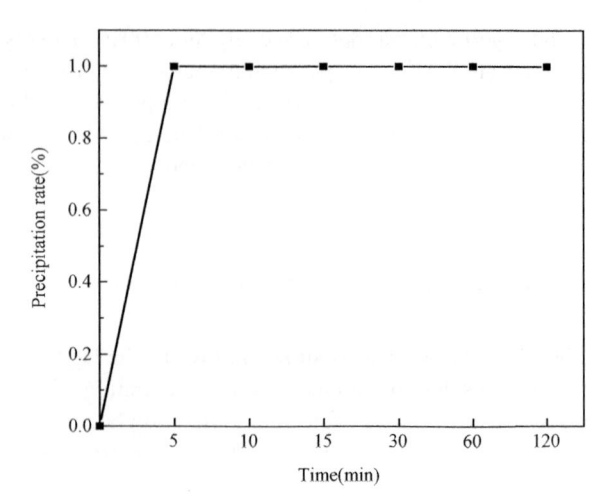

the polymer of molybdenum and iron ($Fe_2(MoO_4)_3$), then the NaOH was added to precipitate Fe(III) completely.

After 2 h of stirring and 5 min of standing, the precipitate settled at the bottom of the container and the supernatant was presumed impurity free. Then the precipitate and the supernatant can be separated by filtration. The supernatant was taken to measure the molybdenum concentration, and the filter cake was roasted at 800 °C for 2 h to obtain the mixture of molybdenum and iron.

Results and Discussion

Effect of Reaction Time

The precipitation kinetics of molybdenum at room temperature was studied by adding 300 mg/L Fe^{3+} to the initial concentration of molybdenum at 100 mg/L, at pH = 5, and the results are shown in Fig. 2. It can be seen that the reaction can be complete in a very short time (5 min) and the precipitation rate of molybdenum can reach 100%.

Effect of pH

The influence of pH on the precipitation transformation of molybdenum was investigated at room temperature when the initial molybdenum concentration was 100 mg/L and the Fe^{3+} amount was 300 mg/L.

The pH had a great influence on the precipitation conversion effect of molybdenum in the solution. It can be seen from Fig. 3 that the precipitation conversion rate of

Fig. 3 The influence of pH on precipitation and transformation of molybdenum. (n(Mo):n(Fe) = 1:5; stirring speed = 150 r/min; 25 °C)

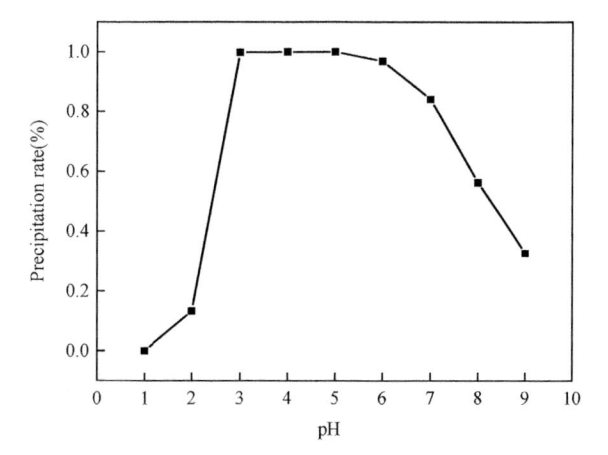

molybdenum was high under the weak acid and neutral conditions, while it was opposite under the strong acid and alkaline conditions. When pH was less than 2, there was no precipitation in the solution, as well as the precipitation conversion rate of molybdenum was close to zero. At the time, as shown in Fig. 4(a), molybdenum in the solution mainly existed in the form of MoO^{2+}, while iron existed in the form of Fe^{3+}, which were free in the solution, so there was no precipitation. In case of the pH > 2, the precipitation conversion rate of molybdenum increased with the increase of pH value. At the time, the molybdenum existed in the solution beginning to form the $MoO_3 \cdot H_2O$, which decreased the concentration of molybdenum. At pH > 3, the precipitation conversion rate of molybdenum was maximal, which can reach to 100%, therefore the reasons for this result may be that Fe^{3+} started to polymerize, forming $Fe(OH)^{2+}$, $Fe(OH)_2^+$ etc. in the pH = 3, as shown in Fig. 4(b), which can react with molybdenum by electrostatic adsorption effect. Moreover, the molybdenum in solution also polymerized into a variety of highly charged polymers that facilitated electrostatic interactions. However, in condition of pH > 5, the precipitation rate of

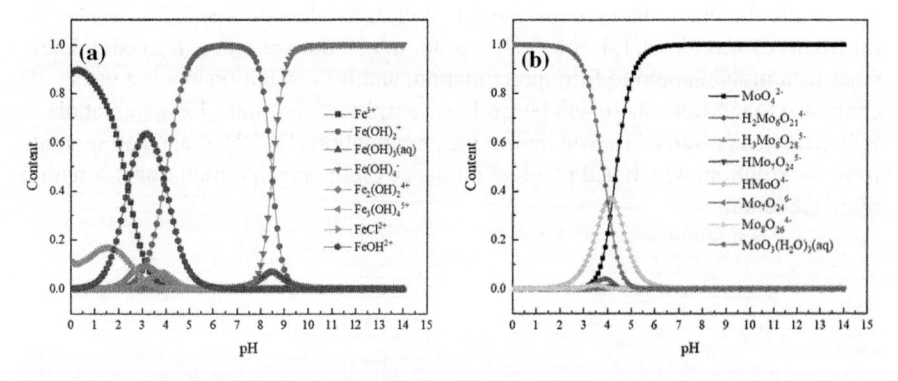

Fig. 4 Mole fraction of Fe and Mo in main species as function of pH value. (Color figure online)

Fig. 5 Effect of Fe(III) addition on the effect of molybdenum precipitation and conversion. (pH = 5; stirring speed = 150 r/min)

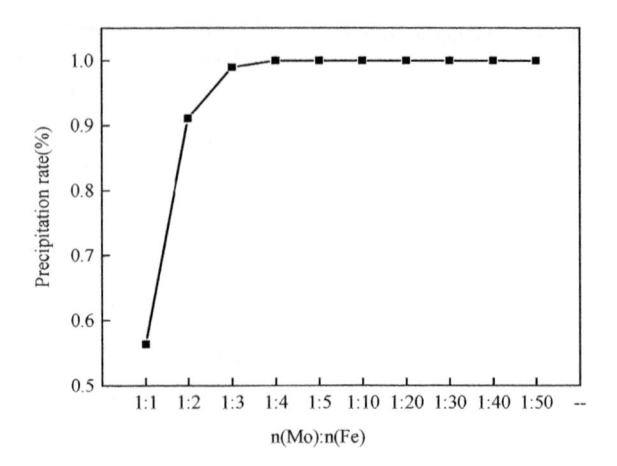

molybdenum decreased with pH increase gradually. This may be due to the fact that when pH was greater than 5, molybdenum mainly existed in the form of MoO_4^{2-}, and the electrostatic interaction decreased between positive and negative ions. Moreover, with the continuous increase of pH value, the polymer of Fe^{3+} gradually presented negative electricity, which inhibited the precipitation transformation of molybdenum.

Effect of Fe(III) Addition

Figure 5 revealed the influence of Fe(III) dosage on the precipitation conversion effect of molybdenum when the initial concentration of molybdenum was 100 mg/L at pH = 5. It can be seen that the addition amount of Fe(III) had a great influence on the precipitation conversion rate of molybdenum. With the increase of the addition amount of Fe(III), the precipitation conversion effect of molybdenum gradually strengthened. When the n(Mo):n(Fe) was 1:4, it reached equilibrium and peak value, meanwhile the precipitation conversion rate of molybdenum was 100%. When the n(Mo):n(Fe) was about 1:4, the Fe(III) addition amount was enough to completely react with molybdenum to form precipitation, and if the addition amount of Fe(III) continued to increase, there will be a relative surplus of iron ions. However, at pH = 5, Fe(III) was presented as iron hydroxide precipitation ($Fe(OH)_3$) and can separate from the solution, which had no great influence on the precipitation transformation of molybdenum.

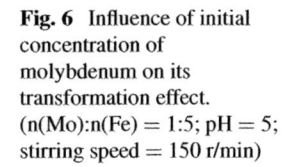

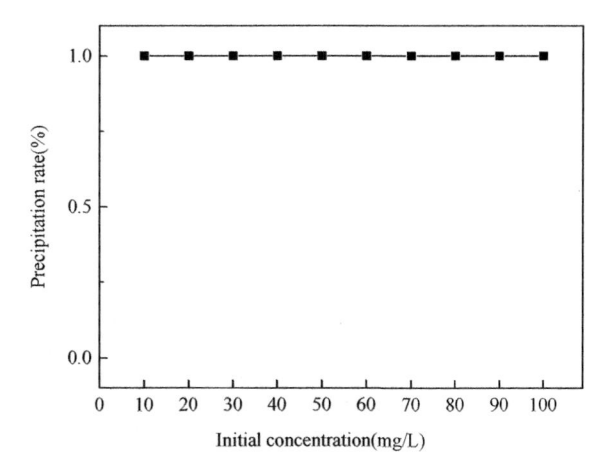

Fig. 6 Influence of initial concentration of molybdenum on its transformation effect. (n(Mo):n(Fe) = 1:5; pH = 5; stirring speed = 150 r/min)

Effect of Initial Molybdenum Concentration

The effect of initial concentration of molybdenum on precipitation was investigated under optimal conditions (pH = 5 and 300 mg/L Fe(III)), and the results are displayed in Fig. 6. It can be seen from the figure that, under the optimal conditions, the initial concentration of molybdenum had no effect on its precipitation conversion effect, and the precipitation rate was close to 100%.

Particle Size Distribution of Precipitation

In this paper, the precipitate, which generated from a solution with initial concentration of molybdenum of 100 mg/L by adding 300 mg/L Fe(III), was characterized by the particle size at 100× magnification with the ratio of Temp 0.706034 micron/pixel with Carl Zeis polarizing microscope in Germany, the results are shown in Fig. 7.

As can be seen, the particle size was mostly concentrated in the range of 0–24 μm; the particle size that accounted for about half of the total was between 6 and 12 μm, which was too small to be separated by filtration; consequently, high-speed centrifugation should be adopted. At the same time, it was found that the sediment can settle to the bottom of the solution after 30 min, and there were no fine particles in the supernatant.

SEM Analysis of the Precipitation

SEM was used to characterize the morphology of molybdenum-iron sediment, and EDS analysis was conducted on the sample to understand the main components of

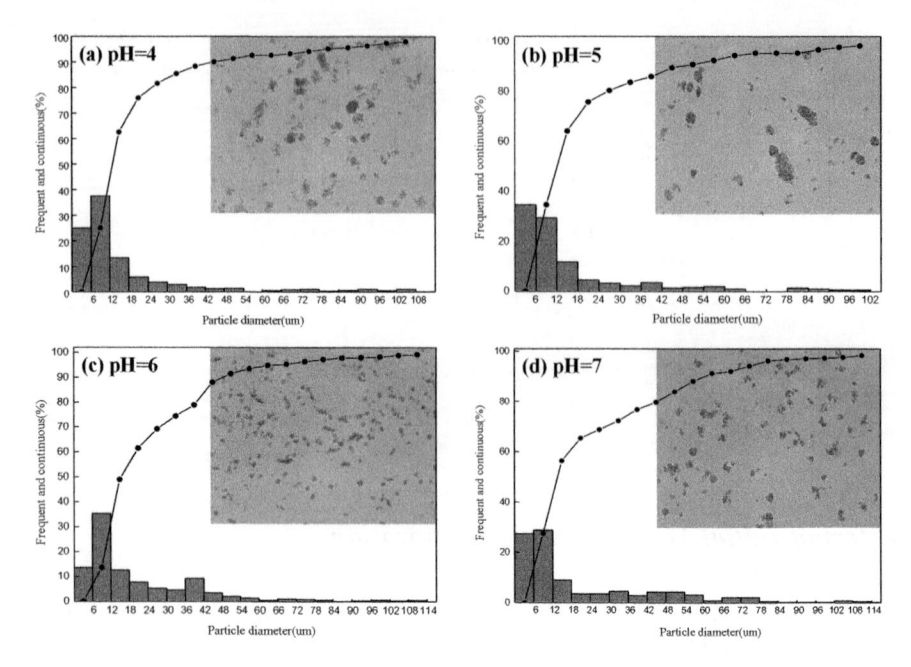

Fig. 7 Particle size analysis of the mixture of molybdenum and iron. (Color figure online)

the sample. The results are shown in Fig. 8. Figure 8a was the SEM diagram of molybdenum-iron sediment at different magnifications, and Fig. 8b was the EDS diagram of molybdenum-iron sediment.

It can be seen that the precipitate was covered with fine particles, and 0.5–1.5 μm similar particles crystallized on the precipitation surface. From the EDS diagram of molybdenum-iron sediment, it can be clearly seen that there were mainly three elements Mo, Fe, and O in the sediment.

Conclusions

In this paper, ferric chloride was used as the precipitant to adsorb and remove molybdenum in the solution, and the process conditions were optimized. Phase analysis and morphology characterization of the products were carried out by SEM to determine the experimental mechanism. At room temperature, in case of the pH = 5, n(Mo):n(Fe) = 1:3 and stirring speed 150 rpm, the precipitation conversion rate of molybdenum reached 100%. At the same time, Fe(III) was also completely removed by precipitation. The precipitation particle size was mostly concentrated between 6 and 12 μm, which was difficult to separate. However, after resting for 5 min, the precipitate settled and the supernatant transparent was colorless. This showed that the sedimentation method can be used to separate the solid and liquid in the

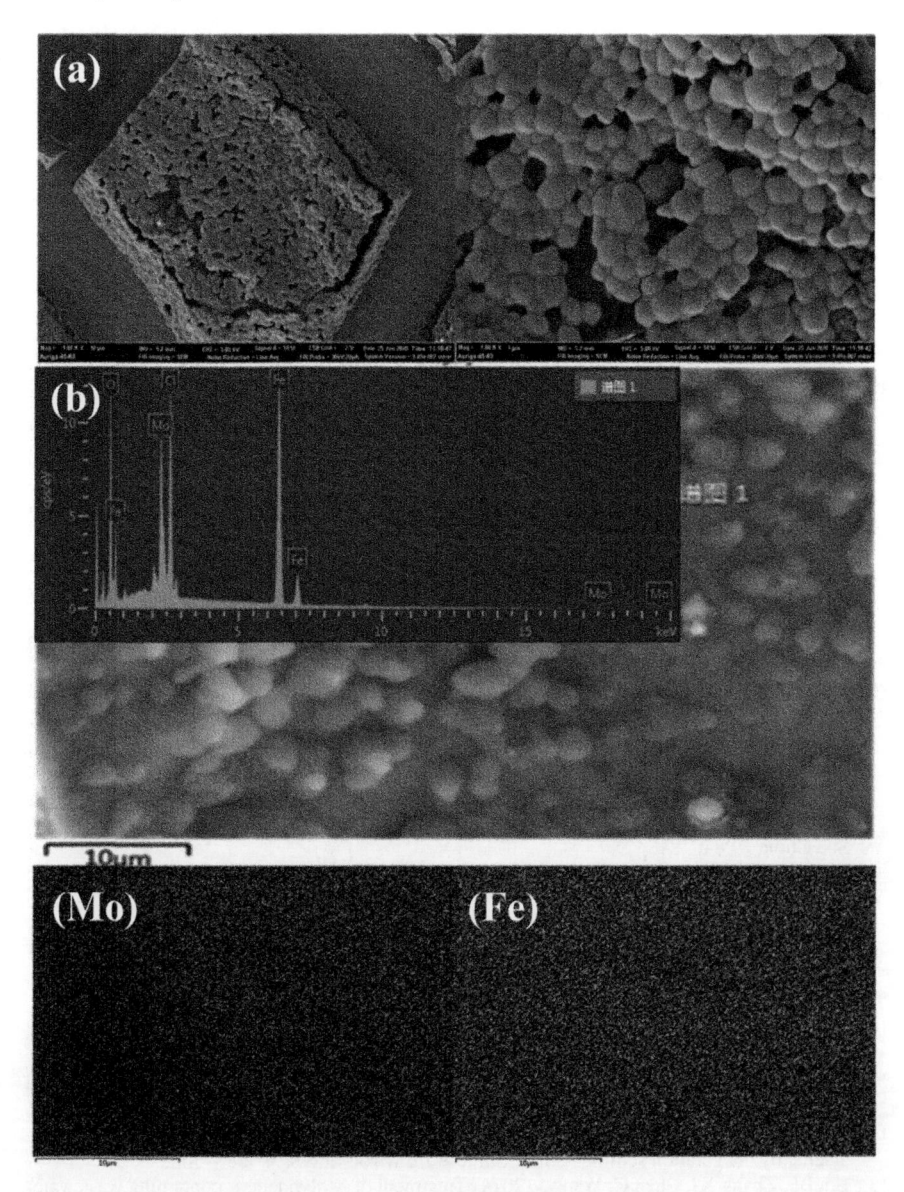

Fig. 8 a SEM image of the molybdenum-iron precipitate; b EDS of the molybdenum-rhenium-iron precipitate. (Color figure online)

actual production process. In conclusion, it was feasible to remove the MoO_4^{2-} from solution with addition of Fe(III), and the method had a high removal rate of molybdenum.

Acknowledgements The authors wish to express their thanks to the China Postdoctoral Science Foundation (Number 2019TQ0289), the Key Scientific Research Project Plan of Henan Colleges and Universities (Number 20A450001), and the Special Support Program for High Level Talents in Henan Province (Number ZYQR201912182) for the financial support.

References

1. Fan Y, Zhou TF, Zhang DY, Yuan F, Ren Z (2014) Spatial and temporal distribution and metallogical background of the chinese molybdenum deposit. J Geol 88(4):784–804
2. Zhang ZH, Li DL, Xing XD, Liu SF, Zhang GX, Lu HH (2020) Application status and development prospect of Molybdenum in alloy steel smelting. J Iron Steel Res 32:1–8
3. Ji Y, Deng HJ, Wang CN, Jiang AL, Jian H (2016) Current status of molybdenum-ore resources in China and screening of national molybdenum-ore geological material data. China Mining 25:139–145
4. Zhang T, Xin WJ, Kang NL (2020) Improvement of determination method of molybdenum content in roasting molybdenum concentrate. China Molybdenum Ind 44:55–57
5. Xiao C, Zeng L, Li YB (2017) Predominance diagram of molybdenite roasting with lime. China Molybdenum Ind 41:40–44
6. Li XL, Wang Z, Wang YF, Zhang BS (2017) Experimental study on oxidation and roasting of molybdenite. China Molybdenum Ind 35:29–31+39.
7. Fan XH, Deng Q, Gan M, Chen XL (2019) Roasting oxidation behaviors of ReS_2 and MoS_2 in powdery rhenium-bearing, low-grade molybdenum concentrate. Trans Nonferrous Metals Soc China 29:840–848
8. Dong H, Cai JJ, Zhang JF, Yu QB (2003) Development and application of an integrated device for dust collecting and Rhenium recovery from roasting tail gas of molybdenum concentrate containing rhenium. China Molybdenum Ind 27(5):19–21
9. Liu HZ, Liu L, Zhang B, Wang W, Cao YH, Liu L, Wang HL (2019) Flow characteristic of rhenium in molybdenum concentrate roasting process by self-heating rotary kiln. Nonferrous Metals 11:42–45
10. Liu ZQ, Wang ZC, Cao X (2018) Iron-Vanadium coprecipitation method to remove vanadium from vanadium-containing wastewater. J Shenyang Univ Chem Technol 32:330–334
11. Song ZY, Sun YM, Hu JL, Zhang X, Ma WJ (2019) Study of pretreatment of glyphosate wastewater by iron salt precipitation. Fine Specialty Chem 27:16–20
12. Yan Q, Gui YG, Zhou NN, Xu J, Yu Y, Luo XP (2014) Treatment of wastewater containing arsenic by coagulation sedimentation. Chinese J Environ Eng 8(9):3683–3688
13. Lin BF, Zhang XJ, Chen C, Wang J (2014) Treatment of Molybdenum-containing wastewater and drinking water. J Tsinghua Univ 54:613–618
14. Zhang JS, Feng X, Cao X (2019) Removal of Molybdenum from Aqueous solution by coprecipitation. J Shenyang Univ Chem Technol 33:97–103
15. Ma Y, Wu W, Han HD (2011) Molybdenum removal in water by chemical sedimentation. Hydrotechny 5:26–27+31
16. Zhang X, Ma J (2016) Removal of Molybdenum ions from water by polyferric chloride. Water Supply Drainage 52:41–45

An Effective Way to Extract Cr from Cr-Containing Tailings

Jie Cheng, Hong-Yi Li, Shuo Shen, Jiang Diao, and Bing Xie

Abstract The high-chromium vanadium slag is treated by magnesium roasting-acid leaching to obtain Cr-containing tailings, which are not the solid waste but important valuable resources for various fields. In order to extract Cr from these tailings, sodium carbonate as annexing agent mixed tailings was roasted and then leached by water. There are several factors that were investigated including roasting temperature, roasting time, the molar ratio of Na/(V+Cr), leaching time, leaching temperature, and the ratio of L/S. The Cr-containing tailings before and after roasting and the residue after leaching were characterized by XRD. It is indicated that Cr in tailings was all entered into leaching solution due to residue not found Cr-containing phase. Under the optimal conditions, the maximum leaching rate of Cr was 92.49%.

Keywords Cr-containing tailings · Extract · Leaching

Introduction

Cr is a precious resource, which is widely used in stainless steel, leather tanning and electroplating, and electrolysis [1–4]. However, the Cr(VI) is very dangerous due to high toxicity [2]. Previous studies have shown that Cr(VI) can enter the human body through the respiratory tract and skin, thereby causing harm to the human body [5]. At present, the main raw material for extracting chromium in China is chromite ore [6] which is roasted in air or oxygen and then leached in water to obtain Cr(VI) [7]. In this process, a part of the chromium enters the wastewater, which will cause harm to the environment [8]. Moreover, the resources of chrome ore are limited in China. Therefore, it is imperative to find a new raw material for extracting Cr(VI).

Cr-containing tailings, which are the residues obtained after the vanadium-chromium slag roasted and leached to extract vanadium [9]. Recently, Cr-containing

---

J. Cheng · H.-Y. Li (✉) · S. Shen · J. Diao · B. Xie
College of Materials Science and Engineering, Chongqing University, Chongqing 400044, China
e-mail: hongyi.li@cqu.edu.cn

Chongqing Key Laboratory of Vanadium-Titanium Metallurgy and New Materials, Chongqing University, Chongqing 400044, China

tailings have attracted more and more attention from researchers because these tailings are a very precious resource that can be used to prepare ferrochrome [10]. However, at present, there was not found an effective way to deal with these tailings.

In this work, sodium carbonate roasting–water leaching method is proposed to effectively extract Cr from Cr-containing tailings. The effects of roasting time, roasting temperature, the molar ratio of Na/(V+Cr), leaching time, leaching temperature, and the ratio of L/S on Cr extraction efficiency are investigated.

Materials and Methods

Materials

Cr-containing tailings are obtained from the vanadium chromium slag of Pan Steel through magnesium oxide roasting—sulfuric acid leaching. Chemicals used (A.R.) were purchased from Chron Chemicals, China. Table 1 and Fig. 1 show the main chemical components and XRD of Cr-containing tailings, respectively.

Table 1 Chemical composition of the leaching residue (wt.%)

Fe_2O_3	V_2O_5	Cr_2O_3	TiO_2	MnO	MgO	CaO	Al_2O_3	SiO_2	P_2O_5
45.49	5.72	9.25	11.59	4.03	8.27	1.09	5.25	25.30	0.096

Note All the iron oxides were counted in the form of Fe_2O_3 for content calculation

Fig. 1 XRD diffraction pattern of Cr-containing tailings. (Color figure online)

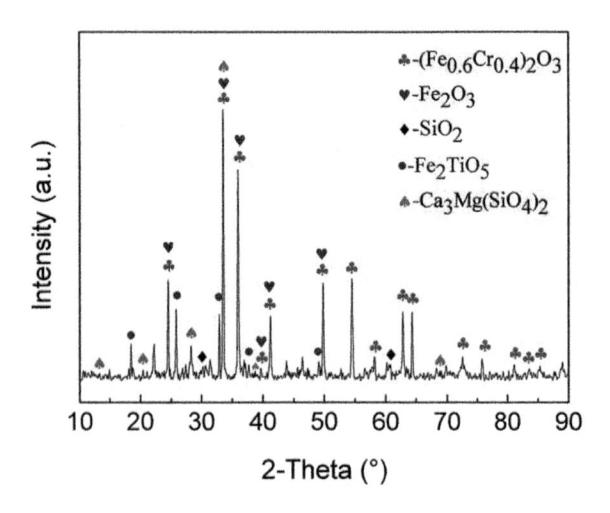

Experimental Procedure

Sodium Carbonate Roasting

Sodium carbonate was mixed with Cr-containing tailings (<74 μm) in the molar ratio of Na/(V + Cr) = 3.0–7.0 and pelletized. Then, the mixture is placed in the muffle furnace and roasted at 1123–1323 K for 30–150 min, and the door of muffle furnace was not completely closed to keep the sufficient oxidizing atmosphere. Then, the roasted slag was cooled down and grounded to power.

Water Leaching

The roasted sample was leached with distilled water in a commercial magnetically stirred water bath with L/S ratios of 4–12 mL/g at 303–343 K for 20–60 min at the stirring speed of 200 r/min. The leaching liquor and residue were separated by filtration. The extraction efficiency of Cr was calculated following Eq. (1):

$$\text{Extraction efficiency } (\%) = C_L V_L / M_{Cr} \times 100\% \qquad (1)$$

where C_L and V_L denote the Cr concentration in leaching liquor and the volume of leaching liquor; M_{Cr} denotes the Cr mass in roasted Cr-containing tailings.

Characterization

The chemical compositions of Cr-containing tailings were characterized by X-ray fluorescence (XRF, Shimadzu XRF-1800, Japan). The phase compositions of Cr-containing tailings before and after roasting and the residue after leaching were characterized by X-ray diffraction analysis (XRD, Rigaku D/MAX 2500PC, Japan).

Results and Discussion

Sodium Carbonate Roasting

Effects of Sodium Carbonate Roasting Conditions on Cr Extraction

Figure 2 shows the effects of roasting temperature, roasting time, and the molar ratio of Na/(V + Cr) on the extraction efficiency of Cr. It can be seen from Fig. 2 that roasting temperature has the biggest effect on the extraction efficiency of Cr.

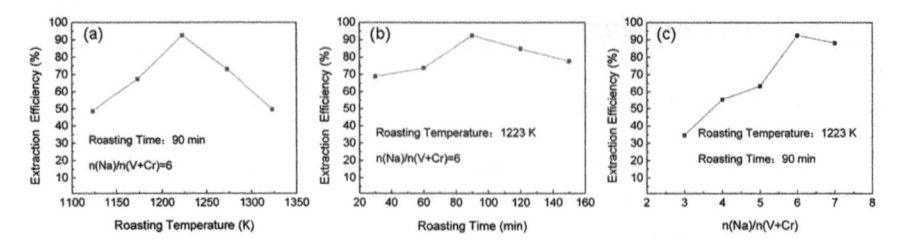

Fig. 2 Influences of sodium carbonate roasting conditions on the extraction efficiencies of Cr **a** roasting temperature; **b** roasting time; **c** Na/(V + Cr) molar ratio. (Color figure online)

As Fig. 2a shows, the extraction efficiency of Cr increased first and then decreased with the increase of temperature, reaching the maximum value of 92.49% at 1223 K. This is that the roasting temperature is not enough, which causes the chromium phase to fail to completely react with the sodium carbonate, and the high roasting temperature can cause the sintering of the roasted sample. Therefore, 1223 K was selected as the optimal roasting temperature.

The effect of roasting time on the extraction efficiency of Cr is shown in Fig. 2b. As roasting time increases from 30 to 90 min, the extraction efficiency of Cr increases from 68.79% to 92.49%. Continuing to extend the roasting time, the extraction efficiency of Cr decreases. This is because the low melting point phases melt and wrap on the surface of the chromium phase, which hindering the diffusion of oxygen and resulting in a decrease in the extraction efficiency of Cr. Therefore, 90 min was determined as the best roasting time.

In Fig. 2c, with the increase of sodium carbonate content, the extraction efficiency of Cr increased first and then reduced, reaching the maximum value of 92.49% at the molar ratio of Na/(V + Cr) = 6.0. The reason for the decrease in the extraction efficiency of Cr is that the unreacted sodium carbonate melts and wraps around the chromium phase, hindering the diffusion of oxygen. So, 6.0 was selected as the best molar ratio.

Water Leaching

Effects of Water Leaching Conditions on Cr Extraction

Figure 3 shows the effects of leaching temperature, leaching time, and L/S ratio on the extraction efficiency of Cr. In Fig. 3a, the leaching efficiency of chromium increases with the increase of temperature, reaching the maximum value of 92.49% at 333 K. This is due to that the solubility of sodium chromate in the roasted sample increases as the leaching temperature increases. Therefore, 333 K was determined as the best leaching temperature. It can be seen from Fig. 3b that the change of leaching time has little effect on the extraction efficiency of chromium. Therefore, in order to save time,

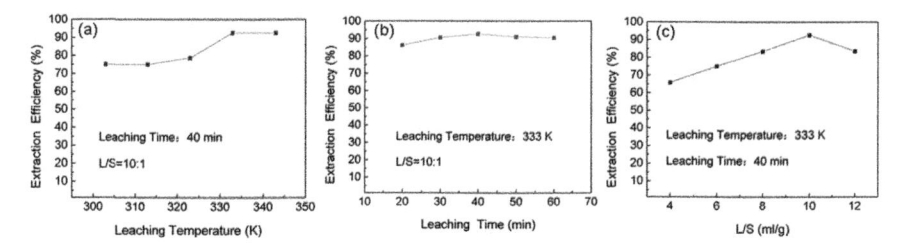

Fig. 3 Influences of water leaching conditions on the extraction efficiencies of Cr. **a** Leaching temperature; **b** Leaching time; **c** L/S ratio. (Color figure online)

the leaching time was controlled to 40 min. In Fig. 3c, at a L/S ratio of 4.0 and 6.0, the extraction efficiency of chromium was 65.66% and 74.92%, indicating that the leaching kinetic conditions are insufficient. The extraction efficiency of chromium reaches maximum value of 92.49% at a L/S ratio of 10.0. Therefore, a L/S ratio of 10.0 was selected as the optimal L/S ratio.

Phase Evolutions Before and After Leaching

Figure 4 shows the phase evolutions before and after leaching of roasted Cr-containing tailings. In Fig. 4b, the major components of leaching residue are Fe_2O_3, $Fe_{2.75}Ti_{0.25}O_4$, $Na_{1.8}Mg_{0.9}Si_{1.1}O_4$, and $NaFeTiO_4$, which can be returned to the blast furnace for the iron-making process after removing sodium. Compared with Fig. 4a, the diffraction peaks of $NaCr_2O_4$ disappeared after leaching, indicating that chromium has been separated successfully.

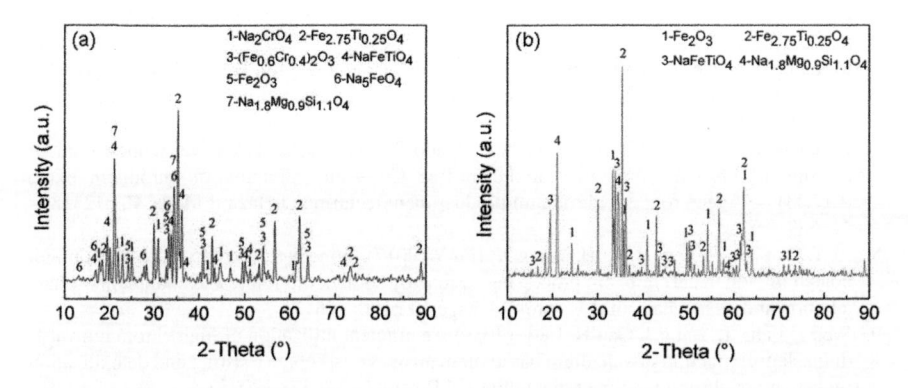

Fig. 4 XRD patterns of **a** roasted Cr-containing tailings; **b** leaching residue. (Color figure online)

Conclusions

In this work, an effective way to extract Cr from Cr-containing tailings by sodium carbonate-water leaching was proposed. When the roasting temperature is 1223 K, the roasting time is 90 min, the molar ratio of Na/(V + Cr) is 6.0, the leaching temperature is 333 K, the leaching time is 40 min, and the L/S ratio is 10.0, the maximum leaching efficiency of chromium can reach 94.29%.

Acknowledgements This work was supported by the National Natural Science Foundation of China [grant number 51674051]; Chongqing Science and Technology Bureau [cstc2019jcyjjqX0006]; Chongqing Talents Plan for Young Talents [CQYC201905050]; Chongqing University [cqu2017hbrc1B08, 2020CDCGCL003].

References

1. Barnhart J (1997) Occurrences, uses, and properties of chromium. Regul Toxicol Pharmacol 26:S3-7
2. Fang D, Liao X, Zhang XF, Teng AJ, Xue XX (2018) A novel resource utilization of the calcium-based semi-dry flue gas desulfurization ash: As a reductant to remove chromium and vanadium from vanadium industrial wastewater. J Hazard Mater 342:436–445
3. Tor A, Arslan G, Muslu H, Celiktas A, Cengeloglu Y, Ersoz M (2009) Facilitated transport of Cr(III) through polymer inclusion membrane with di(2-ethylhexyl)phosphoric acid (DEHPA). J Membr Sci 329:169–174
4. Vellaichamy S, Palanivelu K (2010) Speciation of chromium in aqueous samples by solid phase extraction using multiwall carbon nanotubes impregnated with D2EHPA. Indian J Chem Sect A-Inorgan Bio-Inorgan Phys Theor Anal Chem 49:882–890
5. Sellami F, Kebiche-Senhadji O, Marais S, Colasse L, Fatyeyeva K (2020) Enhanced removal of Cr(VI) by polymer inclusion membrane based on poly (vinylidene fluoride) and Aliquat 336. Separat Purif Technol
6. Moon DH, Wazne M, Koutsospyros A, Christodoulatos C, Gevgilili H, Malik M, Kalyon DM (2009) Evaluation of the treatment of chromite ore processing residue by ferrous sulfate and asphalt. J Hazard Mater 166:27–32
7. Escudero-Castejon L, Taylor J, Sanchez-Segado S, Jha A (2020) A novel reductive alkali roasting of chromite ores for carcinogen-free Cr6+-ion extraction of chromium oxide (Cr2O3)—A clean route to chromium product manufacturing! J Hazard Mater 403:123589–123589
8. Li Y, Cundy AB, Feng J, Fu H, Wang X, Liu Y (2017) Remediation of hexavalent chromium contamination in chromite ore processing residue by sodium dithionite and sodium phosphate addition and its mechanism. J Environ Manage 192:100–106
9. Wen J, Jiang T, Wang J, Gao H, Lu L (2019) An efficient utilization of high chromium vanadium slag: Extraction of vanadium based on manganese carbonate roasting and detoxification processing of chromium-containing tailings. J Hazard Mater 378:120733
10. Wang G, Lin MM, Diao J, Li HY, Xie B, Li G (2019) Novel strategy for green comprehensive utilization of vanadium slag with high-content chromium. ACS Sustain Chem Eng 7:18133–18141

Study on the Enhancement of Iron Removal in the Becher Aeration by a Novel Tubular Reactor

Lei Zhou, Qiuyue Zhao, Mingzhao Zheng, Zimu Zhang, Guozhi Lv, and Tingan Zhang

Abstract An aeration step of the Becher process was carried out in a novel tubular reactor to study the effect of the tubular reactor on the removal of metallic iron. In the tubular reactor, the influence of oxygen flow rate and stirring rate on the removal of metallic iron in reduced ilmenite (RI) were studied and compared with the kettle reactor. The results show that when the reaction system is 2% (w/v) ammonium chloride solution, the optimal condition is that 97% of metallic iron (MFe) can be removed from RI in 3 h. At the same time, the synthetic rutile (SR) obtained in the tubular reactor has less iron oxide precipitated inside its particles. The grade of TiO_2 can reach 87%.

Keywords Becher process · Tubular reactor · Aeration step · Reduced ilmenite · Metallic iron

Introduction

The raw materials in the industrial production of titanium are mainly natural rutile and ilmenite. The small reserves of natural rutile cannot meet the production demand [1]. Therefore, synthetic rutile produced from ilmenite occupies an increasing proportion in the industrial production of titanium. The methods of producing synthetic rutile successfully applied in the industry mainly include Benelite process [2], Austpac process [3], Kataoka process [4], and Becher process [5]. The Benelite process uses carbothermal reduction to convert iron into ferrous state, and then leaches and separates it with hydrochloric acid. The Ausptpac process is to roast ilmenite at high temperature to make it selective magnetization, then magnetic separation to remove gangue, and finally leaching through hydrochloric acid. The Kataoka process is to reduce the iron in ilmenite to ferrous state through reduction roasting, and then separate it through a sulfuric acid leaching step. These three processes all separate impurities through reduction roasting and high-concentration acid leaching, which

---

L. Zhou · Q. Zhao (✉) · M. Zheng · Z. Zhang · G. Lv · T. Zhang
School of Metallurgy, Northeastern University, Shenyang, Liaoning 110000, China
e-mail: zhaoqy@smm.neu.edu.cn

causes high energy consumption and serious pollution. In the Becher process, low-ash coal is used as a fuel and a reducing agent to mix with ilmenite, which is reduced and roasted in a rotary kiln to reduce the iron in the ilmenite to metallic iron. In the NH$_4$Cl solution, metallic iron is oxidized and dissolved into the solution by aeration to form fine iron oxide precipitates. The separated titanium-rich material product undergoes acid washing to remove the remaining iron. Kataoka process employ.

The process of reduction roasting can be explained by the following reaction:

$$4FeTiO_3 + O_2 \rightarrow 2Fe_2O_3 \cdot TiO_2 + 2TiO_2 \tag{1}$$

$$2Fe_2O_3 \cdot TiO_2 + 3CO \rightarrow 2Fe + 2TiO_2 + 3CO_2 \tag{2}$$

$$2FeTiO_3 + CO \rightarrow FeTi_2O_5 + Fe + CO_2 \tag{3}$$

The aeration step is essentially an electrochemical reaction, which can be expressed as:

$$Fe \rightarrow Fe^{2+} + 2e \tag{4}$$

$$O_2 + H_2O + 4e \rightarrow 4OH^- \tag{5}$$

Fe^{2+} diffuses into the solution and further oxidizes to form a precipitate:

$$2Fe^{2+} + 4OH^- + \frac{1}{2}O_2 \rightarrow Fe_2O_3 \cdot H_2O + H_2O \tag{6}$$

The Becher process is carried out in a weakly acidic solution, and the wastewater is neutral, which has environmental advantages [6]. In addition, iron by-products can be used as raw materials for iron making. The disadvantage of this process is that the reaction is slow in the aeration step. In the current industrial production, it may take 20 h to complete the reaction [7]. At present, some literature reports that some catalysts can accelerate the rusting process. These catalysts include: ethylene dichloride, acetic acid, citric acid, sucrose, anthraquinone derivatives, aldehydic compounds, paraquat [7–11, 13, 14]. However, few researchers currently study the influence of different reactors on the aeration process.

In this study, we conducted aeration experiments in a novel tubular reactor to investigate the effect of oxygen flow on the reaction effect. This kind of tank reactor draws on the experience of the continuous production of alumina [14]. This tubular reactor was proposed and designed by the special metallurgy team of Northeastern University. This kind of tubular reactor is still in the stage of testing in the laboratory, we hope it can be applied to industrial production in the future.

Feeding

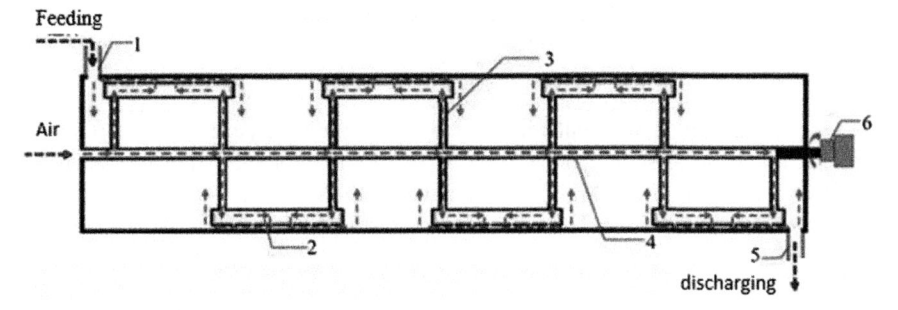

Fig. 1 Diagrammatic sketch of the tubular reactor. (Color figure online)

Experiment

Experimental Materials and Equipment

The reduced ilmenite used in this experiment was taken from a Chinese mining company, and it was obtained by reduction roasting in a rotary kiln at 1100 °C. Other reagents used in the experiment are of analytical grade.

The aeration experiment was carried out in a tubular reactor. The details of the tubular reactor are shown in Fig. 1. The volume of the tubular reactor is 30L. The stirring shaft, connecting rods, and blades of the tubular reactor are all hollow structures, and the walls of the blades are all provided with small vent holes. The gas inlet is connected with the stirring shaft, and the gas enters the reactor through the gas outlet on the blade. The gas generates tiny bubbles in the reactor and is evenly distributed throughout the liquid phase. At the same time, under the action of the rotation of the stirring blade, the solid is uniformly suspended in the tubular reactor and fully contacted with the gas, finally achieving a uniform distribution of gas–liquid–solid three phase.

Experimental Method

Add 8 kg of reduced ilmenite and 16 kg of 2% (w/v) NH_4Cl solution into the reactor and mix, with a liquid–solid mass ratio of 2:1. Oxygen is introduced for aeration reaction. In the experiment, samples were taken every 30 min, and the reaction time was 4 h. Afterward, the synthetic rutile particles and the iron by-products are separated, and the synthetic rutile particles and the iron by-products are respectively suction filtered and dried in an oven at 150 °C for 3 h. The dried samples are tested and analyzed.

The extent of metallic iron removal is obtained by calculation, the formula is as follows:

$$\mu(\%) = \left(1 - \frac{\omega_1 m_1}{\omega_0 m_0}\right) \times 100\% \qquad (7)$$

where ω_0 is the content of metallic iron in reduced ilmenite, m_0 is the mass of reduced ilmenite, ω_1 is the metallic iron content in the product obtained after the reaction, and m_1 is the quality of the product obtained after the reaction.

Analysis Method

X-ray fluorescence spectrometry (ZSXPrimusIV) was used to detect the content of TiO_2, TFe, SiO_2, Al_2O_3, MgO, CaO, and Mn_3O_4 in the sample. The potassium dichromate method was used to determine the content of metallic iron in the sample. X-ray diffractometer (XRD) model D8 ADVANCE was used to characterize the structure of raw materials and products. Use the scanning electron microscope (SEM) of SU-8010 to analyze the surface microtopography of the sample. The BT-9300S laser particle size analyzer was used for particle size analysis of the samples.

Results and Discussion

Sample Characterization

The composition of reduced ilmenite is shown in Table 1. It can be seen that the reduced ilmenite contains 58.47% TiO_2, 31.25% MFe, 2.31% FeO, and the metallization rate is 92.78%. The particle size distribution of the sample is shown in Fig. 2. About 80% of the particles have a particle size in the range of 93–311 μm.

The XRD pattern of reduced ilmenite is shown in Fig. 3. The phase composition of the sample is mainly metallic iron, rutile (TiO_2), and a small amount of pseudobrookite ($FeTi_2O_5$). At the same time, Fig. 4 shows the micromorphology of the raw material. (a) and (b) in the figure are the micromorphology under low magnification and high magnification, respectively. The raw material particle is a spherical ore particle with an irregular surface, which looks like a substance that has been condensed after high-temperature melting.

Table 1 XRF analyses of raw material

Component (wt%)	TiO_2	MFe[a]	FeO	TFe[b]	CaO	MgO	Al_2O_3	SiO_2	Mn
Sample	58.47	31.25	2.31	33.68	0.18	0.44	0.94	0.86	2.68

[a]Metallic iron (MFe); [b]Total iron (TFe)

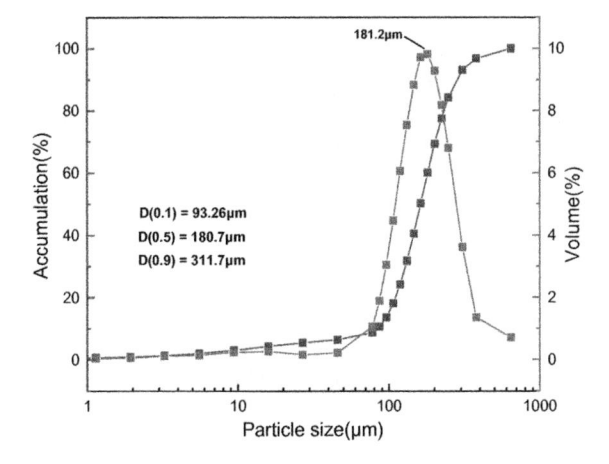

Fig. 2 The particle size distribution of the reduced ilmenite. (Color figure online)

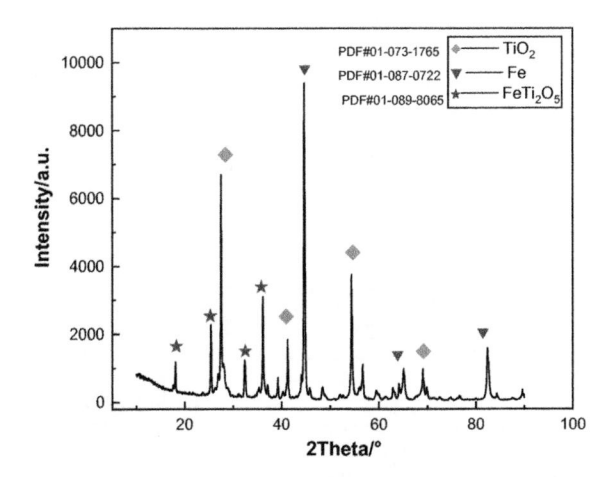

Fig. 3 XRD diagram of reduced ilmenite. (Color figure online)

The Influence of Oxygen Flow Rate

In order to determine the influence of gas flow rate on the reaction rate of the ammonium chloride reaction system, experiments with different gas flow rates were carried out. The oxygen flow rate was set to 0.4 m^3/h, 0.6 m^3/h, and 0.8 m^3/h. The stirring speed is 250 rpm, the liquid-to-solid mass ratio is 2: 1. 2% (w/v) ammonium chloride solution and reduced ilmenite are 16 and 8 kg, respectively, and the reaction time is 4 h.

The result of the change of metallic iron content is shown in Fig. 5. It can be seen from Fig. 5 that as the gas flow rate increases, the time required for the corrosion

Fig. 4 Picture of reduced ilmenite photographed by SEM, **a** Low magnification photos, **b** High magnification photos

Fig. 5 The relationship of metallic iron with oxygen flow and time in the tubular reactor. (Color figure online)

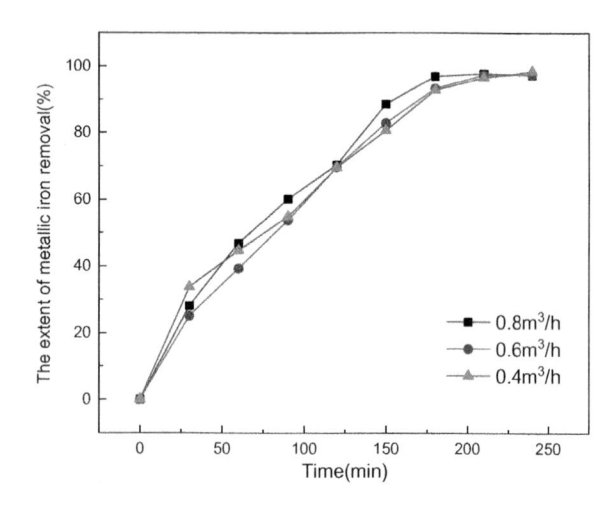

reaction is shortened. When the gas flow rate is 0.8 m³/h, the metal iron content has been reduced to below 1% in only 3 h, and the metal iron removal rate is 97%. According to reports, in a typical Becher process, the removal rate of metallic iron in the middle and late stages of the reaction is very low, only less than 50% [13]. This is because metal iron is oxidized to form a passivation film during the aeration process to block the reaction.

In this experiment, after 4 h of reaction, the content of TiO_2 and TFe is shown in Fig. 6. It can be seen from the figure that after 4 h of reaction, the TFe in the product is reduced to about 4.3%, and the TiO_2 in the product reaches about 86.7%, which shows that there is no obvious passivation phenomenon during the aeration process. This is the reason why the metal iron removal rate in the late stage of the reaction is not as low as reported in the tubular reactor. The stirring effect of the tubular reactor is much stronger than that of the tank reactor, which will strengthen the out-diffusion

Fig. 6 The relationship between the main element content and the oxygen flow rate and time. (Color figure online)

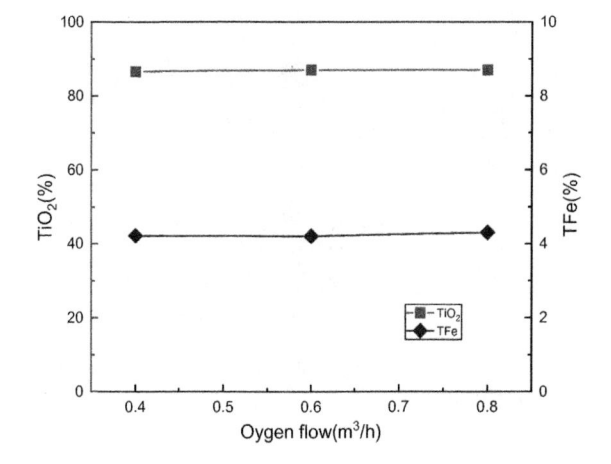

rate of Fe^{2+} in the particles, reduce the residence time of Fe^{2+} inside the particles, and reduce passivation.

Characterization of Product

The XRD pattern of synthetic rutile is shown in Fig. 7. The main phase of synthetic rutile is rutile (TiO_2) with a small amount of pseudobrookite ($FeTi_2O_5$). Figure 8 shows the microscopic surface of synthetic rutile. Compared with reduced ilmenite, synthetic rutile has an irregular surface after melting disappears and becomes a porous structure, which should be caused by the removal of metallic iron in the particles. This also shows that the molten structure on the surface of reduced ilmenite is mainly

Fig. 7 XRD patterns of synthetic rutile. (Color figure online)

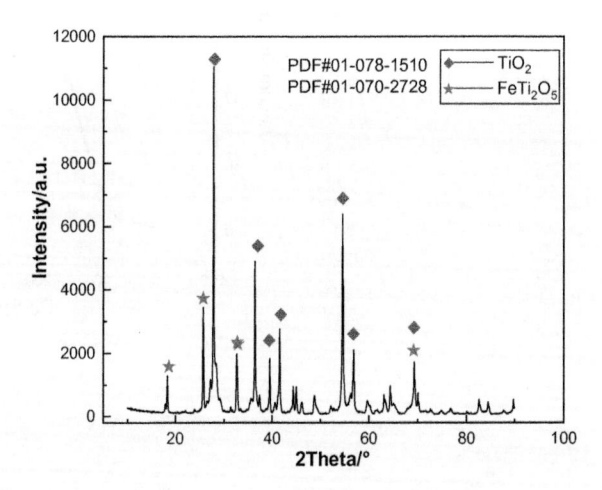

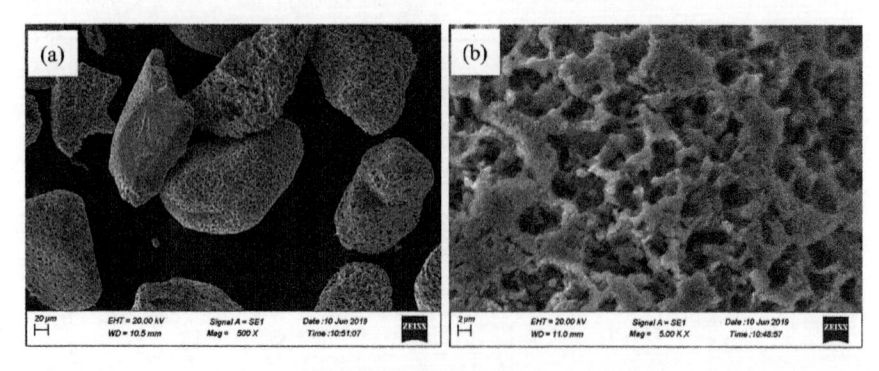

Fig. 8 Picture of synthetic rutile photographed by SEM, **a** Low magnification photos, **b** High magnification photos

metallic iron. The porous structure reduces the mechanical strength of the particles, making them easily broken.

The particle size distribution of synthetic rutile is shown in Fig. 9. It can be seen that the particle size distribution of the titanium-rich material becomes more concentrated after rusting, which is due to the metal iron reaction on the surface of the material and the collision of the stirring paddle with the material to break the particles. The iron by-product produced during the aeration is reddish-brown, and the XRD pattern is shown in Fig. 10. The iron by-product is a mixture of iron oxides (γ-FeOOH, α-Fe$_2$O$_3$).

Fig. 9 Particle size distribution of synthetic rutile. (Color figure online)

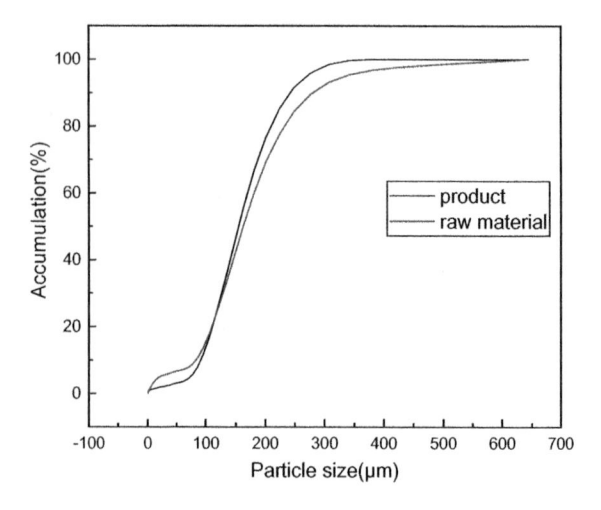

Fig. 10 XRD patterns of iron by-product. (Color figure online)

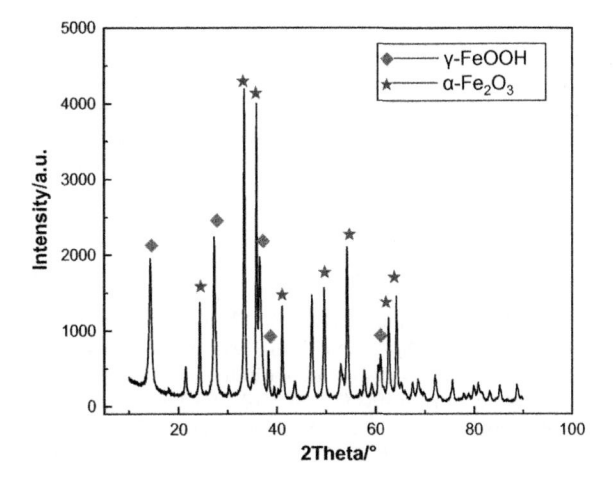

Conclusion

In the tubular reactor, the aeration efficiency is the highest under the conditions of 2% (w/v) NH$_4$Cl aqueous solution, oxygen flow rate of 0.8 m^3/h, and stirring speed of 250 rpm. At 3 h, the removal rate of metallic iron in reduced ilmenite reached 97%. At the same time, the content of TiO$_2$ and TFe reached 86.6% and 4.2%, respectively, without the passivation in the traditional Becher process.

Acknowledgements This research was supported by the National Natural Science Foundation of China (NSFC) (Grant No. 51204040); Fundamental Research Funds for the Central Universities (Grant No. N180725023, N2024005-6).

References

1. U.S. Geological Survey (2019) Mineral commodity summaries, titanium mineral concentrates. https://prd-wret.s3-us-west-2.amazonaws.com/assets/palladium/production/atoms/files/mcs-2019-titan.pdf
2. Zhang W, Zhu Z, Cheng CY (2011) A literature review of titanium metallurgical processes. Hydrometallurgy 108:177–188. https://doi.org/10.1016/j.hydromet.2011.04.005
3. Walpole EA, Winter JD (2002) The Austpac ERMS and EARS processes for the manufacture of high-grade synthetic rutile by the hydrochloride leaching of ilmenite. In: Proceedings of Chloride Metallurgy 2002, 2, Montreal, Canada, pp 401–415
4. Kataoka S, Yamada S (1973) Acid leaching upgrades ilmenite to synthetic rutile. Chem Eng 80(7):92–93
5. Farrow JB, Ritchie IM, Mangano P (1987) The reaction between reduced ilmenite and oxygen in ammonium chloride solutions. Hydrometallurgy 18(1):21–38. https://doi.org/10.1016/0304-386X(87)90014-4

6. Geetha KS, Surender GD (2000) Experimental and modelling studies on the aeration leaching process for metallic iron removal in the manufacture of synthetic rutile. Hydrometallurgy 56(1):41–62. https://doi.org/10.1016/S0304-386X(00)00065-7
7. Bruckard WJ, Calle C, Fletcher S, Horne MD, Sparrow GJ, Urban AJ (2004) The application of anthraquinone redox catalysts for accelerating the aeration step in the becher process. Hydrometallurgy 73(1–2):111–121. https://doi.org/10.1016/j.hydromet.2003.09.003
8. Nguyen TT, Truong TN, Duong BN (2016) Impact of organic acid additions on the formation of precipitated iron compounds. Acta Metall Slovaca 22:259. https://doi.org/10.12776/ams.v22i4.831
9. Nguyen TT, Truong TN, Quoc Dang K (2017) Acetic acid and sodium acetate mixture as an aeration catalyst in the removal of metallic iron in reduced. Ilmenite. Acta Metall. Slovaca. 23(4):371–377. https://doi.org/10.12776/ams.v23i4.1004
10. Ward J, Bailey S, Avraamides J (1999) Use of ethylenediammonium chloride as an aeration catalyst in the removal of metallic iron from reduced ilmenite. Hydrometallurgy 53(3):215–232. https://doi.org/10.1016/S0304-386X(99)00046-8
11. Kumari, E.J., Bhat, K.H., Sasibhushanan, S., Das, P.N. M. (2001) Catalytic removal of iron from reduced ilmenite. Miner. Eng. 14(3): 365-368. DOI: https://doi.org/10.1016/S0892-6875(01)00008-5
12. Fletcher S, Bruckard WJ, Calle C, Constanti-Carey K, Sparrow GJ (2019) Soluble catalysts for the oxygen reduction reaction (orr), and their application to becher aeration. Ind Eng Chem Res 58(24):10190–10198https://doi.org/10.1021/acs.iecr.9b01085
13. Xiang J, Pei G, Lv W, Liu S, Lv X, Qiu G (2019) Preparation of synthetic rutile from reduced ilmenite through the aeration leaching process. Chem Eng Process 147:107774. https://doi.org/10.1016/j.cep.2019.107774
14. Zhao QY, Zhang TA, Cao XC (2008) Application of tubular reactor in alumina production. Chinese J Process Eng. https://doi.org/10.3321/j.issn:1009-606X.2008.z1.067

Author Index

A
Abdulkareem, Aishat Y., 73
Abdul, Mohammed J., 73
Abrenica, Gomer, 217
Adeyemi, Christianah O., 73
Alabi, Abdul G. F., 73
Anderko, Andre, 139
Ayinla, Kuranga I., 73, 239
Azimi, Gisele, 11, 37

B
Baba, Alafara A., 73, 239
Balomenos, Efthymios, 217
Benkoussas, Hana, 163
Benvenuti, Emilia, 277
Bhargava, Suresh, 265

C
Chang, Wei, 291
Chan, Ka Ho, 11, 37
Checketts, Jed, 65
Cheng, Jie, 321
Chivavava, Jemitias, 129
Choubey, Pankaj Kumar, 47, 91, 197
Cun-xiong, Li, 291

D
Dal Santo, Vladimiro, 249
Dambruoso, Paolo, 249, 277
Das, Gaurav, 139
Davris, Panagiotis, 217
Degli Esposti, Alessandra, 277
Diao, Jiang, 321
Diaz, Luis A., 139

E
Eriksen, Dag Øistein, 229

F
Forsberg, Kerstin, 3, 155, 277
Free, Michael L., 211
Fu-li, Tang, 291

G
Gang, Fan, 291
Gardner, James, 3
Ghorbani, Yousef, 173
Girigisu, Sadisu, 73
Gupta, Rajesh, 91

H
Halli, P., 57
Han, Guihong, 311
Haque, Nawshad, 265
Hong, Hyun Seon, 79
Huang, Yuanfang, 311

I
Ibrahim, Abdullah S., 73, 239
Ilyas, Sadia, 303
Islam, Md Khairul, 265

J
Jha, Manis Kumar, 47, 91, 115, 197

K
Kim, Hyunjung, 303

Kim, Tai Gyun, 197
Kumari, Archana, 47, 91, 197

L
Lapinska-Viola, Renata, 277
Lee, Chan-Gi, 79
Lee, Jae-chun, 79, 115, 197
Leleu, David, 163
Lencka, Malgorzata M., 139
Lewis, Alison, 129
Li, Hong-Yi, 321
Lister, Tedd E., 139
Liu, Bingbing, 311
Ludwig, Christian, 187
Lu, Huimin, 29
Lundström, M., 57
Lv, Guozhi, 327

M
Malik, Monu, 11, 37
Marthi, Rajashekhar, 19
Martinez, Ana Maria, 101
Marzocchi, Stefania, 249
Ma, Yiqian, 3
Meagher, Michelle, 139
Mikeli, Eleni, 217
Min-ting, Li, 291

N
Navrotsky, Alexandra, 139
Nazari, Ghazaleh, 217
Neelameggham, Neale R., 29, 65
Nwaila, Glen T., 173

O
Olaoluwa, Daud T., 239
Olsson, Richard T., 3
Osen, Karen Sende, 101

P
Panda, Rekha, 115, 197
Panias, Dimitrios, 217
Park, Jae Ryang, 79
Park, Kyung Soo, 79
Pathak, Devendra Deo, 47, 115
Patil, Ajay B., 187
Patkar, Shailesh, 217
Pfennig, Andreas, 163
Pilihou, Anastasia, 217

Plessis du, Jacolien, 129
Polo, Eleonora, 277
Pownceby, Mark I., 265

R
Raji, Mustapha A., 73, 239
Riman, Richard E., 139
Rinne, M., 57
Rollins, Harry W., 139
Rosenkranz, Jan, 173
Russo, Ornella, 249

S
Sadisu, Girigisu, 239
Sanku, Meher, 155
Sarswat, Prashant K., 211
Satpathy, Swagatika, 163
Shariff, Zaheer Ahmed, 163
Shen, Shuo, 321
Siano, Salvatore, 249
Smith, Dane, 129
Smith, Jody-Lee, 129
Smith, York R., 19
Somerville, Michael, 265
Sommerseth, Camilla, 101
Srivastava, Rajiv Ranjan, 303
Strauss, Mark L., 139
Struis, Rudolf P. W. J., 187
Su, Shengpeng, 311
Svärd, Michael, 3, 155
Swain, Basudev, 79

T
Tang, Kai, 101
Tardio, James, 265
Testino, Andrea, 187
Torreggiani, Armida, 249, 277
Tripathy, Bankim Ch., 239

W
Wilson, B. P., 57

X
Xie, Bing, 321
Xing-bin, Li, 291
Xu, Wen-Qing, 217

Y
Yliniemi, K., 57

Z

Zanelli, Alberto, 277
Zhang, Bei, 311
Zhang, Steven E., 173
Zhang, Tingan, 327
Zhang, Zimu, 327
Zhang, Zongliang, 211
Zhao, Qiuyue, 327
Zheng, Mingzhao, 327
Zhi-gan, Deng, 291
Zhou, Lei, 327

Subject Index

A

Accelerated selective percolation leaching, 179
Acidic organophosphorus feed solution, 156
Advances in deep seabed mining of CRMs, 183
Aeration step, 327, 328
Aluminothermic, 29, 31, 66, 69
Aluminothermic reduction, 29, 31, 66
Aluminum metal powder, 29, 31, 33, 35
Ammonium Metatungstate (AMT), 73, 74, 77, 78
A move toward synchronized process and extractive metallurgy dry laboratories, 182
Analysis, 5, 8, 21, 22, 24, 32–34, 50, 59, 66, 75, 81, 82, 93, 94, 96, 103, 104, 107–109, 142–144, 146, 148, 151, 152, 158, 201, 212, 213, 215, 219, 220, 241, 243, 257–259, 270, 292, 293, 300, 303, 305, 306, 312, 317, 318, 323, 330
Analysis of metal lithium recovery rate, 33
An important metal for the green economy, 229
Arsenazo III solution, 156, 157
Au, 57, 58, 116, 117, 120, 139, 178, 181, 266

B

Batch experiment, 5, 7–9, 129, 132, 135–137, 311
Batch tests, 132, 133, 135
Bauxite residue, 217–220, 222–227, 229
Becher process, 327, 328, 332, 335
Bench-scale simulation of the process, 82
Binary phases diagrams, 6

Bioleaching, 211, 213, 214

C

CARONTE, 256–258
Cathode active material, 11, 13
Ce(IV) extraction with Cyanex 923, 305, 307, 309
Cerium(III) precipitation with oxalic acid, 307
Challenges in extraction design, 163
Characterization, 21, 76, 103, 144, 212, 258, 259, 318, 323, 330, 333
Characterization of product, 318, 333
Characterization studies, 75
Chemical precipitation for Li recovery, 96
Chemical pre-treatment of e-waste, 115, 199
Chemicals and Materials, 12
Chemistry of Scandium in Minerals, The, 230
Chromate compound, 239
Chromatography, 155–160
Chromite, 239–244, 321
Chromite ore dissolution process investigation, 241
Circular economy, 58, 79, 80, 86, 89, 102, 130, 185, 187, 250, 254, 278–280
CNR operations within the EIT RawMaterials framework, 251
Coal Ore Samples crushing, splitting, and screening process, 212
Coal refuse, 212
Cobalt, 7, 8, 11, 38, 47, 48, 50–53, 91, 130, 144, 165–167, 178, 183, 204
Cobalt sulphate ($CoSO_4$), 7, 8, 48, 51, 53
Column conditioner, 156, 157
Column preparation, 158

© The Minerals, Metals & Minerals Society 2021
G. Azimi et al. (eds.), *Rare Metal Technology 2021*, The Minerals, Metals & Materials Series, https://doi.org/10.1007/978-3-030-65489-4

Column preparation and performance, 157
Combustion fly ash, 291
Comparison of recovery to model, 148
Composite-enhanced-extractant polymer resin, 218
Comprehensive experiment for producing lithium, 33
Concept of sustainability, 175
Confirmation of equilibrium, 270
Confirmation of the slag composition, 269
Co-precipitation, 37–42, 44, 45, 98, 150
Co-precipitation method, 38–40
$CoSO_4$, 4–10, 48, 52
Cr-containing tailings, 321–323, 325, 326
Critical materials, 80, 140, 152
Critical raw materials, 80, 130, 173, 261, 262, 278, 282, 286
Crystallization, 3–8, 34, 35, 74, 75, 77, 129, 131–134, 271, 272

D

Definition and changes in assessment over time (2011–2017), 174
Details on selected CNR activities with the EIT RM portfolio, 252
Developed application oriented processes at CSIR-NML, 117
Different solution systems containing REEs, 190
Digital competence, 257
Dissolution mechanism and residual analysis, 243
Dissolution rate determination, 144
Distribution of vanadium in stone coal and fly ash, 298
Downstream extraction of CRMs from primary and secondary sources, 177
Drop quench testing—Phase equilibria determination, 267

E

Education, 249–253, 256, 258, 278, 279, 281, 285, 286, 309
Effect of acid concentration, 306
Effect of extractant concentration, 305
Effect of Fe(III) addition, 316
Effect of Impurities on the Performance of the II-VI SIR resin, 223
Effect of initial molybdenum concentration, 317
Effect of leachant concentration, 76
Effect of Na_2CO_3 Concentration, 242

Effect of Na_2O on the liquidus temperature of quaternary slag system, 273
Effect of oxalic acid addition, 307
Effect of particle size, 76, 242, 243
Effect of pH, 41, 95, 96, 314
Effect of precipitation temperature, 308
Effect of reaction temperature, 76, 243
Effect of reaction time, 314
Effect of sintering conditions on electrochemical performance, 42
Effects of Sodium carbonate roasting conditions on Cr extraction, 323
Effect of synthesis parameters, 41
Effect of time and temperature on dissolution of vanadium in fly ash, 294
Effect of time and temperature on leaching exchangeable fraction, 295
Effect of time and temperature on leaching Fe–Mn oxides fraction, 295
Effect of time and temperature on leaching organic matter fraction, 297
Effects of water leaching conditions on Cr extraction, 324
EIT RawMaterials, The, 249–252, 279, 284
Electrodialysis, 11–16
Energy efficiency, 107, 152
Estimation of the liquidus temperature, 272
Eutectic freeze crystallization, 3–7
E-waste, 79, 80, 86, 115–121, 124, 140, 141, 180, 187–189, 193, 197–200, 202, 204–208, 265, 266, 275
Experimental results and discussion of leaching of davidite, 232
Extract, 19, 29, 31, 33, 35, 121, 122, 164, 165, 200, 203, 205–207, 219, 234, 235, 299, 306, 321, 322, 326
Extraction and stripping isotherms, 306, 309
Extractive metallurgy, 131, 173

F

Farming-based strategies of high-tech and CRMs, 181
Feed solution and analysis, 5
Feed solution preparation and electrodialysis experiment, 13
Fe(III), 230, 231, 312–314, 316–318, 320
Fluidised Bed Reactor, 129, 132–134, 136, 137
Flux, 33, 103–105, 266
FREECATS, 261, 262
Future potential resources, 183

Subject Index

G
Generation of effluent containing Li and Mn, 94, 95

H
H_2SO_4, 5–9, 11–14, 16, 47, 48, 51, 52, 73–78, 93, 94, 118, 122, 141, 202, 204, 206, 207, 218, 219, 221–225, 232, 233, 294, 304, 306, 307, 309
H_3PO_4, 73–78
Heap leaching, 179, 180, 211
High-acid concentration leaching of BR - campaign 1, 223
High-tech CRMs, 173, 174, 176, 179
High-tech raw materials, 173
HNO_3 solution, 156
Hydrometallurgical flow-sheets developed for the recovery of precious metals, 119
Hydrometallurgical processes developed for the recovery of metals from e-waste, 202
Hydrometallurgy, 58, 163, 176, 189

I
Indigenous, 74, 77, 78, 117, 239, 304
Indium, 79–81, 86, 88, 205, 206, 277, 278
Industrial applications, 29, 41, 239, 244
Influence of mineral crystallography and fluid inclusions on CRM extraction route, 176
InnoLOG and InSITE, 258
Influence of oxygen flow rate, The, 327, 331
Innovative techniques and solutions to sustainable access to high-tech CRM, 176
In situ leaching, 179, 180, 181, 185
Integration of process through simulation, 85
Ion-exchange, 19, 115, 118, 124, 197, 198, 205, 206, 217, 219
Italian National Research Council, The, 249, 250
ITO etching, 79–85, 87–89

L
Laboratory experiments equipment, 31
Lab-scale process development, 81
Lantern reactor, 67–71
Leaching, 4, 11, 12, 48–52, 73–76, 79, 85–89, 92–94, 115, 117, 118, 120, 121, 123, 139, 141, 142, 148, 149, 152, 156, 164, 165, 177–181, 185, 197, 198, 202–207, 211–215, 217–227, 232–234, 236, 240–243, 291, 292, 294–299, 321–327
Leaching of cobalt from the obtained precipitate/ cobalt sulphide, 51
Leaching of pre-treated e-waste, 202
Leaching procedure, 49, 74
Leaching studies, 52, 76, 202, 203, 241
Leaching system, 48, 118, 142
Leaching test, 75, 212, 221, 224, 232
Leach liquor, 47–53, 77, 86, 88, 89, 94, 118, 120–123, 203–207, 303, 304
Life Cycle Assessment (LCA), 58
Lifelong learning, 249, 251
Li-ion battery recycling, 3, 4, 9, 10, 91
$LiNi_xMnCo_{1-x-y}O_2$ cathode, 37, 43
Liquidus, 30, 265–267, 271–275
Lithium carbonate, 29–31, 33, 34
Lithium-ion, 3, 4, 9–11, 37–39, 91, 92, 198, 205
Lithium-Ion Batteries (LIBs), 3, 4, 9–16, 37–39, 47–52, 91–94, 99, 198, 204, 205
Lithium (Li), 3, 10, 11, 13, 15, 19–26, 29–35, 37, 38, 42, 47, 48, 51, 52, 91, 198, 204
Low-acid concentration leaching of BR – campaign 2, 224

M
Magnet and steel dissolution studies, 142
Magnets, 130, 139–142, 144, 145, 148, 150, 152, 156, 170, 187–190, 193, 197, 198, 207, 278, 304
Materials, 3–5, 11–13, 19, 22, 29–31, 33–35, 37–42, 44, 45, 47–49, 58, 65, 67, 73, 74, 80, 89, 91–94, 101–105, 112, 116–118, 120, 130, 131, 133, 140–142, 145, 150, 152, 156, 163, 165, 173, 174, 176–179, 181, 184, 188, 189, 197, 199, 200, 202, 204, 205, 211, 212, 218–220, 232, 234–236, 239, 240, 249–254, 256, 258, 259, 261–263, 266, 267, 277–286, 292, 294, 311, 312, 321, 322, 327, 328, 330, 334
Mechanical pre-treatment of e-waste, 199
Metallic iron, 327–335
Metal lithium purity and impurity analysis, 32
Metals circular economy, 58

Methodology, 49, 104, 118, 164, 165, 170, 174, 249, 257, 281, 286
Methods, 3–5, 12, 14, 20, 21, 29–31, 37–41, 43–45, 57–60, 62, 65, 74–76, 93, 101, 102, 117, 120, 121, 130, 132, 140–142, 146, 150, 155, 156, 159, 160, 163, 165, 169, 173, 176, 179–181, 187, 189, 198, 205, 206, 213, 239, 240, 244, 259, 266, 267, 292, 293, 311, 318, 320, 322, 327, 329, 330
Mineral processing, 173
Molten salts, 29, 30, 102, 112
Molybdenum, 178, 181, 311–320

N

Na_2O, 105, 220, 265–267, 269, 271, 273–275
NaOH solution, 16, 95, 157, 214, 235
$NiSO_4$, 5–7, 131

O

Optimal process, 112, 164
Optimum leaching conditions – lower-than-1M H_2SO_4 leaching of BR, 225
Option tree for reactive extraction, 165
Option trees, 163–170
Option trees as support for process design, 164
Oxalate precipitation, 304
Oxidative decomposition, 240

P

Particle size distribution of precipitation, 317
Phase equilibria, 133, 265–267
Phase evolutions before and after leaching, 325
Pilot plant, 218, 220, 226, 227
Pilot-plant demonstration, 85
Platinum-Group Metals (PGM), 101–107, 109, 111, 112, 174, 178–180
Portable Instant Mineral Analysis Systems (PIMAS), 259, 260
Precious metals, 115–122, 124, 125, 163, 198, 205, 206
Precipitation, 4, 12, 16, 40, 41, 45, 47–52, 74, 85, 91–98, 115, 117, 118, 120, 122, 123, 130–132, 141, 143, 145, 146, 149, 150, 165, 190–193, 204, 205, 207, 214, 215, 218, 219, 221, 224, 225, 243, 267, 271, 295, 303–305, 307–309, 313–318
Precipitation and separation process, 313
Precipitation kinetics and mechanism, 308
Precipitation of cobalt from leach liquor, 50
Precipitation procedure, 49
Preliminary leaching campaigns, 221
Preparation principle of lithium metal, 33
Pre-treatment of e-waste, 115, 198, 199
Primary phase identification, 271
Primary resources, 91, 130, 183
Principle of recovery of lithium from spent LIBs utilizing electrodialysis coupled with EDTA, 14
Process design, 163, 164, 167, 170, 184
Process development, 80–82, 85, 87, 117, 129, 163, 164, 169, 188
Processing of wastewater, 97
Purification of recovered powders, 151
Pyrolysis of e-waste, 200
Pyro-metallurgy, 101–108, 111

R

RAISE, 252, 253
Rare earth elements, 129, 130, 139, 140, 150, 155, 174, 181, 187, 188, 211, 212, 214, 215, 229–232, 277
Rare earth metals, 139, 197, 207, 235
Rare earths, 129, 130, 139–141, 150, 155, 174, 178, 181, 187, 188, 197, 198, 207, 208, 211–215, 229–232, 235, 277
Rare metals, 58, 140, 189, 197, 199, 303
Raw materials, 29–31, 33, 34, 65, 74, 117, 118, 120, 130, 163, 173, 174, 177, 188, 239, 249–254, 256, 258, 259, 261–263, 277–280, 282, 284–286, 321, 327, 328, 330
Reactive extraction, 163, 165
Reactor design, 65–67, 69
Recovery, 3, 4, 9, 12, 14–16, 29, 33–35, 48, 50, 52, 57, 58, 60–62, 75, 79, 80, 82–89, 91–93, 96–99, 101, 102, 104–107, 111, 112, 116–124, 129–132, 136, 139–142, 144–152, 155, 160, 170, 176–182, 184, 197–199, 201–205, 207, 208, 213, 217, 219, 221, 225–227, 229, 231, 232, 234, 239, 243, 244, 265, 266, 303, 304, 309, 311–313
Recovery of Ag and Pd from MLCCs, 121
Recovery of Ag, Au, Pd and Pt from ICs, 120

Subject Index

Recovery of Ag from Mylar sheet of keyboards, 122
Recovery of Au from outer surface of PCBs of various electronic equipments, 119
Recovery of Au from waste water/ effluent, 124
Recovery of dissolved REEs, 143
Recovery of metals from leach liquor of metallic concentrate, 203
Recovery of metals from LIBs, 204
Recovery of metals from magnets and tube lights, 207
Recovery of metals from scrap LCD panels, 205
Recovery of precious metals from e-waste, 205
Recovery of REEs from HDDs, 145
Recovery of REEs from HDDs using recycled solution, 150
Recovery of scandium, 227, 232, 234
Recycling, 3–5, 9–12, 48, 80, 88, 89, 91–94, 115–117, 120, 124, 130, 139, 140, 150, 152, 156, 160, 174, 175, 178, 185, 187–189, 193, 198, 203, 204, 207, 208, 227, 250, 262, 266, 279, 311
Red mud, 229, 231, 236
Reduced ilmenite, 327, 329–333, 335
REE solution, 156, 157, 189, 218
Remote asteroid mining of CRMs, 183
Roasting—leaching studies, 241
Roasting process, 239–241

S

Scandium, 130, 131, 181, 217–219, 227, 229–232, 234–236, 304
Schematic diagram of laboratory equipment, 31
School Hands-on Toolkits, 282
Schools, 184, 251–256, 277–286
Science communication by students, 284
Science dissemination, 254, 280, 284
Screening process, 212
Secondary resource and urban mining, 184
Secondary resources, 79, 80, 85, 87, 116, 174, 180, 181, 184, 189
Selective-Ion Recovery (SIR), 217, 219, 220, 222–227
Selective precipitation of Mn from the effluent, 95
SEM analysis of the precipitation, 317
SEM-ED analysis, 269

Separation experiments, 157, 158
Separation of scandium from the other cations present in the PLS, 234
Sequential chemical extraction, 291, 293
Significance and outlook, 158
Significance of the simulated chemistry to the REE recycling from magnets, 193
Simulation method and software, 189
Slag, 35, 57, 102–105, 107, 111, 112, 184, 265–275, 321–323
Smart multi-filtering system to identify specific reagents, 176
Smelting, 30–32, 265, 266, 274, 275
Sodium carbonate roasting, 322–324
Sodium metal, 65
Sodium sulfate, 14, 65, 66, 69, 70, 139, 144, 145, 152
Solid-state, 37–39, 44, 45
Solid-state reaction method, 44
Solvating extraction, 304
Solvent extraction/beneficiation studies, 75, 77
Solvent extraction of pregnant solution from sample CR-A, 214
Sources for scandium, 231
Spectroscopy results, 23
Spent catalyst, 102, 105, 108
Spent LIBs, 12–15, 47–52, 92, 93, 204
Spent lithium-ion Batteries (LIBs), 11–15, 47–52, 92–95, 99, 198, 204, 205, 259, 260, 261
Spray pyrolysis, 37–39, 43–45
Spray pyrolysis, 37–39, 44, 45
Stone coal, 291–293, 298–300
Surface hydroxyls, 19
Sustainability, 62, 130, 155, 173–175, 184, 185, 187, 262, 279
Synthesis and application of layered H_2TiO_3 ion sieve, 20

T

Tellurium, 57–62, 278
Thermochemistry, 65, 66
Thermodynamic modelling, 5, 129, 132, 133, 137, 144
Thermogravimetric analysis, 22
Transformation of vanadium phases in combustion process, 299
Treatment of roasted products, 241
Tubular reactor, 327–329, 332, 335

U
Urban mining, 80, 116, 163, 184, 187, 266

V
Vacuum reduction furnace, 29, 31
Vanadium phases, 291–294

W
Wastewater, 12, 79–85, 87–89, 92, 97, 98, 131, 312, 321, 328
Water leaching, 243, 322–326
Wolframite, 73–78

X
X-ray diffraction, 5, 21, 22, 144, 241, 323